'This is a fascinating book—one built around a conceptual structure and implemented with well-chosen, contrasting case studies. It begins with recognition that small- and medium-size countries do defence planning, but go about it very differently than great powers because "their outlook, ambitions, global recognition, dependencies, and available resources tend to differ substantially from great powers." That said, *how* do they go about it? They encounter dilemmas and contradictions, including some relating to self-image. What happens varies across the cases (Australia, Canada, Finland, Indonesia, Israel, the Netherlands, Oman, Singapore, Slovakia, and the United Arab Emirates). As both scholars and practitioners will appreciate, the differences reflect not only objective matters such as geography, size, and resources, but also the nations' history, culture, personalities, and politics. Commonalities can be found, as can rules of thumb about what factors matter, but changes also occur in response to events and to trends in military technology, authoritarianism, shooting wars, and domestic politics. The authors deserve credit for a very interesting book with much to teach those familiar only with major-power planning'.
  —**Paul K. Davis**, *Senior Principal Researcher (retired) (RAND) and Professor,*
  *Pardee RAND Graduate School, Santa Monica, California*


'Students of international security, strategic and military studies have long neglected what most military and civilian defence practitioners do most of the time: prepare and build the future force through various forms of defence planning. This book fills that important gap in the literature with a series of rich and well-structured national case studies that will be relevant for anyone trying to understand defence planning issues beyond the great powers. With powerful analytical categories for the comparative study of defence planning, the book's general framework and specific findings will be highly useful for both expert practitioners and future research'.
  —**Henrik Breitenbauch**, *Dean, Royal Danish Defence College, Denmark*


'In a world that is increasingly dangerous, small and medium powers face unique challenges when it comes to defence planning and military innovation. This volume offers incisive and long overdue comparative analysis of how smaller states prepare to defend themselves. Essential reading for our times'.
  —**Theo Farrell**, *President, La Trobe University, Australia*

I0031540

# Defence Planning for Small and Middle Powers

This book examines the processes, practices and principles of defence planning in small and middle powers.

Small and middle powers are recalibrating their force postures in this age of disruption. They are adapting their defence planning and military innovation processes to protect the security of their nations. The purpose of this book is to explore defence planning and military innovation in 11 contemporary case studies of small and middle powers in North America, Europe, the Middle East, Asia and Oceania. Employing a structured focused comparison framework, it traces patterns in the choices of small and middle powers across the following themes: (1) alliances, dependencies and national ambitions; (2) approaches, processes, methods and techniques; and (3) military innovation strategies and outcomes. Breaking new theoretical ground, it offers a three-pronged typology distinguishing between the strategic defence planner, the transactional defence planners and the complacent defence planner. The book offers a rich array of insights into cases that fall across different geographies, strategic cultures and governance systems. These insights can help guide discussions on how to structure decision-making structures, arrive at ambition levels, formulate priorities, select partners and design defence planning and military innovation processes.

This book will be of much interest to students of defence studies, security studies, public policy and international relations, as well as to professionals in defence planning.

**Tim Sweijs** is Director of Research at The Hague Centre for Strategic Studies and Senior Research Fellow at the War Studies Research Centre of the Netherlands Defence Academy.

**Saskia van Genugten** is Senior Director at MacroScope Strategies (M2S).

**Frans Osinga** is Professor of War Studies, Leiden University.

# Routledge Advances in Defence Studies

*Routledge Advances in Defence Studies* is a multi-disciplinary series examining innovations, disruptions, counter-culture histories, and unconventional approaches to understanding contemporary forms, challenges, logics, frameworks, and technologies of national defence. This is the first series explicitly dedicated to examining the impact of radical change on national security and the construction of theoretical and imagined disruptions to existing structures, practices, and behaviours in the defence community of practice. The purpose of this series is to establish a first-class intellectual home for conceptually challenging and empirically authoritative studies that offer insight, clarity, and sustained focus.

For more information about this series, please visit: www.routledge.com/Routledge-Advances-in-Defence-Studies/book-series/RAIDS

# Defence Planning for Small and Middle Powers

Rethinking Force Development
in an Age of Disruption

**Edited by Tim Sweijs, Saskia van Genugten
and Frans Osinga**

R Routledge
Taylor & Francis Group

LONDON AND NEW YORK

First published 2025
by Routledge
4 Park Square, Milton Park, Abingdon, Oxon OX14 4RN

and by Routledge
605 Third Avenue, New York, NY 10158

*Routledge is an imprint of the Taylor & Francis Group, an informa business*

*British Library Cataloguing-in-Publication Data*
A catalogue record for this book is available from the British Library

ISBN: 978-1-032-50356-1 (hbk)
ISBN: 978-1-032-50360-8 (pbk)
ISBN: 978-1-003-39815-8 (ebk)

DOI: 10.4324/9781003398158

Typeset in Times New Roman
by Apex CoVantage, LLC

# Contents

# Acronyms and abbreviations

ADF:            Australian Defence Force
AKP:            Justice and Development Party
ANZAC:          Australian and New Zealand Army Corps
AUKUS:          Australia-United Kingdom-United States
AUV:            Autonomous Underwater Vehicles
BAE:            British Aerospace
BAP:            Baltic Air Policing
BAPPENAS:       *Badan Perencanaan Pembangunan Nasional*
                or National Development Planning Agency
BG:             Brigadier General
CAF:            Canadian Armed Forces
CBC:            Canadian Broadcasting Corporation
CBP:            Capability-Based Planning
CD&E:           Concept Development & Experimentation
CDS:            Chief of the Defence Staff
CGS:            Chief of the General Staff
CJOC:           Canadian Joint Operations Command
COS:            Chief of Staff
CSDP:           Common Security and Defence Policy
CUAS:           counter-uncrewed aerial systems
DARPA:          Defense Advanced Research Projects Agency
DART:           Disaster Assistance Response Team
DCDC:           Development, Concepts and Doctrine Centre
DM:             deputy minister
DND:            Department of National Defence
DoA:            Defence of Australia
DPCR:           Defence Planning Capability Review
DPR:            *Dewan Perwakilan Rakyat* or People's
                Representative Council
DRDC:           Defence Research Development Canada
DSR:            defence strategic review
EEZ:            exclusive economic zone
EFP:            Enhanced Forward Presence

| | |
|---|---|
| EI2: | European Intervention Initiative |
| EOP: | enhanced opportunities partner |
| ESSM: | evolved SeaSparrow missile |
| EU: | European Union |
| FDA: | Force Development and Analysis |
| FDF: | Finnish Defence Forces |
| GCC: | Gulf Cooperation Council |
| GDP: | Gross Domestic Product |
| GMLRS-ER: | Extended-Range Guided Multiple Launch Rocket System |
| GOC: | Government of Canada |
| HADR: | Humanitarian Assistance and Disaster Relief |
| HCSS: | Hague Centre for Security Studies |
| Hybrid CoE: | Center of Excellence for Countering Hybrid Threats |
| IAP: | Iceland Air Policing |
| IDEaS: | Innovation for Defence Excellence and Security |
| IDF: | Israel Defense Forces |
| IMET: | International Military Education and Training |
| INTERFET: | International Force East Timor |
| ISAF: | International Security Assistance Force |
| ISED: | Innovation, Science and Economic Development Canada |
| ITAR: | International Traffic in Arms Regulations |
| JASSM: | Joint Air-to-Surface Standoff Missile |
| JEF: | Joint-Expeditionary Force |
| JSF: | Joint Strike Fighter |
| KKIP: | *Komite Kebjakan Industri Pertahanan* or Defense Industry Policy Committee |
| LAV: | Light Armoured Vehicle |
| MaFat: | Hebrew acronym of the Administration for the Development of Weapons and Technological Infrastructure |
| MANPADS: | Man-portable air-defence systems |
| MBT: | Main-battle tanks |
| MBZ: | Sheikh Mohammed bin Zayed Al Nahyan |
| MDA: | Missile Defense Agency |
| MEF: | Minimum Essential Force |
| MOKYS: | Mobilný komunikačný systém (Mobile Communication System) |
| MOOTW: | Military Operations Other Than War |
| MRO: | Maintenance Repair and Overhaul |
| NASAMS-II: | Norwegian Advanced Surface-to-Air Missile System |
| NATO: | North Atlantic Treaty Organization |
| NCO: | Non-Commissioned Officer |
| NDHQ: | National Defence Headquarters |
| NDPP Targets: | NATO Defence Planning Process Targets |
| NORAD: | North American Aerospace Defence Command |
| NORDEFCO: | Nordic Defense Cooperation |

| | |
|---|---|
| NSA: | National Security Agency (US) |
| NSC: | National Security Council |
| NSS: | National Shipbuilding Strategy |
| NVA: | East German Army |
| OEM: | Original Equipment Manufacturer |
| OTRI: | Operational Theory Research Institute |
| OUR: | Urgent Operational Requirements |
| OYTEP: | Ten-Year Procurement Plan |
| PDD: | Project Definition Document |
| PfP: | Partnership for Peace |
| PKK: | Kurdistan Workers Party |
| PPBES: | Planning, Programming, Budgeting and Execution System |
| PSF: | Peninsula Shield Force |
| PSPC: | Public Services and Procurement Canada |
| RAAF: | Royal Australian Air Force |
| RAF: | British Royal Air Force |
| RAFO: | Royal Air Force of Oman |
| RAMSI: | Regional Assistance Mission to Solomon Islands |
| RAN: | Royal Australian Navy |
| RCAF: | Royal Canadian Air Force |
| RCN: | Royal Canadian Navy |
| RFP: | Request for Proposals |
| RMA: | Revolution of Military Affairs |
| RNO: | Royal Navy of Oman |
| SHP: | Strategic Targets Plan |
| SIGINT: | Signal Intelligence |
| SIPRI: | Stockholm International Peace Research Institute |
| SNMG: | Standing NATO Maritime Groups |
| SNMGC: | Standing NATO Mine Countermeasures Group |
| SOF: | Special Operations Forces |
| SSB: | Defence Industry Agency |
| SSE: | Strong, Secured, Engaged |
| SSİK: | Defence Industry Executive Committee |
| SSM: | Undersecretariat for Defence Industries |
| TAF: | Turkish Armed Forces |
| TBCS: | Treasury Board of Canada Secretariat. |
| TCA: | Turkish Court of Accounts |
| TED: | Technology, Entertainment, Design |
| TNI: | *Tentara Nasional Indonesia* or Indonesian National Defence Forces |
| TP-UTVA: | Tasavallan Presidentti-Ulko-ja Turvallisuuspoliittinen ministerivaliokunta |
| TRL: | technology readiness level |
| UAS: | unmanned aircraft system |
| UAV: | unmanned aerial vehicle |

| | |
|---|---|
| UOR: | urgent operational requirements |
| US CENTCOM: | United States Central Command |
| US: | United States |
| VHRJTF: | Very High Readiness Joint Task Force |
| WMD: | weapons of mass destruction |

# Contributors

**Murat Caliskan** is a post-doc researcher at the Université Catholique de Louvain, Belgium, and a former Turkish military officer. He completed his PhD on 'the concept of Hybrid Warfare and its meaning for NATO' from the perspective of strategic theory in 2021. He has published on topics ranging from war, strategy and peace in the following journals: *Union Politics, Journal of Common Market Studies, Contemporary Security Policy, European Journal of Political Research, Cooperation & Conflict, The Nonproliferation Review* (and elsewhere). From 2018 to 2019 he was a junior faculty fellow at Stanford University's Center for International Security and operations, peacebuilding, Belgian Foreign Policy, Turkish Foreign/Domestic Policy and Russia-Ukraine War. His recent research interests are defence policy and planning and the use of Artificial Intelligence in defence. He teaches master's level War, Strategy and Technology course at the Université Catholique de Louvain.

**Andrew Carr** is a senior lecturer in the Strategic and Defence Studies Centre at the Australian National University. His research focuses on Strategy, Middle Powers and Australian Defence Policy. He has published widely in outlets such as *Survival, The Journal of Strategic Studies, Australian Foreign Affairs, The Washington Quarterly, International Theory, Comparative Strategy*, and the literary magazine *Meanjin*. He has an authored book with Melbourne University Press and has edited books with Oxford University Press and Georgetown University Press.

**Paul Dickson** is an expert-level senior strategic analyst with the Canadian Department of National Defence's Defence Research and Development Canada, currently posted as a strategic advisor, strategies and effects, in the Strategic Joint Staff's Director General Strategies, Effects and Readiness. He has taught at Wilfrid Laurier University's Centre for Military, Strategic and Disarmament Studies and Queen's University's International Study Centre (1996–2003) in the United Kingdom. He was a Military History Summer fellow at the United States Military Academy at West Point in 2003 and joined the Department of National Defence in 2004. Defence Department focus areas comprise analysis and advice on strategy and force development, including joint and hybrid warfare and in areas of conflict management, such as deterrence. Select unclassified external publications include chapters in 'Trends in Military Intervention' (McGill-Queens University Press), 'The Pursuit of Relevance: Strategic Analysis in Support of International Policy-making' (Rowman & Littlefield), and 'Analysis in Support of Policy' (Pearson Peacekeeping Cornwallis Group). The monograph *A Thoroughly Canadian General: A Biography of General H.D.G. Crerar* (University of Toronto Press, 2007) was awarded the 2008 C.P. Stacey Award for the best book on the study of conflict and society in Canada published in 2006–2007 and is recommended in the *Guide to Canadian Army Professional Reading*. Professional awards and decorations include the Associate Deputy Minister (S&T) Commendation (2007), RCAF Commendation (2011), Canadian Forces Medallion for Distinguished Service (2014), Meritorious Service Decoration (2016) for service in Afghanistan; and Task Force Commander Commendation (2017, OP ARTEMIS Science Advisor).

**Stephan Frühling** is Professor in the Strategic and Defence Studies Centre at the Australian National University and has widely published on Australian defence policy, defence planning and strategy, nuclear weapons and NATO. Stephan was the Fulbright professional fellow in Australia-US Alliance Studies at Georgetown University in Washington, DC, in 2017. He worked as a 'partner across the globe' research fellow in the Research Division of the NATO Defense College in Rome in 2015, and was a member of the Australian Government's External Panel of Experts on the development of the 2016 Defence White Paper.

**Nikolas Gardner** is Professor of Strategy at the UAE National Defence College. He came to the NDC from the Royal Military College of Canada, in Kingston, Ontario, where he served as Chair of the War Studies graduate program, and the Class of 1965 Chair in Leadership. From 2006-2011 Dr. Gardner taught at the USAF Air War College in Montgomery, Alabama, where he served as the first academic director of the Grand Strategy Program, developing a specialized curriculum designed for senior military officers with advanced academic qualifications. Dr. Gardner is the author of *Trial by Fire: Command and the British Expeditionary Force in 1914* (Praeger, 2003), and *The Siege of Kut-al-Amara: At War in Mesopotamia, 1915–1916* (Indiana University Press, 2014). He has published numerous articles and chapters on a broad range of topics, including military and strategic thought, the First World War, and the role of contractors in British foreign policy in the Middle East. In 2008 he was made a Fellow of the Royal Historical Society.

**Saskia van Genugten** is a director at the Abu Dhabi office of MacroScope Strategies. From 2018–2021, Saskia served as a strategy and policy advisor at the Dutch Ministry of Defense. Prior to joining the Dutch MoD, she was based in the Middle East, working as a senior research fellow at the Anwar Gargash Diplomatic Academy, as a manager for government affairs at Pricewaterhouse-Coopers, and as a political affairs officer at the UN Mission to Libya (UNSMIL). Early in her career, she worked at the Foreign Affairs, Defence and Development Cooperation Committee of the Senate of the Netherlands and at the Dutch

Ministry of Foreign Affairs. Saskia received her PhD in 2012 from Johns Hopkins University's School of Advanced International Studies (SAIS, Washington, DC), in the field of international relations. She also holds an MA degree from SAIS, in addition to an MA in diplomatic history, a minor in journalism, and a BA in Italian language and literature from the University of Utrecht (the Netherlands). She authored and co-edited several articles and books, including the monograph *Libya in Western Foreign Policies* (2016, Palgrave Macmillan) and the co-edited volume *Stabilising the Contemporary Middle East and North Africa* (2020, Palgrave Macmillan).

**Lenny Hazelbag** is currently working as a strategic advisor in the cabinet of the Commander of the Royal Netherlands Army. Before that, he was the head of the Operational Policy branch of the Dutch Ministry of Defence, responsible for the Force Development process, in support of the Director of Operational Policy and Plans in his role as National Capability Director. The Operational Policy branch is also responsible for the coordination of all processes and instruments within NATO and EU with regard to concept development and capability development, inter alia NDPP, NWCC, NFM (NATO) and PESCO, CARD, CDP, EDF (EU). Lenny has an operational background as an infantry officer in air mobile, reconnaissance and mechanized infantry units and was deployed to Bosnia and Afghanistan. Lenny holds a PhD in public administration from Erasmus University in Rotterdam (2016) and a master's degree in political sciences from the University of Leiden (2002). In 2014, he completed the Advanced Command and General Staff course at the Netherlands Defence Academy.

**Hans "Hammer" Klinkenberg** is currently working as a strategic advisor at the Dutch Ministry of Defence, Policy Directorate. Before that, he was the deputy head of the Operational Policy branch of the Dutch Ministry of Defence. Hans holds a master's degree in public administration—crisis and security management from Leiden University (2016). He also holds a professional graduate certificate in nuclear deterrence, from Harvard Extension School (2022). In 2018, he completed the Advanced Command and General Staff course at the Netherlands Defence Academy. Hans has a military background as an F-16 pilot, including multiple combat deployments to Afghanistan. Hans also was the first to lead the military Space Office (now: Defence Space Security Centre) and has worked at a management position at the Defence Cyber Command. Hans is still active as a part-time flying instructor in basic military pilot training, which takes place at Woensdrecht AFB.

**Evan A. Laksmana** is Senior Fellow for Southeast Asia Military Modernisation at the International Institute for Strategic Studies (IISS) based in Singapore. He previously held senior research positions at the Lee Kuan Yew School of Public Policy of the National University of Singapore and the Centre for Strategic and International Studies (CSIS) in Jakarta, Indonesia. He has also held non-resident and research positions with Carnegie China, the Lowy Institute for International Policy, ISEAS-Yusof Ishak Institute, the National Bureau of Asian

Research, Sydney University, German Marshall Fund of the United States, and the S. Rajaratnam School of International Studies. He earned his PhD in political science from the Maxwell School of Citizenship and Public Affairs, Syracuse University, as a Fulbright Presidential Scholar.

**Michal Onderco** is Professor of International Relations at Erasmus University Rotterdam and affiliate at Peace Research Center Prague. He studies international security, with focus on nuclear politics and on domestic politics of foreign policy. He published extensively on foreign and security policy in Central Europe. He authored *Networked Nonproliferation* (Stanford UP, 2021) and *Iran's Nuclear Program and the Global South* (Palgrave, 2016), as well as papers which appeared in *International Studies Quarterly* and *European* Cooperation.

**Frans Osinga** is the Chair in War Studies at the Institute of Governance and Global Affairs of Leiden University. Until his retirement from the Royal Netherlands Air Force in 2023 as Air Commodore, starting in 2010, he was the Chair in War Studies and Head of the War Studies Department of the Netherlands Defence Academy. A former F-16 pilot and graduate of the Netherlands Advanced Staff College, he was the MoD Senior Research Fellow at the Clingendael Institute before a posting at NATO Allied Command Headquarters. He holds a PhD in political science from Leiden University. Author of more than 90 publications, his interests focus on contemporary warfare, strategy and defence policy. His books include S*cience, Strategy and War: the Strategic Theory of John Boyd* (2007, Routledge), *Military Adaptation in Afghanistan* (2013, Stanford University Press, with Theo Farrell and James Russell), and *Deterrence in the 21st Century* (2020, Asser Springer, with Tim Sweijs). He is a member of the government's Advisory Council on International Affairs.

**Michael Raska** is Assistant Professor in the Military Transformations Programme at the S. Rajaratnam School of International Studies, Nanyang Technological University in Singapore. His research and teaching focus on defence and military innovation, specifically how key nation-states strive to maintain or prolong margins of military-technological superiority through defence innovation; the effects of emerging technologies such as AI on force planning and future warfighting concepts; and open source-based intelligence assessments on emerging threats such as next-generation hybrid conflicts and digital warfare. He is the author of *Military Innovation and Small States: Creating Reverse Asymmetry* (Routledge, 2016), co-editor of the AI Wave in Defence Innovation (Routledge, 2023) and *Defence Innovation and the 4th Industrial Revolution* (Routledge, 2022). He has published in journals such as the *Journal of Strategic Studies, Strategic Studies Quarterly, Prism—Journal of Complex Operations, Air Force Journal of Indo-Pacific Affairs, Korea Journal of Defence Analysis, Pointer-Journal of Singapore Armed Forces,* and *Sirius—Journal of Strategic Analyses.*

**Ash Rossiter** is an associate professor of international security at Khalifa University. He received his PhD from the University of Exeter in 2014 after earlier completing an MA in War Studies from King's College London. He is the author of *Security in the Gulf* published in 2020 by Cambridge University Press. His next book, *Warfare in the Robotics Age*, will be published this year (2024, Lynne Rienner). Rossiter is the book series editor for Security and Strategy for University of Exeter Press.

**Eitan Shamir** currently serves as the Managing Director of the Begin-Sadat Center for Strategic Studies (BESA) and holds the position of Associate Professor in the Department of Political Science at Bar Ilan University. Furthermore, he is the academic head of the MA programme in Security and Cyber Security. Before delving into the academic realm, Shamir served as the head of the National Security Doctrine Department at the Office of the Prime Minister (Israel), within the Ministry of Strategic Affairs. He also served as a senior fellow at the Dado Center for Interdisciplinary Military Studies (CIMS) in the Israel Defense Forces (IDF). He has authored numerous articles published in leading academic journals. He published several books, including *Transforming Command: The Pursuit of Mission Command in the US, UK, and Israeli Armies* (Stanford UP, 2011), *Moshe Dayan: The Making of a Strategist* (Modan and Ministry of Defense, 2023 in Hebrew) and *The Art of Military Innovation: Lessons from the IDF* (Cambridge, MA: Harvard UP, 2023) co-authored with Edward Luttwak. He earned his PhD from the Department of War Studies at King's College London, Additionally, he holds a master's degree in organisational strategy from Brigham Young University.

**Olli Pekka Suorsa** is an assistant professor at Rabdan Academy in Abu Dhabi, the UAE. Olli's research focuses on air power, maritime security, defence industry and technology, and smaller power's role in international relations. Before embarking on his academic career, Olli worked in the defence and aerospace industry in Finland.

**Tim Sweijs** is the Director of Research at The Hague Centre for Strategic Studies and Senior Research Fellow at the Netherlands' War Studies Research Centre of the Netherlands Defence Academy. His work is multidisciplinary in nature and straddles political science, strategic studies and war studies. He is a board member at the European Initiative for Security Studies, where he chairs the War, Coercion and Statecraft Working Group. He is also a research affiliate at the Center for International Strategy, Technology and Policy in the Sam Nunn School of International Affairs at the Georgia Institute for Technology. Tim has advised international organisations, governments and defence departments across the globe. He has provided expert testimony to the United Nations Security Council, the European Parliament, the Dutch Parliament, as well as to NATO's Parliamentary Assembly. He teaches an advanced graduate and executive staff command course on the future of war. Tim holds degrees in war studies (PhD, MA), international relations (MSc) and philosophy (BA) from King's College, London, and the University of Amsterdam. His book on state threat behaviour and interstate crisis escalation over the past century, titled *The Use* *and Utility of Ultimata in Coercive Diplomacy*, was published by Palgrave Macmillan in 2023. Visit the book's website at *www.coercivediplomacy.com*. The volume—co-edited with Jeffrey Michaels—titled *Beyond Ukraine: Debating the Future of War* was published by Hurst in March 2024.

**Athol Yates** heads the Institute for International and Civil Security at Khalifa University. His research focuses on the history of the UAE police and armed forces, and his major publications are *The Military and Police Forces of the Trucial States and United Arab Emirates 1951–1980* (2019, Helion and Company) and the *Evolution of the Armed Forces of the United Arab Emirates* (2020, Helion and Company).

# Figures and tables

## Figures

## Tables

# Acknowledgements

We would like to thank the Faculty of Military Sciences at the Netherlands Defence Academy for providing the financial support for the author symposium that allowed for in-depth, face-to-face, and spirited discussion that continues to be vital for the scholarly enterprise. We would also like to extend our gratitude to Geert Kuiper at the Dutch Ministry of Defence for his continuous support and his dedication to this specific topic. He saw the project's potential in its conception stages, helped sustain the position of a research fellow at the Netherlands Defence Academy and generously shared his insights at the author symposium in Amsterdam in March 2023. We would like to thank Peter Viggo Jakobsen for insisting that we double down on our proposed structured focused comparison approach that you will read more about in these pages. We would like to thank Harm Roelant van de Plassche and Alisa Hoenig for their research assistance and support in the preparation of this manuscript. Finally, we would like to express our gratitude to Andrew Humphrys and Devon Harvey: Andrew for greenlighting our book proposal, and Devon and Andrew for their dedication to seeing its results into print. Last but not least, thanks go out to all the authors, who participated in online meetings, travelled to Amsterdam from sometimes far-flung places and contributed the fascinating case studies that are part of an emerging but burgeoning literature in this age of disruption.

# 1 Introduction

*Tim Sweijs, Saskia van Genugten*
*and Frans Osinga*

This book examines defence planning in small and middle powers (SMPs) in North America, Europe, the Middle East, Asia and Oceania. It does so in the context of significant geopolitical and military-strategic developments that are fundamentally disrupting the environment in which SMPs operate. Defence planners in SMPs across the globe face fundamental issues concerning the design and calibration of force postures to safeguard the security of their nations.

In Europe, the changing security environment has exposed the low level of readiness of European forces at a moment when the most important alliances and partnerships underpinning Europe's security are being challenged.[1] The return of large-scale conventional war to the European continent prompted a rapid process of geopolitical reawakening. Especially after Russia's invasion of Ukraine in February 2022, the notion of greater strategic autonomy has entered mainstream policy discourses both in Brussels and in various other European capitals. Yet, even after two years of war and significant increases in defence budgets, European efforts remain limited compared to those of the US, China and various others. European defence planners confront real trade-offs between the renovation and renewal of their armed forces against the background of considerable military-technological change which includes advances in C4ISR (Command, Control, Communications, Computers, Intelligence, Surveillance and Reconnaissance), the proliferation of long-range precision strike capabilities, the emergence of robots and unmanned systems and the progressive introduction of artificial intelligence (AI) in the military domain. From a geopolitical perspective, the war in Ukraine and doubts about the strength of the US commitment to NATO are raising concerns amongst policymakers in European SMPs. They fear that on their own, they are unable to effectuate a credible deterrence strategy and, if necessary, defend European territory. The necessity for change is also felt by other NATO members, including Canada. Repeated NATO communiques have stressed the continuous need for 'significant measures to further enhance NATO's deterrence and defence posture in all domains, including strengthening forward defences and the Alliance's ability to rapidly reinforce any Ally that comes under threat'.[2]

In the Middle East, defence spending among the wealthier states is consistently high in the context of persistent instability, multiple regional rivalries and recurring wars. Given the current outlook, this focus on defence and security will most likely

DOI: 10.4324/9781003398158-1

remain significant. Various states are increasing their military footprint in the region and decision-making around defence-related issues is often bolder and faster than it can be in their European counterparts. The region's position, resources and volatility make it an important area of interest for the world's great powers, not only the US but also China as well as Russia. At the same time, many SMPs in the Middle East region still cope with a lack of national expertise and technological know-how. While defence planners in smaller Gulf states are less bound by financial constraints, they face a host of other challenges, for example with regard to human capital. Consequently, these states tend to focus on building international partnerships geared towards technological transfers and military innovation, to try to be at the forefront of acquiring new technologies and their applications for their armed forces.

Meanwhile, intensifying geopolitical competition in Asia and Oceania, perhaps a prelude to larger-scale militarised conflict, poses a whole different set of challenges in that region. China's rise as a regional power, manifested in the doubling of its defence budget over the past decade and the ongoing establishment of a maritime sphere of influence in its neighbourhood, is a cause of concern for many SMPs in the region that struggle with the question whether to pursue balancing, bandwagoning or hedging (including engagement) strategies to address the security implications associated with China's ascent.[3]

A look back at the steady stream of reports published since 2010 on trends in international security indicates that, while the security environment since 2022 has rapidly deteriorated, the downward trends were already visible since the late 2000s when Russia's aggression against Georgia revealed its revisionist policy, the West was failing in efforts to stabilise and transform the societies of Afghanistan and Iraq, and China started asserting itself as a future peer competitor. An era of great power competition was emerging, with polarising international rhetoric, an increase in economic coercion and the prevalence of hostile state activity below the threshold of large-scale violent conflict. Hard military power, including nuclear weapons, was once again becoming a visible currency in international relations, at least for the major powers. Yet, most SMPs seemed to struggle to adjust to the pace of change in the security environment that emerged and slowly accumulated during the decade.

Against this background, this book aims to better understand the fundamentals of what makes SMPs' defence planning and military innovation different from that of great powers, what common challenges and opportunities these SMPs face, and whether there are patterns in SMP defence planning and military innovation processes and practices that can be distilled from a variety of contemporary case studies.

This introduction is structured as follows: it situates the meaning of defence planning and military innovation within the wider literature, identifies the lacunas in that literature on SMPs and outlines the assumptions around why SMPs are different from great powers. On that basis, it presents a structured focused comparison framework that has been developed to assess similarities and differences between the various SMP case studies in this volume, followed by an explanation of the selection of cases and the structure of the book. The introduction concludes with the contribution this volume has to offer to a small but emerging SMP defence planning literature.

**Defence planning and military innovation**

Defence planning and military innovation as disciplines of academic inquiry are both concerned with the way in which defence organisations prepare to meet future challenges. Defence planning examines how political guidance is translated into concrete choices for force profiles and concomitant force capabilities. Defence planning has garnered relatively little scholarly interest, perhaps even less so than the topic of military innovation. That is somewhat surprising, given that global defence planning involves well over USD 2,200 billion of defence expenditures annually and the extent to which defence organisations leverage military innovation concerns real-world capability choices that are equally critical to war and peace. Defence planning, in the words of the late 'evangelist' of strategy, Colin Gray, involves 'preparations for the defence of a polity in the future (near-, medium- and far-term)'.[4] Paul Davis from the RAND Corporation has provided a slightly more specific description, defining it as the 'deliberate process of planning a nation's future forces, force postures, and force capabilities', which 'is distinct from operations planning on how to employ forces in war'.[5] In a similar vein, Magnus Håkenstad and Kristian Knus-Larsen refer to it as the 'process by which a given state arrives at political decisions regarding the future development of the structure, organisation and capabilities of their armed forces'.[6] Because the future is uncertain and future force choices ultimately evolve around a state's survival, defence planning can also be conceptualised as the 'management of strategic risk', in the words of Stephan Frühling.[7]

Henrik Breitenbauch and André Ken Jakobsson have pointed out that defence planning represents a 'strategic fact' only to be ignored at our own peril.[8] They observe that as a discipline or field of inquiry, defence planning is certainly related to, yet distinct from its more popular cousin strategic studies. Strategic studies are predominantly concerned with the various ways in which *currently available* human and material resources can be applied to achieve political objectives in the context of matters related to war and peace. Defence planning is, however, more concerned with how to ensure the right human and material resources *for the future*. In the words of Stephan De Spiegeleire, defence planning is about 'upstream' matters: the allocation of those means, including the design, derivation, engineering acquisition and management of defence capabilities.[9] Defence planning is typically conducted by states and is necessarily concerned with the future. Nonetheless, because defence planning is by its very nature institutionalised, it is often very much focused on the present and the perpetuation of existing ways and means.[10]

Because of its breadth, defence planning involves a panoply of topics, including the actual methods and processes utilised to translate political guidance into, among others, the procurement of materiel, the training of people and the conclusion and maintenance of national and international partnerships. An important element in interstate military competition resides in the ability to translate scientific and technological progress into timely relevant adjustments in military capabilities, doctrines and structures to meet challenges in the security environment. The field of military innovation studies, with its focus on the ability of defence organisations to incorporate change brought about by scientific and technological developments, is thus equally relevant. Military innovation studies therefore complements the study of defence planning as it seeks to understand and explain the trajectory of change in military organisations. Change, as Theo Farrell has noted and Michael Raska, amongst others, further elaborated, manifests in three archetypical ways: emulation (imitation), adaptation (incremental adjustment) and innovation (actual transformation).[11] Change manifests itself in the adoption of new technologies, doctrines, organisational structures and/or force orientations. Change results from different sets of drivers, including interstate competition, military advantages that accrue from specific military capabilities, levels of national development and organisational structures.[12]

The concept of military innovation has many definitions. Most scholars would agree that it involves significant changes in how militaries fight, relative to existing ways of war, manifesting itself in changes to tactics, doctrine and military organisation.[13] Adam Grissom summarised the general consensus among scholars in a landmark survey of the state of the field of military innovation studies 17 years ago stating that an innovation changes the manner in which military formations function in the field (thus excluding administrative and bureaucratic changes); it is significant in scope and impact; and the objective is to increase the ability of a military community to generate power. It is often associated with technological developments, but political forces, changes in threat perceptions, shifts in national strategy and societal changes may also prompt defence organisations to adapt. Depending on the significance of the change, this may qualify as military innovation.[14]

As Grissom noted, several schools of thought each highlight specific sets of drivers to explain the degree to which defence organisations are able to incorporate change. One perspective suggests that the defence policies of major powers and the innovation processes within their militaries are predominantly informed by external developments in the security environment (e.g., geopolitical developments and threats) even if these are certainly mediated by the perceptions of the senior political and military leadership.[15] Another perspective suggests that defence policy, force orientation and investment priorities are instead heavily shaped by dynamics within ministries of defence, the separate military services and affiliated security organisations, and by the competition between those organisations. This second perspective identifies organisational culture, inter- and intra-service rivalry and the quality of civil-military relations as key factors in explaining why certain changes in policy or force structure are adopted or not; why such changes do not occur; or why they are delayed.[16]

A third perspective, which straddles the previous two perspectives, considers, in the words of Horowitz, the 'adoption capacity' of states to new technologies as a function of 'the financial resources and organisational changes required' in the context of external threats, which shapes the types of choices for force postures that will be made.[17] Other factors include the difference between innovation in wartime versus peacetime and whether the organisational culture of a military fosters experimentation, amongst others, through the stimulation of bottom-up initiatives, senior leadership's openness to novel operational approaches and experiments that may run counter to existing investment plans, forces structures and service identities.[18] Specifically for SMPs, international security organisations are an important factor to be reckoned with. Institutions such as NATO serve as conduits for transmitting best practices, norms concerning operational practices and technical standards. Moreover, such organisations serve as signposts for innovation and investment priorities.[19]

## Literature skewed towards great powers

Over time, processes associated with defence planning and military innovation have become increasingly complex, while the methods applied have grown ever more sophisticated. Prior to the nineteenth century, defence planning was the exclusive concern of the sovereign and the Commander in Chief—in many cases one and the same person.[20] The growth in size and complexity of military organisations and operations, in combination with the availability of new scientific insights, led to the development and take up of more sophisticated planning methods by military organisations.[21] During the twentieth century, great powers became global powers that had to deal with many different types of military operations, often in various corners of the world at the same time. Defence organisations therefore grappled with questions ranging from how to best protect the safe passage of maritime vessels over the Atlantic Ocean during the Second World War, or how to deter nuclear-armed opponents in the Cold War, on to how to win and fight two major wars in the immediate post-Cold War era.[22]

It was predominantly in the twentieth century that great powers and their defence organisations started to formalise and institutionalise defence planning processes and methods. Operational Research methods became instrumental in solving strategically significant issues with important implications for the development of defence capabilities. The use of formal methods in defence planning became more entrenched during the 1950s and 1960s, when analytical techniques from system theory and economics were introduced, which in turn became part and parcel of institutionalised processes, especially in the leading military power of that period, the US.[23] Indeed, the practices and principles of US defence planning have been amply documented and the US-centric literature covers defence planning from a variety of angles, including the portfolio of analytical methods,[24] the impact of ideas and the role of people,[25] the structure and the design of decision-making processes; and the effect of bureaucratic politics.[26] While the literature ranges from descriptive to prescriptive, most authors are concerned with the question of how to get defence planning right, typically with an eye to making the process of defence planning more effective and/or more efficient.

In this context, most of the currently existing defence planning and military innovation literature relies heavily on Interbellum and Cold War case studies. A smaller number of studies look at the post–Cold War period and the way Western states have responded to rapid military technological changes after the First Gulf War (Operation Desert Storm, 1990–91), with most of these studies focusing on the practices of the US and the UK. More recently, following the experiences in counter-insurgency operations in Iraq and Afghanistan, military innovation literature has focused on the question to what extent Western militaries (once again predominantly the US and the UK) managed to adapt to the challenges of fighting violent non-state actors.[27]

In contrast, with some exceptions of course, very few book-length studies are available that home in on the defence planning processes and military innovation dynamics within the militaries of SMPs.[28] Much less is known, therefore, about the practices of defence planning and military innovation in SMPs. This represents a gap in the literature because the insights drawn from case studies on great powers are not necessarily transferable to SMPs. After all, their outlook, ambitions, global recognition, dependencies and available resources tend to differ substantially from great powers. As per the description of Jack Levy, great powers are states that play a major role in international politics and have global security concerns. They possess the capability to project military power beyond their borders and beyond their region. They can do so on their own, are willing to aggressively defend their interests, can initiate and join many alliances and are frequently involved in wars. Finally, other states perceive them as 'great', which is typically codified through their position in international congresses, organisations and/or treaties.[29] When it comes to SMPs, there is a small but burgeoning literature that offers an in-depth and nuanced reflection on what constitutes a small power and what a middle power is and how they are different from great powers. The literature traces it back to objective features (e.g., the limited amount of power they possess), subjective features (e.g., self-identification of decision-makers as a small or middle power that has limited sway in the international system), or ideological features (e.g., concern with international law).[30] Recurring in the literature as a common denominator is a description of SMPs that essentially juxtaposes Levy's great power definition: SMPs do not play a major role in international security-related issues, they have fewer military capabilities, which they cannot deploy outside their region for extended periods of time, they tend to be less aggressive in their foreign policies, they conclude fewer alliances, are less frequently involved in war and are not accorded great power status by other states.[31] Authors have therefore questioned whether great power defence planning methods serve the needs of SMPs, deploring what they consider 'conflated concepts and nomenclature and muddled thinking'.[32]

SMPs face different challenges, confront different trade-offs, often depend on the support of a great(er) power and operate on vastly different scales. This brings both constraints and opportunities. First, from a risk management perspective, as Frühling observes, such powers' 'ability to exercise direct military influence on possible sources of strategic risk' is comparatively smaller which in turn affects not just their risk calculus but also their portfolio of means to manage risks.[33] At the same time, amidst geopolitical competition between the great powers, SMPs can create opportunities as great powers vie for their support. In many ways, the role of domestic politics and alliance considerations may appear more salient to these SMPs than any global strategic outlook.

### Structured focused comparison: a framework

On the basis of the commonalities that set them apart from great powers, as well as the perceived differences between SMPs, we have developed a framework to conduct a structured focused comparison of the defence planning process of SMPs. Within our framework, authors contributing to this volume have analysed what considerations SMPs incorporate in their strategic planning, including the role of allies, what approaches they use in their defence planning processes, what specific challenges they grapple with, and how these in turn shape military innovation outcomes. The framework allows for an analysis of individual country case studies as well as the comparison between those, as it uses a standardised set of questions enabling a 'systematic comparison and [ac]cumulation of the findings of the cases possible' based on a 'focused' analysis of specific features.[34] The framework focuses on three core themes and related sets of questions that derive from the characteristics, constraints and challenges sketched: (1) alliances, dependencies and national ambitions; (2) approaches, processes, methods and techniques; and (3) military innovation strategies and outcomes (see Figure 1.1).

In order to facilitate comparability across cases, for each question a limited set of standardised answer options has been identified, complemented with the possibility to identify another option. While within the individual chapters, authors examine their country's case studies in more depth, the framework allows us to identify patterns across the individual case studies.

The framework suggests a number of commonalities between SMPs and traces the variety within those commonalities, which are further explained subsequently.

| | SMPS ADJUST THEIR APPROACHES BASED ON SYSTEMIC CHANGE OUT OF THEIR CONTROL | SMPS REACT TO CHANGES IN THE RELEVANT SECURITY COMPLEX(ES) THE SMP IS PART OF | | | |
|---|---|---|---|---|---|
| *Alliances, Dependencies and National Ambitions* | | | | | |
| SMPS FACE CONSTRAINTS IN SAFEGUARDING NATIONAL SECURITY | FINANCIAL | DEMOGRAPHIC | CULTURAL | GEOGRAPHICAL | OTHER |
| SMPS NEED TO POSITION THEMSELVES VIS-À-VIS GLOBAL AND REGIONAL POWER CONSTELLATIONS | BALANCING | HEDGING | BANDWAGONING | NON-ALIGNING | OTHER |
| SMPS HAVE TO ENGAGE WITH OTHER STATES TO INCREASE THEIR DEFENCE CAPABILITIES | SECURITY GUARANTEES | TRANSACTIONAL | FORCE INTEGRATION | SPECIALISATION | OTHER |
| SMPS HAVE ARMED FORCES FOR DIFFERENT PURPOSES | INSTRUMENT OF DIPLOMACY | PRESTIGE PROJECT | REDUCE STRATEGIC RISK (DEFEND) | DOMESTIC STABILITY | OTHER |
| *Approaches, Processes, Methods and Techniques* | | | | | |
| SMPS TAKE A MORE CONSTRAINED APPROACH TO DEFENCE PLANNING | THREAT-NET ASSESMENT BASED | PORTFOLIO-BASED | TASK-BASED | MOBILISATION | OTHER |
| SMPS HAVE LIMITED RESOURCES TO DESIGN AND IMPLEMENT DEFENCE PLANNING PROCESSES | TAILORED ANALYTICAL PROCESS | OUTSOURCING STRATEGY | ACCEPT DILUTION / FAIR SHARE | EXTERNAL TEMPLATE | OTHER |
| SMP EMPLOY MULTIPLE METHODS AND TECHNIQUES | COPY GREAT POWERS | INFORMAL CONSENSUS | TAILORED METHODS | AD HOC | OTHER |
| SMPS RELY ON DIFFERENT PROCESSES TO DECIDE ON FORCE POSTURE | ANALYTICAL RESULTS | POLITICS DECIDES | MILITARY DECIDES | GREAT POWER DECIDES | OTHER |
| *Military Innovation* | | | | | |
| SMPS FACE CHALLENGES IN KEEPING PACE WITH TECHNOLOGICAL ADVANCES | BUILD DOMESTIC TECH/INDUSTRY | TECH TRANSFER AGREEMENTS | ACQUIRE READY INNOVATIONS | PARTNER WITH GREAT POWER | OTHER |
| SMPS PURSUE VARIOUS MILITARY INNOVATION PATHS AND PATTERNS | EMULATION / SPECULATION | ADAPTATION / EXPERIMENTATION | INNOVATION / IMPLEMENTATION | OTHER / OTHER | OTHER |
| SMPS FIND THEMSELVES AT DIFFERENT STAGES OF ORGANISATIONAL CHANGE | EXPLORATION | MODERNISATION | TRANSFORMATION | OTHER | OTHER |

*(Left margin labels: GENERAL COMMONALITIES; VARIATIONS WITHIN THESE COMMONALITIES)*

*Figure 1.1* Defence planning of small and middle powers: a structured focused comparison framework.

*Source:* Authors

***Theme 1: alliances, dependencies and national ambitions***

SMPs tend to adjust strategies based on systemic changes out of their control, and in particular on those affecting the regional security complex or complexes the specific SMP considers itself part of. In this context, SMPs suffer from inherent constraints on self-reliance to safeguard their security, which can be predominantly financial, demographic, cultural/historical or geographical in nature. In the context of these constraints, SMPs will need to make a set of choices. To start with, they will need to position themselves vis-à-vis the global distribution of power. In doing so, they rely on different strategies, which include:

- balancing (aligning with (the) existing power(s) against the rising power based on the deliberate strengthening of relationships along important political, military and economic dimensions of state relations);
- bandwagoning (realigning with the rising/resurging power(s) based on the deliberate strengthening of relationships along important political, military and economic dimensions);
- hedging (a mixed strategy to offset risk through the deliberate diversification of relationships along important political, military and economic dimensions of state relations);
- non-aligning (a strategy of refraining from clear alignment of relationships along important political, military and economic dimensions of state relations, deliberately steering an independent course) (see Figure 1.2).[35]

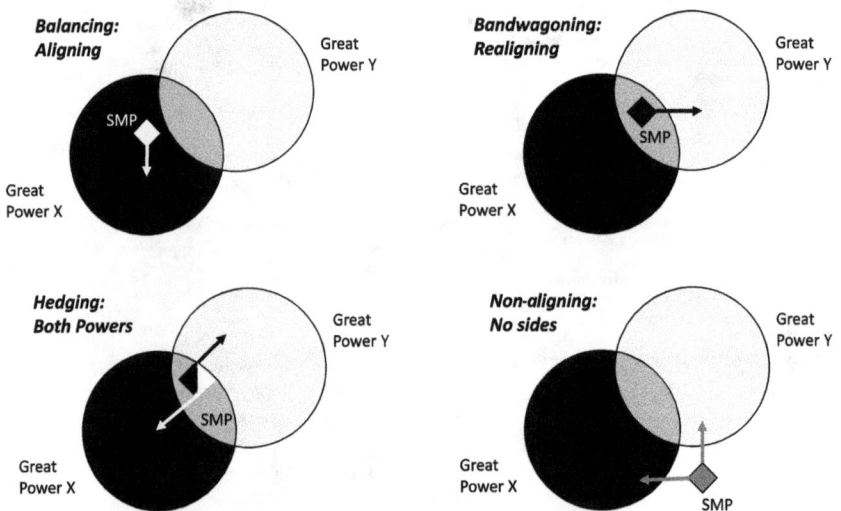

*Figure 1.2* Typology of SMP behaviour.
*Source:* Tim Sweijs et al.[36]

As a corollary to that choice of strategy, SMPs partner with other nations to increase their defence capabilities for different reasons and through different modalities, which may include the provision of security guarantees, force integration and specialisation, as well as other transactional considerations. Finally, in the context of typically limited military ambitions, the principal purpose, written or unwritten, of armed forces varies. The purpose may be predominantly focused on the reduction of strategic risk and self-defence, but it can also be an instrument of diplomacy, a prestige project, or a tool to maintain domestic or regional stability.

### *Theme 2: approaches, processes, methods and techniques*

What approaches, processes, methods and techniques do SMPs use to decide and implement different aspects of defence planning? Defence planning, processes, methods and techniques can be formalised and codified, over time institutionalised through tradition, or perhaps used in a more ad hoc and irregular fashion. SMPs have designed processes, methods and techniques related to the constraints and trade-offs they face. This includes the ways in which political guidance is translated into the procurement of materiel, training and the conclusion and maintenance of national and international partnerships. It also relates to how such processes are structured: top down, bottom up or consultative; open or closed; the bureaucratic actors that are involved in these processes; and the impact of current geopolitical and technological change on processes and methods.

Yet, an important knowledge gap remains regarding the way SMPs go about defence planning activities.[37] The case studies in this volume seek to gain a better understanding of the most important characteristics of the actual defence planning processes of SMPs. In short, what factors drive the trajectory of the evolution of the defence planning policies and force postures of SMPs? This includes but is not limited to the 'responsiveness, relevance, and expertise' involved in strategic analysis to inform the defence planning of SMPs.[38] It also applies to the portfolio of methods and tools employed that should be tailored to the purpose, the needs, as well as the budgetary means of actors in the context of a changing environment. After all, to accommodate change, the portfolio of analytical techniques and methods also needs to adapt, as do the existing processes of planning to novel circumstances.[39] This includes more reflexive approaches in which planning is conceived as an emerging and dynamic process instead of rational, top-down approaches,[40] and attention to the actual implementation of defence planning in terms of processes. Stephan de Spiegeleire et al. lament 'the policy formulation—policy implementation gap' and call for 'more experiential ("design") ways rather than . . . industrial-age bureaucratic ones'.[41]

In an 'ideal' rational analytical process, such as in theory practised in the US Quadrennial defence review process on a regular basis, a threat or the potential utility of a particular new strategy doctrine, tactic or technology would be evaluated in light of existing national interests, emerging security political priorities and the costs and complexity (the impact on the existing organisation) of a range of potential solutions assessed.[42] Solutions to security challenges include concluding alliances with new partners, pressuring existing alliance partners to strengthen their capabilities, enlarging inventories of existing military capabilities and developing and acquiring new technologies and weapon systems. The feasibility and suitability would then be assessed in terms of effectiveness, costs, domestic support and alliance politics. It is likely that such processes are different for SMPs.

Theme 2 therefore scrutinises the defence planning process of SMPs in greater detail. First, it examines what defence planning approaches SMPs use in the context of existing constraints, building on an overview of archetypal approaches as elaborated by Stephan Frühling, another contributor to this volume. Case study assessments distinguish between threat assessment ('to meet the risk of conflict in the present and near future, with one known and understood adversary'); mobilisation ('to meet the risk of conflict in the future, at an uncertain time and from a threat that is yet to develop'); portfolio based ('to plan for different configurations to address multiple risks' at different times); and task based ('the ability to conduct basic military tasks' in the context of uncertainty about future engagements).[43] It then assesses the ways in which defence planning processes are structured, looking at cultural factors, including the role of consensus, rivalry and authority and assessing the type of methods that are employed (tailored analytical process, outsourcing, acceptance of dilution/fair share, or reliance on an external template); techniques (copied from great powers, informal consensus, tailored methods, ad hoc); and decision-making procedures (analytical results, political leadership decides, military leadership and/or individual services decide, great power decides). On this basis, similarities and differences between the defence planning processes of the SMPs are identified.

### Theme 3: military innovation and adaptation

As Williamson Murray and Allan Reed Millet concluded in their survey of the dynamics of military innovation in the Interbellum, military innovation is most likely to succeed if both the political establishment and military organisation are interested in solving tangible strategic or operational problems, willing to experiment with potential solutions and open towards options even if these may significantly alter the (dominant) role of existing military capabilities, structures and practices as well as the roles of different services.[44]

In reality, as both Grissom and Horowitz have observed, such processes are affected by the fact that the military is not a monolithic organisation but a fragmented one consisting of a diverse set of services and branches, each with its own organisational interests, authority, responsibilities and outlook on pressing problems and potential solutions. Inter-service and intra-service rivalry not only influence the trajectory of a process (with one service attempting to block an innovation process advocated by another service because it will take up too large a share of the defence budget) but also colour the perception of threats and potential solutions to those threats.[45] The nature of the civil-military relationship—that is, the relationship between the senior military leadership on the one hand and the civilian leadership of the Ministry of Defence and the political leadership on the other—will shape the degree of autonomy of the military in planning processes. Implementation of innovation may falter if, for instance, responses to changes in the international security environment require significant additional financial expenditures. The introduction of novel military capabilities may also be deemed inappropriate in the domestic political arena due to existing strategic culture, or result in the establishment of novel military organisational structures.

Through this lens, the third theme explores how SMPs engage in military innovation and adaptation in times of rapid change and in the context of limited resources, financial and otherwise. These limited resources, in particular in combination with a growing number of threat scenarios and/or long periods of budget cuts, inevitably force them to make capability choices. This applies both to their planning departments which typically have more limited resources in terms of human capital and to their industrial base which will most likely not be sufficient to generate complete innovative capability packages that are cost-effective. Instead, SMPs are dependent on external suppliers, will rarely be first movers in the development of sophisticated heavy platforms and are likely to seek industrial relations with those external suppliers in the form of co-development, co-production or niche-product development. At the same time, innovation may also be inhibited by a preference to at least protect a small but struggling national military industry. This may drive decisions to not acquire state-of-the-art foreign-produced military systems but instead procure what capabilities the national industry can furnish. Furthermore, as SMPs' security outlook and defence policy development are shaped by more regionalised security outlooks alongside alliance politics and domestic strategic culture, capability development and acquisition decisions are often shaped by other factors. These include domestic politics, status and reputation within an alliance, interoperability concerns, affordability issues, as well as inter-service rivalry, as opposed to the single-minded pursuit of operational solutions to pressing security challenges informed by an appreciation of the emerging character of war and warfare.

To account for this, the various case studies in this volume assess the strategies deployed by SMPs to deal with challenges in keeping pace with technological advances; they analyse whether SMPs rely on indigenously produced technology by domestic industry, whether they establish technology transfer agreements, buy off the shelf, and/or partner with a great power, in the development and life-time support and maintenance of military platforms. Building on this, the case studies examine the nature and extent of military innovation pursued by SMPs using a framework developed by Michael Raska, one of the contributors to this book. The framework can be broken down into two dimensions: conceptual paths on the one hand and technological patterns on the other, with each consisting of three levels representing progressive degrees of military innovation through the magnitude of organisational change (see Textbox 1.1 and Figure 1.3 for a longer description).

---

**Textbox 1.1    Military Innovation: Conceptual Paths, Technological Patterns, Magnitude of Change**

**Conceptual Paths:**

- *emulation* (which amounts to 'importing of new tools and ways of war through imitation of other military organisations');
- *adaptation* (which involves a gradual change by 'adjustments of existing military means and methods') and
- *innovation* (in which military change is effectuated by 'developing novel military technologies, tactics, strategies, and structures').

**Technological Patterns:**

- *speculation* (in which a military actively explores 'new ways for solving existing operational problems or acknowledging the potential of emerging technologies');
- *experimentation* ('with new concepts, force structures, weapons technologies, and warfare methods'); or scale up to
- *implementation* (by adopting, adapting and refining 'selected experimental operational concepts, combat tactics, organisational force structures, or new generations of weapons systems and technologies').[46]

The extent or 'diffusion of military innovation' can be gauged by considering the magnitude of organisational change which can be seen as the product of the two dimensions of conceptual paths and technological patterns, again yielding a three-pronged typology to compare SMPs:

- *exploration* which is the pairing of speculation and emulation and largely involves attempts to develop novel areas of (technological) expertise while retaining existing force structures.
- *modernisation* is a slightly more advanced approach that follows from the pairing of experimentation and adaptation and pertains to the actual improvement of existing force structures.
- *transformation* is the most far-reaching element in which a military goes through steps of speculation and experimentation with novel technologies and/or modus operandi, which it actually implements, and thus achieves substantial innovation in the way its military operates.

---

Finally, the individual chapters also consider force development. While acknowledging that defence planning involves more than force development (including but not limited to financial management, personnel policies, outsourcing-insourcing-resourcing, quality of training and organisation, governance and oversight and

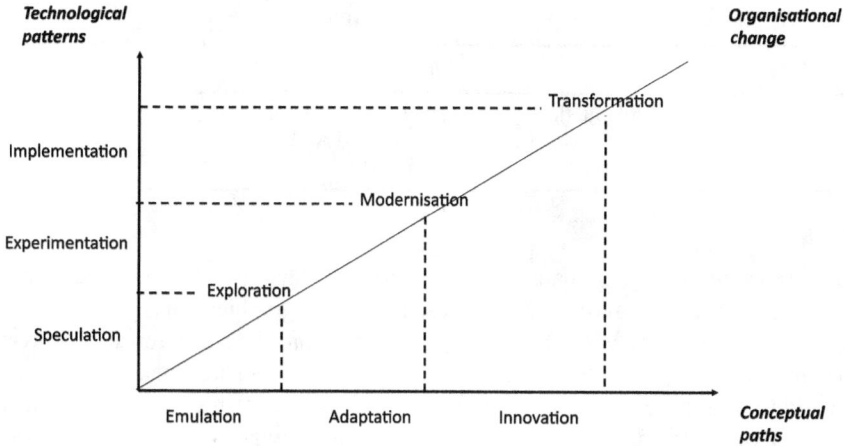

*Figure 1.3* Military innovation trajectories.

*Source:* Figure created by authors based on Michael Raska, Military Innovation of Small States—Creating a Reverse Asymmetry (Routledge, 2016)[47]

industrial relations), an overall focus on force development is warranted because it is an important reflection of policy choices concerning the types of missions the military forces of a nation should be preparing for, which threats are deemed relevant and where future operations are expected to occur. Force development also is the outcome of inter-service and intra-service organisational dynamics as it details the kind of investment priorities among the different branches. As such, it offers a vista into the institutional ability to learn and innovate and the organisational and political factors that drive or hamper that ability. Force development is therefore a real and tangible parameter to consider the actual outcome of practices and principles of the defence planning process. For this reason, it will recur across the different themes, as a common thread in that each chapter will reflect in the conclusion on the relationship between the different core themes even if it has not been parametrised in the structured focused comparison framework.

## Case study selection and structured focused comparison

The book deploys the structured focused comparison framework to conduct in-depth case studies of 11 SMPs. The cases included in this volume all represent SMPs that in their own way are active military actors, while none of them are major military powers. Within the category of SMPs, the book has deliberately sought to include the widest possible variation, in order to not only distil similarities—and in that differentiation from great powers—but also better understand differences amongst SMPs. The variation is first and foremost reflected in the various geographies of the case studies with the inclusion of SMPs from North America (Canada), Europe (Finland, the Netherlands, Slovakia), the Middle East (Israel, Oman, Turkey, the United Arab Emirates) and Asia and Oceania

*Table 1.1* Small and middle powers surveyed in the volume

| North America | Europe | Middle East | Asia and Oceania |
|---|---|---|---|
| Canada | Finland, the Netherlands, Slovakia | Israel, Oman, Turkey, the United Arab Emirates | Australia, Indonesia, Singapore |

(Australia, Indonesia, Singapore) (see Table 1.1). Thereby, the risk of extrapolating conclusions regarding SMPs based on a geographical bias is mitigated. Geographic diversity will, for example, also translate into differences in threat perspectives and potential cultural variables regarding decision-making practices. In addition to the geographical spread, the case studies also comprise a number of other variations, including diverging historical pathways and strategic culture, security environment, regime type, socio-economic development and embedment within alliances and in defence organisations' purchasing power (from the relatively large budgets of the Gulf States to the more modest spending power of a country such as Slovakia).

**Structure of the book**

The volume presents 11 case studies. Each chapter presents the results of the structured focused comparison approach based on the three themes and questions. The comparative analysis of the findings presented in the country case studies reflects on the trade-offs SMPs must incorporate in their strategic planning given the constraints they are under, the principal purposes of their armed forces in the context of the more general (non-)alignment strategies they pursue, the internal dynamics that shape defence planning and military innovation processes, the specific challenges that they grapple with, as well as the opportunities they can reap, the methods, techniques and decision-making procedures they employ in defence planning and the type of military innovation this results in. The concluding chapter of the book compares the similarities and differences found within the various case studies. On this basis, the conclusion offers suggestions for the re-conceptualisation of the actual practice of defence planning in SMPs themselves by outlining a three-pronged typology of the defence planning processes of SMPs.

As such, this book seeks to fill an important gap in the literature generating insights that are also of immediate relevance to defence planners. Most of the insights in the literature are based on and biased towards great power defence planning and military innovation. The knowledge base of defence planning and military innovation processes of SMPs is small in comparison. By mapping the principles and practices of such processes in SMPs, the book provides the empirical evidence base for scholars of SMP defence planning and military innovation to draw on. The book's structured comparative framework allows for the further development, refinement and testing of typologies and theories on a topic that has largely been ignored by the existing literature.

The book's findings will also prove useful to defence planners. In today's context of increasing interstate competition and rising defence spending, the need to better understand SMPs' defence planning is growing, from both a scholarly and a practitioner's perspective. The overarching security situation is getting more fluid, complex and less stable. Geopolitical change, military-technological developments and societal trends are bound to leave their impact on the ability of defence organisations of SMPs to execute their tasks. An in-depth look at the defence planning processes of SMPs helps to understand how well they are positioned to adjust to current and future changes taking place in their security environment.

## Notes

1  Hugo Meijer and Stephen G. Brooks, 'Illusions of Autonomy: Why Europe Cannot Provide for Its Security If the United States Pulls Back', *International Security* 45, no. 4 (20 April 2021): 7–43, https://doi.org/10.1162/isec_a_00405; Stephen G. Brooks and Hugo Meijer, 'Europe Cannot Defend Itself: The Challenge of Pooling Military Power', *Survival* 63, no. 1 (2 January 2021): 33–40, https://doi.org/10.1080/00396338.2021.1881251.
2  NATO Heads of State and Government, 'Vilnius Summit Communiqué', NATO, 11 July 2023, www.nato.int/cps/en/natohq/official_texts_217320.htm.
3  Mingjiang Li, 'The Belt and Road Initiative: Geo-Economics and Indo-Pacific Security Competition', *International Affairs* 96, no. 1 (1 January 2020): 169–87, https://doi.org/10.1093/ia/iiz240; TV Paul, 'When Balance of Power Meets Globalization: China, India and the Small States of South Asia', *Politics* 39, no. 1 (1 February 2019): 50–63, https://doi.org/10.1177/0263395718779930.
4  Lawrence Freedman, 'Strategy's Evangelist', *Naval War College Review* 74, no. 1 (25 February 2021): Article 4; Colin S. Gray, *Strategy and Defence Planning: Meeting the Challenge of Uncertainty* (Oxford: Oxford University Press, 2021), 4.
5  Paul K. Davis, 'Defense Planning When Major Changes Are Needed', *Defence Studies* 18, no. 3 (3 July 2018): 375, https://doi.org/10.1080/14702436.2018.1497444.
6  Magnus Håkenstad and Kristian Knus Larsen, 'Long-Term Defence Planning: A Comparative Study of Seven Countries', *Oslo Files on Defence and Security* 5, no. 1 (2012): 12, https://fhs.brage.unit.no/fhs-xmlui/handle/11250/99805.
7  Stephan Frühling, *Defence Planning and Uncertainty: Preparing for the next Asia-Pacific War*, Routledge Security in Asia Pacific Series 27 (Abingdon and New York: Routledge, 2014), chap. Introduction.
8  Henrik Breitenbauch and André Ken Jakobsson, 'Defence Planning as Strategic Fact: Introduction', *Defence Studies* 18, no. 3 (3 July 2018): 256, https://doi.org/10.1080/14702436.2018.1497443.
9  Stephan De Spiegeleire, 'Taking the Battle Upstream: Towards a Benchmarking Role for NATO' (Center for Technology and National Security Policy, Institute for National Strategic Studies, National Defense University, September 2012), 9, https://ndupress.ndu.edu/Portals/68/Documents/DefenseTechnologyPapers/DTP-098.pdf?ver=2017-06-22-143047-360.
10  D.N. Nelson argues, for instance, for 'defence de-planning', saying that 'less is more'. D.N. Nelson, 'Beyond Defence Planning', *Defence Studies* 1, no. 3 (1 September 2001): 28, https://doi.org/10.1080/714000042.
11  Theo Farrell and Terry Terriff, *The Sources of Military Change: Culture, Politics, Technology* (Lynne Rienner Publishers, 2002), www.rienner.com/title/The_Sources_of_Military_Change_Culture_Politics_Technology.; Michael Raska, *Military Innovation in Small States: Creating a Reverse Asymmetry*, Cass Military Studies (London: Routledge, Taylor & Francis Group, 2016).

12  See, for example, Emily O. Goldman and Richard B. Andres, 'Systemic Effects of Military Innovation and Diffusion', *Security Studies* 8, no. 4 (1 June 1999): 79–125, https://doi.org/10.1080/09636419908429387.

13  Michael C. Horowitz and Shira Pindyck, 'What Is a Military Innovation and Why It Matters', *Journal of Strategic Studies* 46, no. 1 (2 January 2023): 5, https://doi.org/10.1080/01402390.2022.2038572.

14  Adam Grissom, 'The Future of Military Innovation Studies', *Journal of Strategic Studies* 29, no. 5 (1 October 2006): 907, https://doi.org/10.1080/01402390600901067.

15  Stephen Peter Rosen, *Winning the Next War: Innovation and the Modern Military*, Cornell Studies in Security Affairs (Cornell University Press, 1991), www.jstor.org/stable/10.7591/j.ctv2n7k6j.

16  Seminal studies include Stephen Peter Rosen, 'New Ways of War: Understanding Military Innovation', *International Security* 13, no. 1 (1988): 134–68, https://doi.org/10.2307/2538898; Rosen, *Winning the Next War*; Deborah D. Avant, 'The Institutional Sources of Military Doctrine: Hegemons in Peripheral Wars', *International Studies Quarterly* 37, no. 4 (1993): 409–30, https://doi.org/10.2307/2600839.

17  Michael C. Horowitz, *The Diffusion of Military Power: Causes and Consequences for International Politics* (Princeton, NJ: Princeton University Press, 2010), chap. Preface, https://doi.org/10.2307/j.ctt7sqwd.

18  Williamson Murray and Allan R. Millett, 'Introduction', in *Military Innovation in the Interwar Period*, ed. Allan R. Millett and Williamson R. Murray (Cambridge: Cambridge University Press, 1996), 1–5, https://doi.org/10.1017/CBO9780511601019.001.

19  Terry Terrif, Frans Osinga, and Theo Farrell, eds., *A Transformation Gap?: American Innovations and European Military Change* (Stanford, CA: Stanford University Press, 2010).

20  Justin Kelly and Michael Brennan, 'Alien: How Operational Art Devoured Strategy', *Monographs, Collaborative Studies, & IRPs* 620, 1 September 2009, https://press.army-warcollege.edu/monographs/620.

21  Lawrence Freedman, 'The Meaning of Strategy: Part I: The Origin Story', *Texas National Security Review* 1, no. 1 (December 2017): 93; William H. McNeill, *The Pursuit of Power: Technology, Armed Force, and Society since A.D. 1000* (Chicago, IL: University of Chicago Press, 1984), 158, https://press.uchicago.edu/ucp/books/book/chicago/P/bo5975947.html; Martin van Creveld, *Supplying War: Logistics from Wallenstein to Patton* (Cambridge: Cambridge University Press, 1977).

22  Owen Cote, 'The Third Battle: Innovation in the U.S. Navy's Silent Cold War Struggle with Soviet Submarines', *The Newport Papers* 16 (1 January 2003): 7–10, https://digital-commons.usnwc.edu/newport-papers/38; Lawrence Freedman and Jeffrey Michaels, *The Evolution of Nuclear Strategy: New, Updated and Completely Revised* (London: Palgrave Macmillan UK, 2019), https://doi.org/10.1057/978-1-137-57350-6; Hal Brands and Evan Braden Montgomery, 'One War Is Not Enough: Strategy and Force Planning for Great-Power Competition', *Texas National Security Review* 2, no. 3 (2020), https://tnsr.org/2020/03/one-war-is-not-enough-strategy-and-force-planning-for-great-power-competition/.

23  Herman Kahn and Irwin Mann, 'Techniques of Systems Analysis' (Santa Monica, CA: RAND Corporation, 1 January 1956), www.rand.org/pubs/research_memoranda/RM1829-1.html; Charles J. Hitch and Roland N. McKean, *The Economics of Defense in the Nuclear Age* (Cambridge, MA: Harvard University Press, 1960).

24  Paul K. Davis et al., *New Challenges for Defence Planning: Rethinking How Much Is Enough*, ed. Paul K. Davis (Santa Monica, CA: RAND Corporation, 1994), www.rand.org/pubs/monograph_reports/MR400.html.

25  Benjamin M. Jensen, 'The Role of Ideas in Defence Planning: Revisiting the Revolution in Military Affairs', *Defence Studies* 18, no. 3 (3 July 2018): 302–17, https://doi.org/10.1080/14702436.2018.1497928.

26 Raphael S. Cohen, 'The History and Politics of Defence Reviews' (Santa Monica, CA: RAND Corporation, 25 April 2018), www.rand.org/pubs/research_reports/RR2278. html; Jordan Tama, 'Tradeoffs in Defense Strategic Planning: Lessons from the U.S. Quadrennial Defense Review', *Defence Studies* 18, no. 3 (3 July 2018): 282, https://doi. org/10.1080/14702436.2018.1497442.

27 Ben Barry, *Harsh Lessons: Iraq, Afghanistan and the Changing Character of War* (International Institute for Strategic Studies, 2016), www.iiss.org/publications/ adelphi/2017/harsh-lessons-iraq-afghanistan-and-the-changing-character-of-war/; Theo Farrell, James A. Russell, and Frans Osinga, eds., 'Conclusion: Military Adaptation and the War in Afghanistan', in *Military Adaptation in Afghanistan* (Stanford, CA: Stanford University Press, 2013); Frank G. Hoffman, *Mars Adapting: Military Change During War* (Annapolis, MD: Naval Institute Press, 2021).

28 See for instance Terrif, Osinga, and Farrell, *A Transformation Gap?*; Hugo Meijer and Marco Wyss, eds., *The Handbook of European Defence Policies and Armed Forces* (Oxford: Oxford University Press, 2018).

29 Jack S. Levy, *War in the Modern Great Power System: 1495–1975* (University Press of Kentucky, 1983), 10–19, https://doi.org/10.2307/j.ctt130jjmm.

30 Annette Baker Fox, *The Power of Small States: Diplomacy in World War II* (Chicago, IL: University of Chicago Press, 1959), https://press.uchicago.edu/ucp/books/book/ chicago/P/bo8928706.html; Robert L. Rothstein, *Alliances and Small Powers* (New York: Columbia University Press, 1968); Robert O. Keohane, '"Lilliputians" Dilemmas: Small States in International Politics', *International Organization* 23, no. 2 (1969): 291–310; David Vital, *The Inequality of States: A Study of the Small in International Relations* (Oxford: Clarendon Press, 1967); Trygve Mathisen, *The Functions of Small States in the Strategies of the Great Powers*, Scandinavian University Books (Oslo: Universitetsforlaget, 1971); Peter R. Baehr, 'Small States: A Tool for Analysis', ed. Edward E. Azar and Marshall R. Singer, *World Politics* 27, no. 3 (1975): 456–66, https://doi. org/10.2307/2010129; Håkan Wiberg, 'The Security of Small Nations: Challenges and Defences', *Journal of Peace Research* 24, no. 4 (1 December 1987): 339–63, https://doi. org/10.1177/002234338702400403; Jeanne A. K. Hey, *Small States in World Politics: Explaining Foreign Policy Behavior* (Boulder, CO: Lynne Rienner Publishers, 2003); Eduard Jordaan, 'The Concept of a Middle Power in International Relations: Distinguishing between Emerging and Traditional Middle Powers', *Politikon* 30, no. 1 (1 November 2003): 165–81, https://doi.org/10.1080/0258934032000147282; T. Sweijs, 'The Role of Small Powers in the Outbreak of Great Power War', Occasional Paper Series (The Centre of Small State Studies, The Institute of International Affairs, University of Island, 2010), www.semanticscholar.org/paper/The-role-of-small-powers-in-the-outbreak-of-great-Sweijs/2f2db7d8fedf06c829891b0a1e58e279343e741a; Sverrir Steinsson and Baldur Thorhallsson, 'Small State Foreign Policy', in *Oxford Research Encyclopedia of Politics*, ed. Cameron Thies (Oxford: Oxford University Press, 2017); Edström Håkan, Dennis Gyllensporre, and Jacob Westberg, *Military Strategy of Small States: Responding to External Shocks of the 21st Century* (Abingdon, OX: Routledge, 2019), www. routledge.com/Military-Strategy-of-Small-States-Responding-to-External-Shocks-of-the/Edstrom-Gyllensporre-Westberg/p/book/9780367529598; Håkan Edström and Jacob Westberg, *Military Strategy of Middle Powers: Competing for Security, Influence, and Status in the 21st Century* (Abingdon, OX: Routledge, 2020); Samuël Kruizinga, ed., *The Politics of Smallness in Modern Europe: Size, Identity and International Relations Since 1800* (London: Bloomsbury Academic, 2022); Meltem Müftüler Baç, 'Middle Power', *Encyclopedia Britannica*, 14 November 2023, www.britannica.com/topic/middle-power.

31 For a discussion, see Sweijs, 'The Role of Small Powers in the Outbreak of Great Power War', 3–5.

32 Thomas-Durell Young, 'Questioning the "Sanctity" of Long-Term Defence Planning as Practiced in Central and Eastern Europe', *Defence Studies* 18, no. 3 (3 July 2018): 358, https://doi.org/10.1080/14702436.2018.1497445.

33  Frühling, *Defence Planning and Uncertainty*, 19.
34  Alexander L. George and Andrew Bennett, *Case Studies and Theory Development in the Social Sciences* (Cambridge, MA: The MIT Press, 2005), 67, https://mitpress.mit.edu/9780262572224/case-studies-and-theory-development-in-the-social-sciences/.
35  Edström Håkan, Dennis Gyllensporre, and Jacob Westberg, *Military Strategy of Small States*; Eric J. Labs, 'Do Weak States Bandwagon?', *Security Studies* 1, no. 3 (1 March 1992): 383–416, https://doi.org/10.1080/09636419209347476; Erich Reiter and Heinz Gärtner, eds., *Small States and Alliances* (Heidelberg: Physica Heidelberg, 2001), http://link.springer.com/10.1007/978-3-662-13000-1; Tim Sweijs et al., 'Why Are Pivot States So Pivotal? The Role of Pivot States in Regional and Global Security' (The Hague: The Hague Centre for Strategic Studies, 9 July 2014), https://hcss.nl/wp-content/uploads/2014/07/Why_are_Pivot_States_so_Pivotal__The_Role_of_Pivot_States_in_Regional_and_Global_Security_C.pdf.
36  Adapted from Tim Sweijs et al., 'Why Are Pivot States So Pivotal? The Role of Pivot States in Regional and Global Security'. With permission from authors.
37  An exception is, for instance, a critical study of European NATO militaries concerning their success in implementing key elements of NATO's military transformation initiative. Terrif, Osinga, and Farrell, *A Transformation Gap?*
38  Michael Roi and Paul Dickson, 'Balancing Responsiveness, Relevance, and Expertise', in *Strategic Analysis in Support of International Policy Making: Case Studies in Achieving Analytical Relevance*, ed. Thomas Juneau (Washington, DC: Rowman & Littlefield, 2017).
39  Davis, 'Defence Planning When Major Changes Are Needed'; Henrik Breitenbauch and André Ken Jakobsson, 'Coda: Exploring Defence Planning in Future Research', *Defence Studies* 18, no. 3 (3 July 2018): 391, https://doi.org/10.1080/14702436.2018.1497447.
40  Magnus Christiansson, 'Defence Planning beyond Rationalism: The Third Offset Strategy as a Case of Metagovernance', *Defence Studies* 18, no. 3 (3 July 2018): 262–78, https://doi.org/10.1080/14702436.2017.1335581.
41  Stephan De Spiegeleire et al., 'Implementing Defence Policy: A Benchmark-"Lite"', *Defence & Security Analysis* 35, no. 1 (2 January 2019): 59, https://doi.org/10.1080/14751798.2019.1565365.
42  P. Liotta and Richmond Lloyd, 'From Here to There—The Strategy and Force Planning Framework', *Naval War College Review* 58, no. 2 (5 April 2018): 122–37.
43  Frühling, *Defence Planning and Uncertainty*, 3.
44  Williamson Murray and Allan Reed Millett, eds., *Military Innovation in the Interwar Period* (Cambridge and New York: Cambridge University Press, 1996), https://doi.org/10.1017/CBO9780511601019.
45  Horowitz, *The Diffusion of Military Power*; Grissom, 'The Future of Military Innovation Studies'.
46  Michael Raska, 'RMA Diffusion Paths and Patterns in South Korea's Military Modernisation', *Korean Journal of Defence Analysis* 23 (1 January 2011): 371–72; Raska, *Military Innovation in Small States*, 168–69. Raska draws inspiration from Thomas G. Mahnken, 'Uncovering Foreign Military Innovation', *Journal of Strategic Studies* 22, no. 4 (1 December 1999): 26–54, https://doi.org/10.1080/01402399908437768; Thomas Mahnken, 'Uncovering Foreign Military Innovation', *Journal of Strategic Studies* 22, no. 4 (1999); 26–54; Farrell and Terriff, *The Sources of Military Change*; Andrew L. Ross, 'On Military Innovation: Toward an Analytical Framework', *SITC* 2010, no. Policy Brief 1 (2010), https://escholarship.org/uc/item/3d0795p8.
47  Michael Raska, "A Structured-Phased Evolution: The Third Generation Force Transformation of the Singapore Armed Forces", in *Military Innovation in Small States: Creating a Reverse Asymmetry*, ed. M. Raska (New York: Routledge, 2016), 130–62.

## Bibliography

Avant, Deborah D. 'The Institutional Sources of Military Doctrine: Hegemons in Peripheral Wars'. *International Studies Quarterly* 37, no. 4 (1993): 409–30. https://doi.org/10.2307/2600839.

Baehr, Peter R. 'Small States: A Tool for Analysis'. Edited by Edward E. Azar and Marshall R. Singer. *World Politics* 27, no. 3 (1975): 456–66. https://doi.org/10.2307/2010129.

Barry, Ben. *Harsh Lessons: Iraq, Afghanistan and the Changing Character of War*. International Institute for Strategic Studies, 2016. www.iiss.org/publications/adelphi/2017/harsh-lessons-iraq-afghanistan-and-the-changing-character-of-war/.

Brands, Hal, and Evan Braden Montgomery. 'One War Is Not Enough: Strategy and Force Planning for Great-Power Competition'. *Texas National Security Review* 2, no. 3 (2020). https://tnsr.org/2020/03/one-war-is-not-enough-strategy-and-force-planning-for-great-power-competition/.

Breitenbauch, Henrik, and André Ken Jakobsson. 'Coda: Exploring Defence Planning in Future Research'. *Defence Studies* 18, no. 3 (3 July 2018a): 391–94. https://doi.org/10.1080/14702436.2018.1497447.

———. 'Defence Planning as Strategic Fact: Introduction'. *Defence Studies* 18, no. 3 (3 July 2018b): 253–61. https://doi.org/10.1080/14702436.2018.1497443.

Brooks, Stephen G., and Hugo Meijer. 'Europe Cannot Defend Itself: The Challenge of Pooling Military Power'. *Survival* 63, no. 1 (2 January 2021): 33–40. https://doi.org/10.1080/00396338.2021.1881251.

Christiansson, Magnus. 'Defence Planning beyond Rationalism: The Third Offset Strategy as a Case of Metagovernance'. *Defence Studies* 18, no. 3 (3 July 2018): 262–78. https://doi.org/10.1080/14702436.2017.1335581.

Cohen, Raphael S. 'The History and Politics of Defence Reviews'. Santa Monica, CA: RAND Corporation, 25 April 2018. www.rand.org/pubs/research_reports/RR2278.html.

Cote, Owen. 'The Third Battle: Innovation in the U.S. Navy's Silent Cold War Struggle with Soviet Submarines'. *The Newport Papers* 16 (1 January 2003). https://digital-commons.usnwc.edu/newport-papers/38.

Creveld, Martin van. *Supplying War: Logistics From Wallenstein to Patton*. Cambridge: Cambridge University Press, 1977.

Davis, Paul K. 'Defence Planning When Major Changes Are Needed'. *Defence Studies* 18, no. 3 (3 July 2018): 374–90. https://doi.org/10.1080/14702436.2018.1497444.

———, Richard Mesic, Charles T. Kelley, Christopher J. Bowie, Glenn Buchan, David Kassing, Bruce W. Bennett, et al. *New Challenges for Defence Planning: Rethinking How Much Is Enough*. Edited by Paul K. Davis. Santa Monica, CA: RAND Corporation, 1994. www.rand.org/pubs/monograph_reports/MR400.html.

De Spiegeleire, Stephan. *Taking the Battle Upstream: Towards a Benchmarking Role for NATO*. Centre for Technology and National Security Policy, Institute for National Strategic Studies, National Defence University, September 2012. https://ndupress.ndu.edu/Portals/68/Documents/DefenseTechnologyPapers/DTP-098.pdf?ver=2017-06-22-143047-360.

———, Karlijn Jans, Mischa Sibbel, Khrystyna Holynska, and Deborah Lassche. 'Implementing Defence Policy: A Benchmark-"Lite"'. *Defence & Security Analysis* 35, no. 1 (2 January 2019): 59–81. https://doi.org/10.1080/14751798.2019.1565365.

Edström, Håkan, Dennis Gyllensporre, and Jacob Westberg. *Military Strategy of Small States: Responding to External Shocks of the 21st Century*. Routledge, 2019. www.routledge.com/Military-Strategy-of-Small-States-Responding-to-External-Shocks-of-the/Edstrom-Gyllensporre-Westberg/p/book/9780367529598.

Edström, Håkan, and Jacob Westberg. *Military Strategy of Middle Powers: Competing for Security, Influence, and Status in the 21st Century*. Routledge, 2020.

Farrell, Theo, James A. Russell, and Frans Osinga, eds. 'Conclusion: Military Adaptation and the War in Afghanistan'. In *Military Adaptation in Afghanistan*. Stanford, CA: Stanford University Press, 2013.

Farrell, Theo, and Terry Terriff. *The Sources of Military Change: Culture, Politics, Technology*. Lynne Rienner Publishers, 2002. www.rienner.com/title/The_Sources_of_Military_Change_Culture_Politics_Technology.

Fox, Annette Baker. *The Power of Small States: Diplomacy in World War II*. Chicago, IL: University of Chicago Press, 1959. https://press.uchicago.edu/ucp/books/book/chicago/P/bo8928706.html.

Freedman, Lawrence. 'The Meaning of Strategy: Part I: The Origin Story'. *Texas National Security Review* 1, no. 1 (December 2017): 90–105.

———. 'Strategy's Evangelist'. *Naval War College Review* 74, no. 1 (25 February 2021): Article 4.

———, and Jeffrey Michaels. *The Evolution of Nuclear Strategy: New, Updated and Completely Revised*. London: Palgrave Macmillan, 2019. https://doi.org/10.1057/978-1-137-57350-6.

Frühling, Stephan. *Defence Planning and Uncertainty: Preparing for the Next Asia-Pacific War*. Routledge Security in Asia Pacific Series 27. Abingdon and New York: Routledge, 2014.

George, Alexander L., and Andrew Bennett. *Case Studies and Theory Development in the Social Sciences*. Cambridge, MA: The MIT Press, 2005. https://mitpress.mit.edu/9780262572224/case-studies-and-theory-development-in-the-social-sciences/.

Goldman, Emily O., and Richard B. Andres. 'Systemic Effects of Military Innovation and Diffusion'. *Security Studies* 8, no. 4 (1 June 1999): 79–125. https://doi.org/10.1080/09636419908429387.

Gray, Colin S. *Strategy and Defence Planning: Meeting the Challenge of Uncertainty*. Oxford: Oxford University Press, 2014.

Grissom, Adam. 'The Future of Military Innovation Studies'. *Journal of Strategic Studies* 29, no. 5 (1 October 2006): 905–34. https://doi.org/10.1080/01402390600901067.

Håkenstad, Magnus, and Kristian Knus Larsen. 'Long-Term Defence Planning: A Comparative Study of Seven Countries'. *Oslo Files on Defence and Security* 5, no. 1 (2012). https://fhs.brage.unit.no/fhs-xmlui/handle/11250/99805.

Hey, Jeanne A. K. *Small States in World Politics: Explaining Foreign Policy Behavior*. Boulder, CO: Lynne Rienner Publishers, 2003.

Hitch, Charles J., and Roland N. McKean. *The Economics of Defence in the Nuclear Age*. Cambridge, MA: Harvard University Press, 1960.

Hoffman, Frank G. *Mars Adapting: Military Change During War*. Annapolis, MD: Naval Institute Press, 2021.

Horowitz, Michael C. *The Diffusion of Military Power: Causes and Consequences for International Politics*. Princeton, NJ: Princeton University Press, 2010. https://doi.org/10.2307/j.ctt7sqwd.

———, and Shira Pindyck. 'What Is a Military Innovation and Why It Matters'. *Journal of Strategic Studies* 46, no. 1 (2 January 2023): 85–114. https://doi.org/10.1080/01402390.2022.2038572.

Jensen, Benjamin M. 'The Role of Ideas in Defence Planning: Revisiting the Revolution in Military Affairs'. *Defence Studies* 18, no. 3 (3 July 2018): 302–17. https://doi.org/10.1080/14702436.2018.1497928.

Jordaan, Eduard. 'The Concept of a Middle Power in International Relations: Distinguishing between Emerging and Traditional Middle Powers'. *Politikon* 30, no. 1 (1 November 2003): 165–81. https://doi.org/10.1080/0258934032000147282.

Kahn, Herman, and Irwin Mann. *Techniques of Systems Analysis*. Santa Monica, CA: RAND Corporation, 1 January 1956. www.rand.org/pubs/research_memoranda/RM1829-1.html.

Kelly, Justin, and Michael Brennan. 'Alien: How Operational Art Devoured Strategy'. Monographs, Collaborative Studies, & IRPs 620, 1 September 2009. https://press.armywarcollege.edu/monographs/620.

Keohane, Robert O. 'Lilliputians' Dilemmas: Small States in International Politics'. *International Organization* 23, no. 2 (1969): 291–310.

Kruizinga, Samuël, ed. *The Politics of Smallness in Modern Europe: Size, Identity and International Relations Since 1800.* London: Bloomsbury Academic, 2022.

Labs, Eric J. 'Do Weak States Bandwagon?' *Security Studies* 1, no. 3 (1 March 1992): 383–416. https://doi.org/10.1080/09636419209347476.

Levy, Jack S. *War in the Modern Great Power System: 1495–1975.* University Press of Kentucky, 1983. https://doi.org/10.2307/j.ctt130jjmm.

Li, Mingjiang. 'The Belt and Road Initiative: Geo-Economics and Indo-Pacific Security Competition'. *International Affairs* 96, no. 1 (1 January 2020): 169–87. https://doi.org/10.1093/ia/iiz240.

Liotta, P., and Richmond Lloyd. 'From Here to There—The Strategy and Force Planning Framework'. *Naval War College Review* 58, no. 2 (5 April 2018): 122–37.

Mahnken, Thomas G. 'Uncovering Foreign Military Innovation'. *Journal of Strategic Studies* 22, no. 4 (1 December 1999): 26–54. https://doi.org/10.1080/01402399908437768.

Mathisen, Trygve. *The Functions of Small States in the Strategies of the Great Powers.* Scandinavian University Books. Oslo: Universitetsforlaget, 1971.

McNeill, William H. *The Pursuit of Power: Technology, Armed Force, and Society Since A.D. 1000.* Chicago, IL: University of Chicago Press, 1984. https://press.uchicago.edu/ucp/books/book/chicago/P/bo5975947.html.

Meijer, Hugo, and Stephen G. Brooks. 'Illusions of Autonomy: Why Europe Cannot Provide for Its Security If the United States Pulls Back'. *International Security* 45, no. 4 (20 April 2021): 7–43. https://doi.org/10.1162/isec_a_00405.

Meijer, Hugo, and Marco Wyss, eds. *The Handbook of European Defence Policies and Armed Forces.* Oxford: Oxford University Press, 2018.

Müftüler Baç, Meltem. 'Middle Power'. *Encyclopedia Britannica.* 14 November 2023. www.britannica.com/topic/middle-power.

Murray, Williamson, and Allan Reed Millett. 'Introduction'. In *Military Innovation in the Interwar Period,* edited by Allan R. Millett and Williamson R. Murray, 1–5. Cambridge: Cambridge University Press, 1996a. https://doi.org/10.1017/CBO9780511601019.001.

———, eds. *Military Innovation in the Interwar Period.* Cambridge and New York: Cambridge University Press, 1996b. https://doi.org/10.1017/CBO9780511601019.

NATO Heads of State and Government participating in the meeting of the North Atlantic Council in Madrid 29 June 2022. 'Madrid Summit Declaration'. NATO, 29 June 2022. www.nato.int/cps/en/natohq/official_texts_196951.htm.

Nelson, D. N. 'Beyond Defence Planning'. *Defence Studies* 1, no. 3 (1 September 2001): 25–36. https://doi.org/10.1080/714000042.

Paul, T. V. 'When Balance of Power Meets Globalisation: China, India and the Small States of South Asia'. *Politics* 39, no. 1 (1 February 2019): 50–63. https://doi.org/10.1177/0263395718779930.

Raska, Michael. 'RMA Diffusion Paths and Patterns in South Korea's Military Modernisation'. *Korean Journal of Defence Analysis* 23 (1 January 2011): 369–85.

———. *Military Innovation in Small States: Creating a Reverse Asymmetry.* Cass Military Studies. London: Routledge, Taylor & Francis Group, 2016.

Reiter, Erich, and Heinz Gärtner, eds. *Small States and Alliances.* Heidelberg: Physica Heidelberg, 2001. http://link.springer.com/10.1007/978-3-662-13000-1.

Roi, Michael, and Paul Dickson. 'Balancing Responsiveness, Relevance, and Expertise'. In *Strategic Analysis in Support of International Policy Making: Case Studies in Achieving Analytical Relevance,* edited by Thomas Juneau. Washington, DC: Rowman & Littlefield, 2017.

Rosen, Stephen Peter. 'New Ways of War: Understanding Military Innovation'. *International Security* 13, no. 1 (1988): 134–68. https://doi.org/10.2307/2538898.

———. *Winning the Next War: Innovation and the Modern Military.* Cornell Studies in Security Affairs. Cornell University Press, 1991. www.jstor.org/stable/10.7591/j.ctv2n7k6j.

Ross, Andrew L. 'On Military Innovation: Toward an Analytical Framework'. *SITC* 2010, no. Policy Brief 1 (2010). https://escholarship.org/uc/item/3d0795p8.

Rothstein, Robert L. *Alliances and Small Powers*. New York: Columbia University Press, 1968.

Steinsson, Sverrir, and Baldur Thorhallsson. 'Small State Foreign Policy'. In *Oxford Research Encyclopedia of Politics*, edited by Cameron Thies. Oxford: Oxford University Press, 2017.

Sweijs, Tim. *The Role of Small Powers in the Outbreak of Great Power War*. Occasional Paper Series. The Centre of Small State Studies, The Institute of International Affairs, University of Island, 2010. www.semanticscholar.org/paper/The-role-of-small-powers-in-the-outbreak-of-great-Sweijs/2f2db7d8fedf06c829891b0a1e58e279343e741a.

———, Willem Theo Oosterveld, Emily Knowles, and Menno Schellekens. *Why Are Pivot States So Pivotal? The Role of Pivot States in Regional and Global Security*. The Hague: The Hague Centre for Strategic Studies, 9 July 2014. https://hcss.nl/wp-content/uploads/2014/07/Why_are_Pivot_States_so_Pivotal__The_Role_of_Pivot_States_in_Regional_and_Global_Security_C.pdf.

Tama, Jordan. 'Tradeoffs in Defence Strategic Planning: Lessons from the U.S. Quadrennial Defence Review'. *Defence Studies* 18, no. 3 (3 July 2018): 279–301. https://doi.org/10.1080/14702436.2018.1497442.

Terrif, Terry, Frans Osinga, and Theo Farrell, eds. *A Transformation Gap?: American Innovations and European Military Change*. Stanford, CA: Stanford University Press, 2010.

Vital, David. *The Inequality of States: A Study of the Small in International Relations*. Oxford: Clarendon P, 1967.

Wiberg, Håkan. 'The Security of Small Nations: Challenges and Defences'. *Journal of Peace Research* 24, no. 4 (1 December 1987): 339–63. https://doi.org/10.1177/002234338702400403.

Young, Thomas-Durell. 'Questioning the "Sanctity" of Long-Term Defence Planning as Practiced in Central and Eastern Europe'. *Defence Studies* 18, no. 3 (3 July 2018): 357–73. https://doi.org/10.1080/14702436.2018.1497445.

# 2 Australia

## The limits of pragmatism

*Andrew Carr and Stephan Frühling*

### Introduction: Australian defence planning across three major disruptions

As an archetypical 'middle power', Australia is in many ways an ideal case study for the defence planning of medium-sized countries.[1] A population of 25 million and a GDP that is the 13th largest in the world place it squarely amongst the 'middle' range of industrialised countries globally. Militarily and economically, Australia remains a significant power relative to other countries in Southeast Asia and the Southwest Pacific. Australia can also access significant external support—in the form of technology, materiel, intelligence and training—through its US alliance and historic links to the UK, which have had a profound effect on the experience of generations of Australian service personnel. But despite these close links, Australia is not integrated into institutional structures for defence planning with its partners.[2]

This chapter traces Australia's defence planning in terms of, first, alliances and national ambitions; second, defence planning processes and methods; and third, approaches to innovation and adaptation, through three distinct eras since the 1970s. Australia's area of defence interest is vast. It spans from Antarctica to the South China Sea and from the Indian Ocean to the island states of the Pacific. Australia is conscious that its security depends on a stable regional order that it needs to support, but it has also experienced the limits of allied support when the United States was not willing to support it with combat forces in conflicts and crises with its immediate neighbour, Indonesia. Hence, the question of where to focus Australia's defence effort, and how to structure for it, has been a perennial challenge. The need to focus on limited resources means that the 'disruption' that led to a significant change in Australian defence planning was always a combination of geostrategic change, with a change in how Australia itself conceives of its own security, its international role and defence priorities.

The early 1970s saw the effective end of the Cold War as the main focus of Australian defence policy. Instead, it sought to develop approaches and forces to defend its own continent against a regional threat, without relying on allied combat troops. By the mid-1980s, an external review of defence capabilities and the 1987 Defence White Paper established a policy framework that led to an era of relative stability for Australian defence planning. The Australian Defence Force (ADF) transformed

DOI: 10.4324/9781003398158-2

into a genuinely joint force designed to meet a limited threat in Australia's particular geographic circumstances. With a focus on territorial security and increasingly during the 1990s low-level regional contingencies such as peacekeeping and stability operations, Australia was less affected by the collapse of the Soviet Union than many countries in the Northern Hemisphere and seen as an exemplar of post–Cold War defence planning by some Australian and US commentators.[3]

This relatively stable framework was disrupted in the late 1990s by the combination of regional instability requiring major Australian deployments, especially the 1999 INTERFET (International Force East Timor) operation in East Timor, as well as the reverberations of the 9/11 and the 2002 Bali terrorist attacks and the subsequent global war on terror in Afghanistan and Iraq. Australia reconceived its security role in a more global, expeditionary and coalition-based approach, which meant that some elements of the ADF now had to emulate US capabilities closely so as to be able to seamlessly integrate Australian task forces into US-led operations.

By the late 2010s, this new framework was again disrupted by an increasingly urgent and singular focus on the possibility of conflict with China, as Beijing continued to expand its military to dominate Australia's wider region, became more aggressive under Xi Jinping and economically coerced Australia. Australia is now moving towards closer force posture integration with the United States in an explicitly balancing role within the Indo-Pacific, while seeking to transform the ADF into a force that can prevail in defending Australia in a high-intensity conflict against a major power.

This chapter argues that Australia's approach to defence planning across all three of these eras arose from a relatively pragmatic policymaking culture and has been defined and coordinated by sets of ideas and key concepts, rather than formal processes. Ideas such as 'Defence of Australia (or DoA)', 'core force', 'self-reliance' and 'low-level conflict' aligned efforts across the defence organisation and shaped internal arguments.[4] Australia never developed systematic, formal processes for force development as they exist in the United States or the NATO defence planning process. Instead, Australian governments have developed and published 'Defence White Papers' when the political leadership believed policy direction and defence capacity had moved too far out of alignment. These papers follow neither a standard process nor schedule and generally are accompanied by capability programmes that prevail until the next major review.

For many decades, this approach has sufficed because the Australian policymaking system is able to make pragmatic expert judgements, is relatively small and flexible and because Australia's strategic environment was stable, meaning that adjustments were only necessary to the margins of established policy traditions. As such, governments have generally allowed the system to manage and develop as it sees fit, with the political leash for force development only occasionally given a sharp tug at moments of strategic concern. But at the same time, an underlying concern with the nation's capacity for effective defence planning, especially to deal with major disruption, is also an enduring characteristic. The Department of Defence has faced over 35 major reviews since the 1970s.[5] Tellingly, at two crucial turning points—in 1985–1986 and 2022–2023—the Australian Government sought recommendations on force structure and defence planning from outside advisors rather than its most senior officials.

## The legacy of 'imperial defence' and 'forward defence'

Britain's Australian colonies federated in 1901 to form a Dominion within the British Commonwealth, but well into the 1950s Australian defence planning was not truly national. Emancipating the organisation of the armed forces from the force structures, doctrines, logistics and cultures from their British origins, was a decades-long process.[6] Australia is often portrayed during these Imperial (1901–1940s) and Forward Defence (1940s–1960s) eras as 'dependent'.[7] While there is some truth to this at moments—such as the confidence in the Singapore Strategy in the 1930s— this is not the case more broadly. The nation had a curious mix of independent conception of its strategic interests, and passivity in developing distinct institutional structures for achieving those interests. If it could work through or with partners, this was always the preference. Australia's challenge in the early Cold War was therefore not whether to join US-led containment efforts, but reconciling its own focus on Southeast Asia with the peripheral nature of that theatre to the United States.[8]

Well into the 1960s, the culture and doctrine of the Australian services reflected close cooperation with their British sister services. In the late 1950s, as it increasingly focused on Southeast Asia and the struggle with communism, Australia slowly began to adopt American military equipment and practices. This was a reluctant marriage in two ways. First, Australia was still keen to retain a significant British presence in Southeast Asia for as long as possible and the shift away from the tight 'kinship' with Britain to a more transactional US was challenging not only in strategic terms but also emotionally and culturally.[9] Second, the Americans proved difficult partners. The US Joint Chiefs of Staff remained firmly opposed to joint planning with Australia and New Zealand.[10] Although Australia would embrace the US-led Southeast Asian Treaty Organisation, move in 1957 to purchase American military equipment and briefly experiment with adopting the US 'Pentomic' structure for the Army, Australian and US defence planning proceeded in almost complete isolation.

Two patterns can be identified during this period, which are important for the arguments to come. First is a pragmatic culture of policy refinement and military adaption. Australian scholars have long pointed to an essential 'pragmatism' within the nation's strategic culture,[11] which also includes a relative disdain for intellectual leadership on the one hand and formal processes and structures on the other. A culture focused on tactical efforts and adaptation helped to distinguish Australian forces during conflicts such as Vietnam, where their contribution was ultimately not of strategic significance.[12]

Second, Australia was slow in developing an effective central strategic policy structure. Until 1972, each of the three military services was administered by their own department and service board and had their own Minister representing them in the Cabinet. The Department of Defence was a small affair that concentrated power in the Secretary as a chief advisor to the government but had little capacity to determine force structure priorities or even coordinate defence planning. Each military service has therefore powerfully shaped its role and capabilities, often with an eye to emulating larger US and UK peers,[13] with whom they had been in continuous combat operations for two decades from Korea, to the Malay Emergency, 'Confrontation' with Indonesia, to Vietnam. The realities of being a small country that relied extensively on the logistics of overseas partners for major operations limited the need for a coherent national approach to force design and sustainment.[14]

With the reduction of Cold War tensions in Asia in the early 1970s, Australia confronted the need to develop policy, planning approaches and forces for the defence of the continent itself. The country thus found itself, as a senior defence official commented in the mid-1980s, with traditions that raised 'organizational, structural and perceptual barriers to implementation of the defence of Australia itself', 'almost totally deficient in a capacity for strategic intelligence and assessment', 'commanders [who] were good tacticians, but by their roles were precluded from strategy', 'lacking in experience and knowledge of how to develop a national strategy that was more meaningful than shibboleths', 'no joint force doctrine' and no experience operating in and from Australia itself.[15] The way it confronted these challenges established traditions and planning cultures that persist to this day.

### National ambitions, alliances and dependencies

Australia's primary ambition for its military is to have a 'self-reliant' capacity for territorial self-defence. Self-reliance does not mean self-sufficiency, as Australia still assumes access to US non-combat support (such as logistics, resupply and intelligence).[16] But the ability to defend the continent without relying on US combat forces is explicit across Defence White Papers from the 1970s to the 2020s and born from the experience that allies either could not (such as Britain in 1941) or did not want to (such as the US in repeated crises with Indonesia) support Australia militarily. This goal provides a common understanding of the purpose of Australian defence planning, one that has bi-partisan political support and has been internalised across the public service and armed forces.

Its second ambition, and one whose form and significance shifts considerably over time, is the use of Australian forces to support regional and global order. This task flows partly from the nation's strategic culture, born of and comfortable within Western liberalism and partly from the expectations of its alliance with the United States. When these two ambitions come into tension—as they have at several times in Australian history—it is the former, territorial concerns, which usually takes priority in arguments about force structure, even as the defence organisation's attention is typically focused on the latter, supporting overseas operations.

#### *The 1980s: global balancing but regional autonomy: compensating for the limits of the US alliance*

The 1980s were the era of 'self-reliance within the alliance'. At the global level, Australia politically remained committed to the US alliance and the West. The end of the Vietnam War, the US Guam doctrine and the defeat of Communism in maritime Southeast Asia, had brought to an end both the strategic need and political support for military engagement overseas. The conservative Fraser government's 1976 White Paper enshrined both 'self-reliance' and 'Defence of Australia' as the key principles of Australian defence policy. Both were embraced by the Hawke Labor government, which sought to emancipate Australia from the last vestiges of its colonial relationship with the UK and shape Australia's economic and security future as a part of Asia. By the late 1980s, Australia started again to make contributions to overseas UN- and US-operations, but these forces were drawn from a force structure designed for the 'Defence of Australia'.[17] Australia made relatively minor contributions to UN operations in Namibia, Cambodia, Rwanda as well as the first Gulf War, which nonetheless were significant politically insofar as they demonstrated a return to overseas operations 15 years after the Vietnam War.

How to translate 'self-reliance' into force structure priorities however remained very contentious for many years. In 1986, the *Review of Australia's Defence Capabilities* led by Paul Dibb developed a threat-based planning framework, which required Australia to structure and posture its forces to meet the maximum level of effort Indonesia could mount against Australia from its existing forces, even if there was no political indication of such a conflict.[18] The 1987 Defence White Paper that followed established a capability programme consistent with the review that would essentially remain in place until the late 1990s. Key elements of this programme were the shift of one of Australia's three Army brigades to Darwin, completion of a series of 'bare base' airfields across the north, the development and deployment of an Over-the-Horizon Radar system and development of new strategic and operational headquarters for joint operations. While the F-111 and new Collins-class submarines provided potent strike forces, new armoured vehicles for the Army (Bushmaster) and new ANZAC (Australian and New Zealand Army Corps) frigates were relatively light forces commensurate with the potential threat from regional countries.

### The 9/11 era: hedging as a 'security provider'

In 1999, Australia unexpectedly found itself the reluctant leader of an international coalition to stabilise the newly independent country of East Timor, an operation that seriously stretched the capabilities of the ADF. At the same time, the conservative Howard Government was more inclined to align with the US and its partners in support of global security. The 2000 Defence White Paper took account of these developments by adopting a portfolio-based planning approach: in addition to being able to defend Australia, the Army was now required to be able to mount one large and one smaller concurrent stabilisation operation in Australia's region, while the Royal Australian Air Force (RAAF) and the Royal Australian Navy (RAN) had to be able to make contributions to US-led operations in the Asia-Pacific and beyond even in high-intensity conflicts where they would be exposed to adversary capabilities that were beyond those found in Australia's immediate neighbourhood.[19]

The 9/11 attacks, which Howard experienced while visiting Washington DC, and the 2002 Bali bombings, reinforced the ambition of the Howard government to reposition Australia as a key US alliance partner and a wider belief that the post–Cold War era was a very favourable regional order that Australia should seek to protect through political as well as militarily means. Australia assumed a greater regional 'security provider' role for stability in the South Pacific, including in leading the RAMSI (Regional Assistance Mission to the Solomon Islands) operation in the Solomon Islands and in counterterrorism efforts in Southeast Asia and contributed to operations in both Iraq and Afghanistan.[20]

But while concerns about US capacity and China's rise emerged during this period and Australia developed new security relationships, especially with Japan, it also resisted explicit balancing. Australia took a decidedly reluctant approach to the first iteration of the 'Quad', and to US suggestions for closer force structure cooperation in the Indo-Pacific. Despite close cooperation in the Middle East, the alliance with the United States remained bereft of political-military institutions or agreements focused on the Indo-Pacific.[21] 'Avoiding a choice' between the United States as Australia's main security partner, and China as its most important economic partner, was the dominating prism of policy and public debates on the country's strategic situation.

While major capability programmes during the 2000s and 2010s—such as the acquisition of JSF (Joint Strike Fighters), Air Warfare Destroyers and large amphibious ships—accorded with the plans of the 2000 White Paper, continuous deployments to the Middle East started to change the ADF in other ways. The army's plans for a 'hardened and networked army' heavy in armoured forces reflected very different priorities. Logistic and enabling capabilities that had been found wanting in Timor were again shrunk as overseas deployments benefited from coalition logistics. In that sense, in regard to acquisitions tied to operational requirements, such as for Special Forces in Afghanistan and Iraq, defence planning also assumed elements of capability-based planning.

### The 2020s: a balancing power in a more integrated alliance

By 2020, the increasingly assertive behaviour of China in the region and its economic and political coercion of Australia profoundly disrupted the assumptions underpinning Australian policy. Public opinion polls show a sharp rise in the perception of a security threat from China between 2018 and 2021.[22] Australia thus moved towards an increasingly explicit threat-based planning approach against China, in the context of an explicit balancing strategy alongside the United States, Japan and other like-minded countries.

In the 2020 Defence Strategic Update, the conservative Coalition government made explicit that Australia could not assume several years' warning before the major conflict.[23] The same year, Australia and the United States announced an agreement on a classified 'Statement of Principles on Alliance Defense Cooperation and Force Posture Priorities in the Indo-Pacific', with the aim to 'deter coercive acts and the use of force': the first ever mention of deterrence in the context of their alliance cooperation.[24] An initial practical sign of this increased cooperation was Australian-led contracts for infrastructure to host four tanker aircraft at its Tindal air base south of Darwin,[25] as well as US investment in military fuel storage in the port of Darwin.[26] In 2021, in addition to the AUKUS (Australia-United Kingdom-United States) partnership on nuclear submarines with the UK, Australia and the US also agreed to create 'a combined logistics, sustainment, and maintenance enterprise to support high-end warfighting and combined military operations in the region'.[27]

The new Labor government, in power since 2022, shared the security assessment of its predecessor and called for Australian forces to be 'interchangeable' with American ones.[28] The same year, Australia and the United States agreed to expand the Australian air base at Tindal to enable the hosting of six B-52 bombers,[29] and both countries announced plans for further joint enhancement of Australian bases, fuel and ordnance storage sites to enable operations by US air and land forces.[30] In 2023, this was followed by the announcement that as part of AUKUS, the US would establish a 'Submarine Rotational Force West' of up to four Virginia-class submarines in Perth from 2027.[31]

While Australia is increasingly contributing to regional balancing through closer force posture integration with the United States, its national planning too continues to shift towards a more threat-based approach. The new government's 2023 Defence Strategic Review (DSR) called for the pursuit of a 'focused force' comprising Anti-Access, Area-Denial systems and increased littoral manoeuvre for the army, and stated that 'The future Defence planning framework must be based on building force structure, force posture and accelerating preparedness on the basis of a net assessment planning process to ensure it is focused on the levels of risk in our current strategic circumstances'.[32]

## Approaches, processes, methods and techniques

That the 2023 DSR calls for 'a much more focused force structure based on net assessment . . . and realistic scenarios agreed to by the Government'[33] is notable because it implies a far more structured approach to force design than was traditionally the case.[34] Australia's defence planning approach historically has been defined by *ideas rather than processes*. It also functioned as a largely *pragmatic* system that is competent in operations and at meeting government policy direction, though not necessarily implementation. Both aspects have facilitated a planning culture that is relatively low cost and informal, with less systematic top-down control and formal processes and approaches to planning than in the United States or NATO. Hence, Australia never developed a process for whole-of-force design, or used scenarios to genuinely determine force structure priorities and drive capability development; capability scenarios established in the 2000s served to test forces and concepts at the margins but ultimately became arguments against rather than for significant change. In practice, the system became dependent on ad hoc 'White Paper' processes to establish defence planning priorities and overall policy guidance, and in the 2000s even for increasingly minor adjustments to investment plans.

### The 1980s: 'defence planning without a threat'

Australia entered the 1970s with a defence planning system that had existed since the 1940s. At its centre were 'Strategic Basis' papers, which were produced every three years by an inter-departmental 'Defence Committee', which included the Service Chiefs as well as Secretaries of the Departments of Defence, Foreign Affairs, Treasury and Prime Minister and Cabinet. Strategic Basis papers were submitted to Government for comment and endorsement and contained judgments about the strategic environment and broad guidance on defence policy principles and priorities.[35] Interpreting what these should mean for force structure was however largely up to the three Services. Despite the formal creation of the ADF as a joint force in the 1970s, the Services retained significant autonomy; there was no joint military strategy, and central policy areas in the Department of Defence were largely limited to challenging the merits of proposals put forward by the Services.[36]

By 1985, the Australian Defence organisation was thus riven by a virtual 'civil war' over how to interpret the guidance as set down in the 1983 Strategic Basis paper,[37] especially on how to interpret the capability implications of the concept of 'warning time'. Some senior civilian officials disputed the need for capabilities for more than constabulary operations. In contrast, senior military leaders argued for preparing for operations against forces more akin to those of the Soviet bloc than those that existed in the region. In the absence of a process that was able to resolve these differences and impose central direction, the services planned for very different scenarios and operations; the Army for mobilising into a multi-divisional force to defeat an adversary landing in force on the continent; the RAAF for long-range interdiction through strike, while the Navy remained focused on replacement of its light aircraft carrier and re-establishing a fixed air wing.[38]

In 1985, a new Defence Minister, Kim Beazley, commissioned an external review of defence capability priorities led by Paul Dibb, a professor at the Australian National University and long-time intelligence official, to break this internal logjam. With access to Defence files and personnel but completely separate from the Department, the landmark 1986 Dibb Report embraced the concepts of 'low-level' and 'escalated low-level conflict' tied to extant regional capabilities and a military strategy based on 'denial', as the key foundations for how Defence should plan for the 'Defence of Australia'. These concepts formed the basis for the 1987 Defence White Paper[39] and established a framework that remained in place until the end of the century.[40] The underlying approach to 'planning without a threat' was even seen by some as a model of defence planning for the post-Cold War era in Europe.[41]

But the White Paper also marked the end of the traditional approach to the development of defence policy guidance. The inter-departmental defence committee and Strategic Basis paper system faded away. Over the late 1980s and 1990s, the Chief of the Defence Force assumed increasing authority over the ADF and its services, including through the development of joint operational concepts and commands.[42] But even as—and perhaps because—the Department and ADF were focused on implementing the 1987 White Paper programme in challenging financial circumstances, no formal methodology for developing and evolving defence planning emerged to replace the old. Instead, from the early 1990s, in public debate but also within the department, those who saw a need to revisit policy and force structure priorities called for 'a new white paper' as the way to achieve this. Without a regular and established process to revise policy settings once set down by the government in a public White Paper, defence planning increasingly came to rely on ad hoc and direct intervention by the government to create new White Papers.

### The 2000s: pragmatism under strain

The 2000 White Paper, with its portfolio-based approach discussed earlier, added the ability to conduct expeditionary operations to the established framework for the 'Defence of Australia'. But how to balance between the two became increasingly contentious and difficult in the context of the post-9/11 war on terror. Many at the time, including a Parliamentary Defence Committee, argued for a new White Paper that would move away from a priority on the 'Defence of Australia', but they were unable to convince the Prime Minister.[43] Instead, the government published 'Defence Updates' in 2003 and 2005, whose narratives focused on global threats, terrorism and WMD (weapons of mass destruction) and made little attempt at reconciling these with the 2000 White Paper framework. While 'self-reliance' and 'Defence of Australia' still retained currency in policy statements, the intellectual coherence that they previously provided was lost in practice.

In addition, the ability of the organisation to maintain an institutional focus on force development eroded over time. In the late 1990s, the central Force Development and Analysis (FDA) division, which had traditionally provided the internal challenge of major capability proposals, was amalgamated with its acquisition-focused service counterparts to encourage greater 'jointery'.[44] However, this eliminated the only major area that was solely focused on the link between strategic guidance and overall force structure priorities and ultimately led to less institutional attention on force design and its links to policy. At the same time, senior commands proliferated with the establishment of a new three-star position of Chief of Joint Operations, and a three-star position for capability acquisition. While the Vice Chief (also three stars) retained responsibility for future force development, the proliferation of additional three-star positions arguably diffused both the responsibility and ability of senior leadership to focus on force structuring, at a time when the attention of the organisation was dominated by managing the heavy operational tempo of the time.

Hence, there was a widespread sense that earlier discipline in force structuring had been lost,[45] as the ADF sought to acquire, for example, armoured capabilities of a kind that were hard to reconcile with purported policy guidance.[46] In 2009, 2013 and 2016, new governments published three 'White Papers' in rapid succession, all of which were developed through ad hoc processes and with separate, parallel 'force structure reviews' in 2009 and 2016. All of them were in part justified by the argument that previous force structure plans had become unaffordable and could not be rectified without a new White Paper process. But given the ad hoc nature of both Force Structure Reviews, and in the absence of an institutional capacity (and arguably willingness) to internally challenge existing proposals, their impact on existing plans remained limited. Towards the end of this period, in 2015, an external 'First Principles Review' of the Department thus explicitly called for strengthening the 'strategic centre' within Defence, re-establishing internal contestability of equipment programmes and establishing a permanent force design capability to improve the effectiveness and coherence of Australian defence planning.[47]

### The 2020s: towards a new approach for Australian defence planning

While the reforms recommended by the First Principles Review were largely implemented,[48] questions remain about whether they have been sufficient to enable Defence to react in a new era where the speed of transformation is becoming a key challenge. Although the 2016 Defence White Paper emphasised that great power competition, rather than stabilisation operations or counterterrorism, should be the main focus of Defence, it was not infused with a clear sense of urgency. When the government launched the 2020 Defence Strategic Update, the rhetoric was notably different. Prime Minister Scott Morrison said, 'We have not seen the conflation of global economic and strategic uncertainty now being experienced here in Australia, in our region, since the existential threat we faced when the global and regional order collapsed in the 1930s and 1940s'.[49] However, the update itself conformed to the established practice of conducting a major, ad hoc review to update the capability plan. While it launched a number of new programmes, the discrepancy between the policy emphasis on no warning time of major conflict and a capability plan that remained dominated by major shipbuilding programmes, which would lead to increases in the fleet size decades later, remained.

Hence, the defence concerns of the new Labor government that entered office in 2022 were not the overall policy direction, but the alignment of the defence capability plan with that direction. Given the same concerns had been associated with Defence-led White Papers and Reviews in 2009, 2013, 2016 and 2020, it decided not to ask the defence organisation to prepare yet another White Paper, but—as Kim Beazley did in 1985—commissioned an external review. The 2023 DSR was conducted by a former Chief of the Defence Force, Angus Houston and former Defence Minister, Stephen Smith, who like Paul Dibb in 1986 had unfettered access to Defence documents and personnel, but provided a review directly to the minister.

The government accepted all the recommendations of the review, released in April 2023, which would lead to fundamental changes to Defence's approach to planning. A bi-annual 'National Defence Strategy' will be produced to provide 'a more holistic approach' and 'ensure strategic consistency and coordination of national policy implementation'.[50] The paper explicitly directs that Defence should develop 'a net assessment planning process' and, for the first time, base it on 'realistic scenarios agreed to by the Government'.[51] The form of Net Assessment established by the DSR is very different to the US approach, serving primarily as a vehicle for disciplining future force design.[52] If implemented, this would be a significant change that could strengthen the historically tenuous translation from strategic risks to force structure,[53] as it would open up Defence justification of force structure priorities to much greater direct government scrutiny than before.

The intent behind these changes to process represents a fundamental shift in Australia's approach to defence planning. The pragmatic, responsive and sometimes ad hoc approach of previous decades has been overturned. The DSR is an explicit recognition that neither broad concepts as in the relatively stable 1980s and 1990s, nor reliance on internal Defence pragmatism as in the 2000s and 2010s, are sufficient to ensure the speed and rigour required for this new era.[54] Hence, these changes will depend on significant cultural change and at the time of writing it is still too early to speculate on the likelihood of success.

## Military innovation and adaptation

A small local defence science and industry capacity has meant that Australia's technology and modernisation needs tend to draw on allied, especially US technology, organisations and examples. This is reinforced by service cultures that are drawn towards operations alongside major ally forces. Australia's main challenges for innovation and adaptation have thus always been to distinguish its own, and sometimes unique requirements in light of its geographic situation and national ambitions, from the pull exercised by those available from the United States. At home, it had access to a very limited technology and industry base, large parts of which passed to ownership by overseas companies in the early 2000s. This means Australia has variably pursued innovation, adaptation and emulation across the three eras discussed in this chapter.

Applying the 'Conceptualizing Military Innovation Trajectories' framework of Michael Raska, we find Australia traditionally and comfortably operates within the zone of modernisation, combining adaption and experimentation. Two consistent challenges for Australia, which at times have been achieved for select projects, are at location T1 to move up the framework, from modernisation into wider transformation and at location T2 to move across the framework from adaptation to genuine innovation. However, the particular challenges that Australia sought to address through these moves differed significantly with the change in strategic circumstances from the 1980s to the present. In Figure 2.1 we identify Australia's position within Michael Raska's 'Conceptualizing Military Innovation Trajectories' framework. While there is variety across the capabilities already implemented or under acquisition currently for the Australian Defence Force, a common experience can be identified. Australian technological challenges and capacity emphasise adaptation, importing military equipment, especially from the United Kingdom (prior to the 1950s) and United States (ever since) and then seeking to modify it to fit Australia's unusual geography and force structure needs. Australia has a capable, if small Defence Science and Technology organisation and a modest, though adept defence industry. Transformational efforts are rare, though the RAAF's wholescale approach to incorporating Fifth Generation F-35s is a notable example. In general, as in the case of the AUKUS Pillar two projects (see later), Australia seeks to obtain

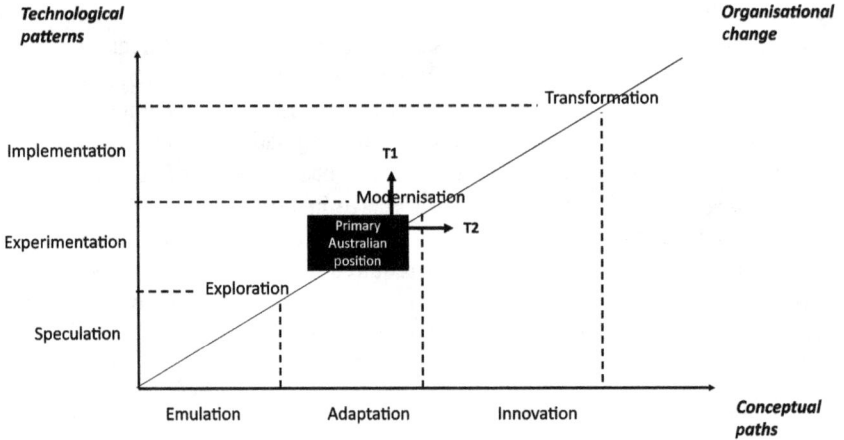

*Figure 2.1* Australia's military innovation trajectories.

*Source:* Figure created by authors based on Michael Raska, Military Innovation of Small States—Creating a Reverse Asymmetry (Routledge, 2016)[55]

the benefits of capabilities that have already been trialled and implemented elsewhere, or work in cooperation with partners to ensure that the costs and risks are shared rather than exclusive to Australia.

### The 1980s: adapting a force to moderate technological ambitions

In the 1970s and 1980s, the quality of Australia's forces easily outclassed those of any other regional country south of Japan and east of India. Contrary to most other countries, adapting the ADF's air and maritime forces to regional circumstances thus involved a deliberate moderation of capabilities; as aircraft and major warships were 'fitted for not with' their most potent armaments. For example, the ANZAC frigates originally were not equipped with Harpoon or SM-2 missiles, nor with sophisticated electronic self-defence systems. Following the 1986 Dibb Review, the Army also adapted the equipment of its medium brigades in particular to the particular geographic challenges of defending the North of the continent against notional, lightly armed incursions, which, for example, led to the development of road-mobile, mine-protected 'Bushmaster' vehicles capable of long-range self-deployment as well as tactical manoeuvre.

However, the defence of Australia also raised specific challenges that Australia had to develop new technology to address. Hence, Australia's defence research and technology effort during this period tended to focus on the adaptation, emulation or innovation of specific systems or sub-systems in light of Australia's particular geographic conditions. In some cases, such as anechoic tiles for submarines or radar warning receivers for Australia's fighter aircraft, Australia emulated US technology that its ally was, at the time, not willing to share.[56] To solve the particular challenge of maintaining surveillance of Australia's vast air and maritime approaches, Australia turned to genuine innovation. The prime example is the development of the Jindalee Over-the-Horizon radar system.[57] Under project Vigilare, Australia developed the ability to network a wide range of civilian sensors of opportunity into its military air surveillance system, leading to the development of particular excellence and expertise in radar track integration. A non-technical innovation was the establishment of Regional Surveillance Units, drawing on reservists from remote First Nations communities to maintain situational awareness in remote littoral regions.[58]

Overall, Australia's approach to innovation and adaptation during this era was thus deliberately top-down, with particular priorities directly derived from the policy of 'Defence of Australia' and the operational challenges that arose from it. The Defence Science and Technology Organization played a major role, in cooperation with industry partners. While this era also saw the corporatisation and then privatisation of the formerly government-owned defence industry, defence industry policy was driven primarily by economic considerations rather than its influence on the ability to innovate.

### The 2000s: operational demands drive bottom-up adaption

The 2000s were an era of significant change for Australian defence acquisition and research. The Howard government came to office with deep concerns about the costs and capability of domestic industry and acquisition. It had a strong free-market ethos and embraced ideas of 'new public management' that affected the entire public service and government policy towards science and industry.[59] The Howard government also encouraged the adoption of US technology and, in line with their cost-cutting concerns, an embrace of Military-off-the-Shelf acquisitions from overseas.

The force structure of the ADF did not change significantly during this period. The broader strategic framework, outlined in the 2000 Defence White Paper, retained a primary focus on defending the Australian continent as the core logic driving force acquisition. But given the rise of more capable regional adversaries, including but not only China, Australian modernisation started to entail acquiring air and naval systems of a similar level of capability as those operated by its allies; most notably the JSF; Air Warfare Destroyers with Aegis combat system and SM-2 missiles and through the implementation of a significant self-defence update to the ANZAC frigates. Australia also adopted the same combat system used in US submarines for its Collins-class boats, bought into the US wideband satellite programme, and acquired Growler electronic warfare aircraft.

In addition to this significant use of off-the-shelf purchases for modernisation, innovation also assumed a more bottom-up approach during this era; which was most pronounced in relation to the Army's need to meet a number of urgent operational requirements for ongoing conflicts in Afghanistan and the Middle East. However, two notable examples of unique Australian innovation in high technology also emerged. One is hypersonics research supported by the Defence Research and

Technology Organization at the University of Queensland, which received significant funding primarily because it was technology that Australia could contribute
to US technology research.[60] Second, the company CEA developed world-leading
solid-state radar technology, which was supported at critical stages by Navy leaders
who recognised the potential of the new technology (which was then implemented
on the ANZAC frigates).[61] While these are very different cases, what they have in
common is that given a much broader and sophisticated threat landscape facing the
ADF, the feasibility and merit of imposing narrow priorities from policy-first principles had eroded compared to the 1980s and 1990s. Hence, the 2016 White Paper
sought to acknowledge and encourage these developments through the establishment of an 'Innovation Hub' providing support to domestic industry innovation.

### The 2020s: becoming a first-tier technological power

While it is too early to fully assess the implications of the move towards a threat-
based policy and planning approach in the 2020s for Australian defence innovation, there are signs that the relatively 'laissez faire' approach to innovation
that characterised the 2000s and 2010s is starting to give way to a more focused
approach once more that reflects the operational challenges (and opportunities) that
arise from conflict with China and a more closely integrated US alliance. Today,
Australia's challenge is not to modernise by directly adapting or emulating US
technology but to work alongside the United States to foster innovation that can
be implemented by the ADF and US military at the same time. Given Australia's
limitations of size and resources as a middle power, this can only be pursued in
close cooperation with the US (and UK), and in specific areas of high priority and/
or particular Australian expertise.

The prime example of this new approach is represented by the AUKUS programme. Announced in September 2021, AUKUS has two distinct elements.
AUKUS 'Pillar One' is the decision to acquire a fleet of nuclear-powered submarines, using a mix of purchasing three to five Virginia class submarines from the
United States, and building a new design with the United Kingdom (in partnership with the US), beginning in the 2030s. This is a case of emulation, albeit one
that given the organisational, technological and regulatory challenges of creating a
nuclear-powered submarine force in a country with almost no extant nuclear reac-
tor expertise, will have a transformative effect across many parts of the Defence
organisation and beyond. AUKUS 'Pillar Two' involves a series of priority areas
for rapid development—hypersonics, counter-hypersonics, undersea warfare,
quantum, artificial intelligence and cyber—which the allies identified as crucial
technologies for future high-intensity conflict. Here, the ambition is for a close
integration of all three countries' research and technology efforts in order to accelerate the fielding of new technology.

Compared to the 1980s, however, this top-down identification of research priorities is complemented by the need for broader experimentation and ultimately
transformation of military forces for a major war against a technologically innovative great power. In this regard, the RAAF is probably the most advanced, as

it already operates one of the most modern fleets in the world and is thus seen as among the 'best small to medium-sized air forces in the world'.[62] The RAAF under 'Plan Jericho' has epitomised this, encouraging bottom-up initiative and rapid experimentation across the force. As a continuation of its efforts at modernisation, the RAAF sponsored the development of the 'Loyal Wingman' as an unmanned, attritable capability to accompany major fighter aircraft.[63] But while developed in Australia, the 'Loyal Wingman' is a Boeing programme. In the undersea warfare area, Australia similarly sponsored the US company Anduril to establish a presence in Australia to develop large unmanned underwater vehicles in the country.[64] Australia's ability to use Australian-based US companies that can tap into broader US technology networks or to tap directly into the US defence industry to build new relationships is thus a key enabler for Australian innovation in this new era: Australia defines defence industrial sovereignty in terms of the ability to undertake activities on Australian territory, not in terms of being International Traffic in Arms Regulations (ITAR)-free.

## Conclusion

In Figure 2.2 we identify Australia's reactions to systemic change, highlighting the most important factors. Australia's decision-making is driven by government—comprising both ministers and the Department of Defence—though calls for greater parliamentary involvement are emerging. Within the department, the military (via the Services) has often had a very significant influence. Constraints, primarily tied to financial and demographic limits, are acutely felt through this

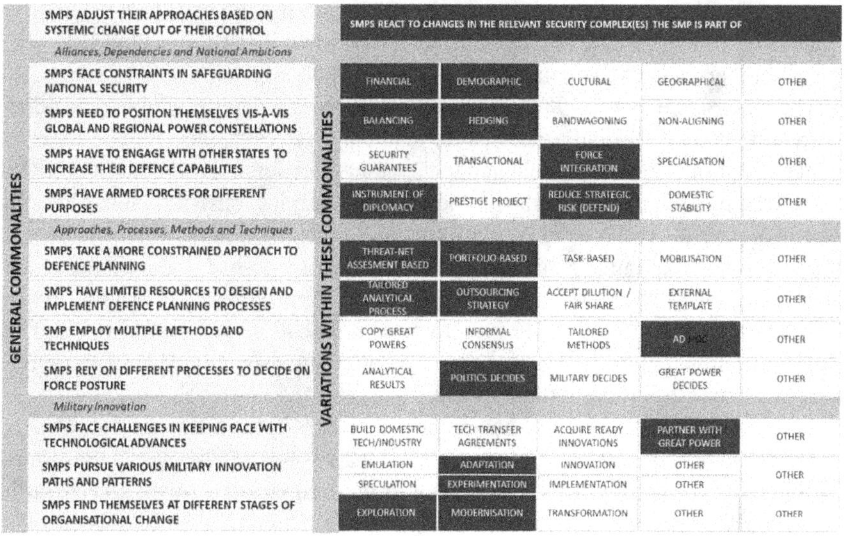

| | SMPS ADJUST THEIR APPROACHES BASED ON SYSTEMIC CHANGE OUT OF THEIR CONTROL | SMPS REACT TO CHANGES IN THE RELEVANT SECURITY COMPLEX(ES) THE SMP IS PART OF | | | | |
|---|---|---|---|---|---|---|
| | *Alliances, Dependencies and National Ambitions* | | | | | |
| | SMPS FACE CONSTRAINTS IN SAFEGUARDING NATIONAL SECURITY | FINANCIAL | DEMOGRAPHIC | CULTURAL | GEOGRAPHICAL | OTHER |
| | SMPS NEED TO POSITION THEMSELVES VIS-À-VIS GLOBAL AND REGIONAL POWER CONSTELLATIONS | BALANCING | HEDGING | BANDWAGONING | NON-ALIGNING | OTHER |
| | SMPS HAVE TO ENGAGE WITH OTHER STATES TO INCREASE THEIR DEFENCE CAPABILITIES | SECURITY GUARANTEES | TRANSACTIONAL | FORCE INTEGRATION | SPECIALISATION | OTHER |
| | SMPS HAVE ARMED FORCES FOR DIFFERENT PURPOSES | INSTRUMENT OF DIPLOMACY | PRESTIGE PROJECT | REDUCE STRATEGIC RISK (DEFEND) | DOMESTIC STABILITY | OTHER |
| | *Approaches, Processes, Methods and Techniques* | | | | | |
| | SMPS TAKE A MORE CONSTRAINED APPROACH TO DEFENCE PLANNING | THREAT-NET ASSESSMENT BASED | PORTFOLIO-BASED | TASK-BASED | MOBILISATION | OTHER |
| | SMPS HAVE LIMITED RESOURCES TO DESIGN AND IMPLEMENT DEFENCE PLANNING PROCESSES | TAILORED ANALYTICAL PROCESS | OUTSOURCING STRATEGY | ACCEPT DILUTION / FAIR SHARE | EXTERNAL TEMPLATE | OTHER |
| | SMP EMPLOY MULTIPLE METHODS AND TECHNIQUES | COPY GREAT POWERS | INFORMAL CONSENSUS | TAILORED METHODS | AD | OTHER |
| | SMPS RELY ON DIFFERENT PROCESSES TO DECIDE ON FORCE POSTURE | ANALYTICAL RESULTS | POLITICS DECIDES | MILITARY DECIDES | GREAT POWER DECIDES | OTHER |
| | *Military Innovation* | | | | | |
| | SMPS FACE CHALLENGES IN KEEPING PACE WITH TECHNOLOGICAL ADVANCES | BUILD DOMESTIC TECH/INDUSTRY | TECH TRANSFER AGREEMENTS | ACQUIRE READY INNOVATIONS | PARTNER WITH GREAT POWER | OTHER |
| | SMPS PURSUE VARIOUS MILITARY INNOVATION PATHS AND PATTERNS | EMULATION / SPECULATION | ADAPTATION / EXPERIMENTATION | INNOVATION / IMPLEMENTATION | OTHER / OTHER | OTHER |
| | SMPS FIND THEMSELVES AT DIFFERENT STAGES OF ORGANISATIONAL CHANGE | EXPLORATION | MODERNISATION | TRANSFORMATION | OTHER | OTHER |

(Left margin labels: GENERAL COMMONALITIES; VARIATIONS WITHIN THESE COMMONALITIES)

*Figure 2.2* Structured focused comparison framework for Australia across 1980s–2020s.

*Source:* Andrew Carr and Stephan Frühling

process. Australian governments portray their forces as primarily defensive—tied to hedging and, more recently, balancing in the Indo-Pacific and emphasise the contribution to diplomacy, despite a heavy operational tempo over the past half-century. Political primacy over choices in defence planning is thus clear, even if the underlying approaches have reflected an overall ad hoc approach. Australia has experienced significant change in its defence planning across 40 years, despite being located in a part of the world which has seen less radical and rapid geostrategic change than many areas in the Northern Hemisphere. The amount of change raises questions about whether the Australian case holds any generalisable lessons on approaches to defence planning by middle powers. Across all three periods, it was the theme of national ambitions, and especially the choices Australia made in response to the policies of its alliance partners, that set the challenges to which processes and innovation had to respond. This is as true of the 1970s when Australia responded to declining US interest and engagement in Southeast Asia after Vietnam as it was when Australia decided to support US engagement in the Middle East after 9/11 or to balance against China through closer integration with the United States today.

Although ad hoc, the 1987 White Paper framework was remarkably effective in providing a core set of ideas that helped bring coherence to both planning and innovation until the turn of the century. But in the 1980s and 1990s, when Australia defined its security concerns in a purely regional context, the moniker of 'middle power' may not be entirely appropriate: relative to the then relatively insubstantial air and maritime forces of Indonesia, let alone the island nations of the South Pacific, Australia was a great power that did not have to rely on direct support from others to ensure its own defence and security. Its struggles to develop a coherent defence planning framework in the early 1980s had more to do with the relative immaturity of its strategic and defence communities and institutions than the need to come to terms with the limitations of being a middle power.

The experience of the 1980s and 1990s also demonstrates that coherence does not require detailed formal processes for planning. But when the strategic environment and national ambitions changed in the 2000s and defence planning was pulled in many directions simultaneously, the limits of relying on pragmatism became readily apparent. At the level of platforms and systems, a pragmatic approach to innovation and making use of access to US technology helped to raise the capability of the ADF at the unit level. But given limited resources as a middle power, Australia struggled to reconcile the practical implications of its regional and its global ambitions, as concerns about the urgent challenges crowded out those that were more long-term, but ultimately more important.[65]

Given the strong reliance on government direction in the form of White Papers, and the reality that urgent commitments ultimately reflected government priorities of the day, this is not to say that more formalised processes would necessarily have helped achieve greater coherence in defence planning. But the fact remains that Australia failed to chew gum and walk at the same time and entered the 2020s with a force that, in the harsh judgment of the 2023 DSR, is 'not fit for purpose'.[66]

Today, the return to threat-based planning promises a return to greater focus in Australian defence planning, and the DSR is explicit about the need to increase the formalisation of planning processes to achieve this. But unlike the 1980s, Australia is also much more directly confronted with the limits of being a middle power in a region threatened by one of the world's two great powers. Hence, in national ambition and defence strategy, in its approach to innovation, and—at least in the context of force posture integration and the future submarine force—in its defence planning, the country is increasingly integrating into wider US alliance structures. The more Australia embraces that integration, it is quite possible that the overall coherence of the ADF force structure as Australian services may also, 50 years after the Vietnam War, lose some of its relevance as Australian services start operating more along their US counterparts than as part of a joint force once more.

## Notes

1 Andrew Carr, 'Is Australia a middle power? A systemic impact approach', *Australian Journal of International Affairs* 68, no. 1 (2014): 70–84; Sarah Teo, 'Middle Powers Amid Sino-U.S. Rivalry: Assessing the "Good Regional Citizenship" of Australia and Indonesia', *The Pacific Review* 35, no. 6 (2022): 1135–61.

2 Stephan Frühling, 'Is ANZUS Really an Alliance? Aligning the US and Australia', *Survival* 60, no. 5 (2018): 199–218.

3 Thomas-Durell Young, 'Capabilities-Based Defence Planning: The Australian Experience', *Armed Forces & Society* 21, no. 3 (1995): 349–69.

4 Stephan Frühling, 'The Fuzzy Limits of Self-Reliance: US Extended Deterrence and Australian Strategic Policy', *Australian Journal of International Affairs* 67, no. 1 (2013): 18–34.

5 Australian Department of Defence, *First Principles Review: Creating One Defence* (Canberra: Commonwealth of Australia, 2015), 13.

6 Albert Palazzo, *The Australian Army: A History of Its Organisation 1901–2001* (Melbourne: Oxford University Press, 2001); Stephan Frühling, ed., *A History of Australian Strategic Policy since 1945* (Canberra: Commonwealth of Australia, 2009).

7 Classic expressions of this view include Alan Renouf, *The Frightened Country* (London: Macmillan, 1979); Allan Gyngell, *Fear of Abandonment: Australia in the World since 1942* (Melbourne: La Trobe University Press, 2017).

8 Stephan Frühling, 'Australian Strategic Policy in the Global Context of the Cold War, 1945–65', in *Fighting Australia's Cold War: The Nexus of Strategy and Operations in a Multipolar Asia, 1945–1965*, ed. Peter Dean and Tristan Moss (Canberra: ANU Press, 2021), 11–34.

9 James Curran and Stuart Ward, *The Unknown Nation: Australia after Empire* (Melbourne: Melbourne University Press, 2010).

10 Dougal Robinson, 'A Sustained Tantrum: How the Joint Chiefs of Staff Shaped the ANZUS Treaty', *Australian Journal of International Affairs* 74, no. 5 (2020): 495–510.

11 Michael Evans, *The Tyranny of Dissonance: Australia's Strategic Culture and Way of War 1901–2005*, Land Warfare Study Paper no. 306 (Canberra: Land Warfare Studies Centre, 2005); Brendan Taylor, 'Is Australia's Indo-Pacific Strategy an Illusion?', *International Affairs* 96, no. 1 (2020): 95–109; Peter J. Dean, 'The Alliance, Australia's Strategic Culture and Way of War', in *Australia's American Alliance*, ed. Peter J. Dean, Stephan Frühling and Brendan Taylor (Melbourne: Melbourne University Press, 2016), 224–50.

12 Peter Edwards, *Australia and the Vietnam War* (Sydney: New South Publishing, 2014).

13  Although focused on the US military, the classic study of the distinct 'personalities' of military services is Carl H. Builder, *The Masks of War: American Military Styles in Strategy and Analysis* (Baltimore: RAND Corporation, The Johns Hopkins University Press, 1989).
14  Arthur Tange, *Defence Policy-Making: A Close Up View 1950–1980,* ed. Peter Edwards (Canberra: ANU Press, 2008); Eric Andrews, *The Department of Defence*, *The Australian Centenary History of Defence. Volume V* (Melbourne: Oxford University Press, 2001).
15  Alan Thompson, *Defence Down Under: Evolution and Revolution, 1971–88*, Working Paper (London: University of London Sir Robert Menzies Centre for Australian Studies, 1988), 5–6.
16  Stephan Frühling, 'Australian Defence Policy and the Concept of Self-Reliance', *Australian Journal of International Affairs* 68, no. 5 (2014): 531–47.
17  David Kilcullen, 'Australian Statecraft: The Challenge of Aligning Policy with Strategic Culture', *Security Challenges* 3, no. 4 (2007): 53.
18  Paul Dibb and Richard Brabin-Smith, 'Indonesia in Australian Defence Planning', *Security Challenges* 3, no. 4 (2007): 67–93; Stephan Frühling, *Defence Planning and Uncertainty* (New York: Routledge, 2014), 52–60.
19  Frühling, *Defence Planning and Uncertainty*, 98–106.
20  James Cotton and John Ravenhill (eds.), *Trading on Alliance Security: Australia in World Affairs, 2001–2005* (Oxford: Oxford University Press, 2007).
21  Frühling, 'Is ANZUS Really an Alliance?'.
22  Lowy Institute, *Lowy Institute Poll 2023: Security and Defence* (Sydney: Lowy Institute, 2023), https://poll.lowyinstitute.org/themes/security-and-defence (accessed 26 July 2023).
23  Australian Department of Defence, *Defence Strategic Update* (Canberra: Commonwealth of Australia, 2020), 14.
24  Australian Department of Foreign Affairs and Trade, 'Joint Statement: Australia-US Ministerial Consultations (AUSMIN) 2020', 28 July 2020, www.dfat.gov.au/geo/united-states-of-america/ausmin/joint-statement-ausmin-2020.
25  Australian Department of Defence, 'United States Force Posture Initiative Airfield Works', www.defence.gov.au/about/locations-property/infrastructure-projects/united-states-force-posture-initiatives-airfield-works (accessed 26 July 2023).
26  'East Arm Fuel Storage Works Gets Underway', *Australian Defence Magazine*, 28 January 2022, www.australiandefence.com.au/defence/estate/east-arm-fuel-storage-works-get-underway.
27  Australian Department of Foreign Affairs and Trade, 'Joint Statement: Australia-US Ministerial Consultations (AUSMIN) 2021', 16 September 2021, www.dfat.gov.au/geo/united-states-of-america/ausmin/joint-statement-australia-us-ministerial-consultations-ausmin-2021.
28  Richard Marles, 'Address: Center for Strategic and International Studies' (speech, Washington, DC, 12 July 2022), www.minister.defence.gov.au/speeches/2022-07-12/address-center-strategic-and-international-studies-csis.
29  Thomas Newdick, 'Australian Airbase Gets Upgrades for American Bomber Deployments', *The Drive*, 31 October 2022, www.thedrive.com/the-war-zone/australian-airbase-gets-upgrades-for-american-bomber-deployments.
30  Australian Department of Foreign Affairs and Trade, 'Joint Statement: Australia-US Ministerial Consultations (AUSMIN) 2022', 6 December 2022, www.dfat.gov.au/international-relations/joint-statement-consultations-australia-us-ministerial-consultations-ausmin-2022.
31  Australian Department of Defence, 'Submarine Rotational Force—West', www.defence.gov.au/about/taskforces/aukus/submarine-rotational-force-west#:~:text=From%20as%20early%20as%202027,West%20(SRF%2DWest).
32  Australian Department of Defence, *National Defence: Defence Strategic Review* (Canberra: Commonwealth of Australia, 2023), 50.

33 Australian Department of Defence, *National Defence*, 53.
34 Stephan Frühling, 'The Defence Strategic Review: A Revolution in Australian Defence Planning?', *The Strategist*, 26 April 2023, www.aspistrategist.org.au/the-defence-strategic-review-a-revolution-in-australian-defence-planning/.
35 Frühling, *A History of Australian Strategic*, 1–50.
36 Tange, *Defence Policy-Making*.
37 Andrews, *The Department of Defence*.
38 A deal was in place to acquire the HMS Invincible from the United Kingdom, though this was cancelled due to the Falklands War, and was abandoned by the incoming Hawke government in 1983.
39 Paul Dibb, *Review of Australia's Defence Capabilities* (Canberra: Commonwealth of Australia, Department of Defence, 1986); Australian Department of Defence, *The Defence of Australia* (Canberra: Commonwealth of Australia, 1987).
40 The one major addition to the 1987 framework was the concept of a 'capability edge' in the 1994 White Paper, which emerged to frame national ambitions with regards to technological superiority over slowly modernising Southeast Asian air and maritime forces.
41 Young, 'Capabilities-Based Defence Planning', 349–69.
42 David Horner, *Making the Australian Defence Force, the Australian Centenary History of Defence. Vol. IV* (Melbourne: Oxford University Press, 2001).
43 Joint Standing Committee on Foreign Affairs, Defence and Trade Defence Subcommittee, *Australia's Maritime Strategy* (Canberra: Commonwealth of Australia, 2004), 69.
44 Andrew Davies, *Let's Test That Idea—Contestability of Advice in the Department of Defence*, Policy Analysis no. 54 (Canberra: ASPI, 2010).
45 Paul Dibb, 'Is Strategic Geography Relevant to Australia's Current Defence Policy?', *Australian Journal of International Affairs* 60, no. 2 (2006): 248.
46 Davies, *Let's Test That Idea*.
47 Australian Department of Defence, *First Principles Review*.
48 Australian National Audit Office, 'Defence's Implementation of the First Principles Review', 17 April 2018, www.anao.gov.au/work/performance-audit/defences-implementation-the-first-principles-review.
49 Peter Hartcher, 'Scott Morrison Is Not Going to Duck This Crisis', *Sydney Morning Herald*, 4 July 2020, www.smh.com.au/national/scott-morrison-is-not-going-to-duck-this-crisis-20200703-p558w5.html
50 Australian Department of Defence, *National Defence*, 99.
51 Australian Department of Defence, *National Defence*, 50.
52 Andrew Carr, 'Australia's Archipelagic Deterrence', *Survival* 65, no. 4 (2023): 88.
53 Frühling, 'The Defence Strategic Review'.
54 While the Department of Defence has come under sustained, and increasing criticism for project management, especially for naval shipbuilding, the Australian political system also has often had a significant—and at times damaging—impact. For instance, though there was a need by 2000 to begin thinking about the next Australian submarine, no government decision to proceed was taken until 2009. In 2013 a new government, finding little clear structure in place, sought to acquire a Japanese build. However, domestic political instability led to a rushed competitive tender in 2014 with a new emphasis on a local build. A decision to acquire French submarines was finally taken in 2016. However in September 2021, Australia abandoned that contract, announcing it would pursue nuclear submarines, a capability that had been ruled out for political reasons over the past twenty-plus years. In 2023, details of the agreement were finally established, now involving two partners, a new design, building across two countries and requiring the Navy to staff up to three different classes of submarine.
55 Michael Raska, 'A Structured-Phased Evolution: The Third Generation Force Transformation of the Singapore Armed Forces', in *Military Innovation in Small States: Creating a Reverse Asymmetry*, ed. M. Raska (New York: Routledge, 2016), 130–62.

56  Although there was also an element of adaptation to these efforts: Australia's submarines operate in much warmer waters than the North Atlantic, for example, and Australia needed its radar warning receivers to also alert against Western-made aircraft and radars operated by Southeast Asian countries.
57  D.H. Sinnott, *The Development of the Over-the-Horizon Radar in Australia*, DSTO Bicentennial History series (Canberra: Commonwealth of Australia, 1988).
58  John Blaxland, *The Australian Army: From Whitlam to Howard* (Melbourne: Cambridge University Press, 2013).
59  Chris Aulich and Janine O'Flynn, 'John Howard: The Great Privatiser?', *Australian Journal of Political Science* 42, no. 2 (2007): 365–81.
60  Australian Department of Defence, 'Hypersonic Precinct to Supercharge Research', *Press Release*, 11 February 2022, www.defence.gov.au/news-events/news/2022-02-11/hypersonic-precinct-supercharge-research.
61  Max Blenkin, 'CEA and DST Group Sign Collaboration Agreement', *Australian Defence Business Review*, 15 February 2018, https://adbr.com.au/cea-and-dst-group-sign-collaboration-agreement/.
62  Robbin Laird, *Joint by Design: The Evolution of Australian Defence Strategy* (Pannsauken: BookBaby, 2021), 1.
63  Inder Singh Bisht, 'Australia Announces $317 Million Loyal Wingman Drone Investment', *The Defence Post*, 19 May 2022, www.thedefensepost.com/2022/05/19/australia-loyal-wingman-drone-investment/.
64  Julian Kerr, 'Anduril Progresses with Sydney R&D Facility', *Australian Defence Magazine*, 22 September 2022, www.australiandefence.com.au/defence/sea/anduril-progresses-with-sydney-randd-facility.
65  See, for instance, the critiques of Hugh White, former Deputy Secretary for Strategy in the Department of Defence and Emeritus Professor at the Australian National University, *Power Shift: Australia's Future Between Washington and Beijing*, Quarterly Essay, no. 39 (Collingwood: Blac Inc., 2010), 1–74.
66  Australian Department of Defence, *National Defence*, 53.

## Bibliography

Andrews, Eric. *The Department of Defence*, The Australian Centenary History of Defence. Vol. V (Melbourne: Oxford University Press, 2001).
Aulich, Chris, and Janine O'Flynn. 'John Howard: The Great Privatiser?', *Australian Journal of Political Science* 42, no. 2 (2007): 365–381.
Australian Department of Defence *The Defence of Australia* (Canberra: Commonwealth of Australia, 1987).
———. *First Principles Review: Creating One Defence* (Canberra: Commonwealth of Australia, 2015).
———. *Defence Strategic Update* (Canberra: Commonwealth of Australia, 2020).
———. 'Hypersonic precinct to supercharge research', *Press Release*, 11 February 2022, www.defence.gov.au/news-events/news/2022-02-11/hypersonic-precinct-supercharge-research.
———. *National Defence: Defence Strategic Review* (Canberra: Commonwealth of Australia, 2023).
———. 'United States force posture initiative airfield works', www.defence.gov.au/about/locations-property/infrastructure-projects/united-states-force-posture-initiatives-airfield-works (accessed 26 July 2023).
———. 'Submarine rotational force—West', www.defence.gov.au/about/taskforces/aukus/submarine-rotational-force-west#:~:text=From%20as%20early%20as%202027,West%20(SRF%2DWest).

Australian Department of Foreign Affairs and Trade. 'Joint Statement: Australia-US Ministerial Consultations (AUSMIN) 2020', 28 July 2020, www.dfat.gov.au/geo/united-states-of-america/ausmin/joint-statement-ausmin-2020.

———. 'Joint Statement: Australia-US Ministerial Consultations (AUSMIN) 2021', 16 September 2021, www.dfat.gov.au/geo/united-states-of-america/ausmin/joint-statement-australia-us-ministerial-consultations-ausmin-2021.

———. 'Joint Statement: Australia-US Ministerial Consultations (AUSMIN) 2022', 6 December 2022, www.dfat.gov.au/international-relations/joint-statement-australia-us-ministerial-consultations-ausmin-2022.

Australian National Audit Office. *Defence's Implementation of the First Principles Review*, 17 April 2018, www.anao.gov.au/work/performance-audit/defences-implementation-the-first-principles-review.

Bisht, Inder Singh. 'Australia Announces $317 Million Loyal Wingman Drone Investment', *The Defence Post*, 19 May 2022, www.thedefensepost.com/2022/05/19/australia-loyal-wingman-drone-investment/.

Blaxland, John. *The Australian Army: From Whitlam to Howard* (Melbourne: Cambridge University Press, 2013).

Blenkin, Max. 'CEA and DST Group Sign Collaboration Agreement', *Australian Defence Business Review*, 15 February 2018, https://adbr.com.au/cea-and-dst-group-sign-collaboration-agreement/.

Builder, Carl H. *The Masks of War: American Military Styles in Strategy and Analysis* (Baltimore, MD: RAND Corporation, Johns Hopkins University Press, 1989).

Carr, Andrew. 'Is Australia a Middle Power? A Systemic Impact Approach', *Australian Journal of International Affairs* 68, no. 1 (2014): 70–84.

———. 'Australia's Archipelagic Deterrence', *Survival*, 65, no. 4 (2023): 79–100.

Cotton, James, and John Ravenhill (eds.), *Trading on Alliance Security: Australia in World Affairs, 2001–2005* (Oxford: Oxford University Press, 2007).

Curran, James, and Stuart Ward. *The Unknown Nation: Australia after Empire* (Melbourne: Melbourne University Press, 2010).

Davies, Andrew. *Let's Test That Idea—Contestability of Advice in the Department of Defence*, Policy Analysis No. 54 (Canberra: ASPI, 2010).

Dean, Peter J. 'The Alliance, Australia's Strategic Culture and Way of War', in Peter J. Dean, Stephan Frühling and Brendan Taylor (eds.), *Australia's American Alliance* (Melbourne: Melbourne University Press, 2016), 224–250.

Dibb, Paul. *Review of Australia's Defence Capabilities* (Canberra: Commonwealth of Australia, Department of Defence, 1986).

———. 'Is Strategic Geography Relevant to Australia's Current Defence Policy?', *Australian Journal of International Affairs* 60, no. 2 (2006): 247–264.

———, and Richard Brabin-Smith. 'Indonesia in Australian Defence Planning', *Security Challenges* 3, no. 4 (2007): 67–93.

'East Arm fuel storage works gets underway', *Australian Defence Magazine*, 28 January 2022, www.australiandefence.com.au/defence/estate/east-arm-fuel-storage-works-get-underway.

Edwards, Peter. *Australia and the Vietnam War* (Sydney: NewSouth Publishing, 2014).

Evans, Michael. *The Tyranny of Dissonance: Australia's Strategic Culture and Way of War 1901–2005*, Land Warfare Study Paper no. 306 (Canberra: Land Warfare Studies Centre, 2005).

Frühling, Stephan, ed. *A History of Australian Strategic Policy Since 1945* (Canberra: Commonwealth of Australia, 2009).

———. 'The Fuzzy Limits of Self-Reliance: US Extended Deterrence and Australian Strategic Policy', *Australian Journal of International Affairs* 67, no. 1 (2013): 18–34.

———. 'Australian Defence Policy and the Concept of Self-Reliance', *Australian Journal of International Affairs* 68, no. 5 (2014a): 531–547.

————. *Defence Planning and Uncertainty* (New York: Routledge, 2014b).

————. 'Is ANZUS Really an Alliance? Aligning the US and Australia', *Survival* 60, no. 5 (2018): 199–218.

————. 'Australian Strategic Policy in the Global Context of the Cold War, 1945–65', in Peter Dean and Tristan Moss (eds.), *Fighting Australia's Cold War: The Nexus of Strategy and Operations in a Multipolar Asia, 1945–1965* (Canberra: ANU Press, 2021), 11–34.

————. 'The defence strategic review: A revolution in Australian defence planning?', *The Strategist*, 26 April 2023, www.aspistrategist.org.au/the-defence-strategic-review-a-revolution-in-australian-defence-planning/.

Gyngell, Allan. *Fear of Abandonment: Australia in the World Since 1942* (Melbourne: La Trobe University Press, 2017).

Hartcher, Peter. 'Scott Morrison Is Not Going to Duck This Crisis', *Sydney Morning Herald*, 4 July 2020, www.smh.com.au/national/scott-morrison-is-not-going-to-duck-this-crisis-20200703-p558w5.html.

Horner, David. *Making the Australian Defence Force*, The Australian Centenary History of Defence. Vol. IV (Melbourne: Oxford University Press, 2001).

Joint Standing Committee on Foreign Affairs, Defence and Trade Defence Subcommittee. *Australia's Maritime Strategy* (Canberra: Commonwealth of Australia, 2004).

Kerr, Julian. 'Anduril Progresses with Sydney R&D Facility', *Australian Defence Magazine*, 22 September 2022, www.australiandefence.com.au/defence/sea/anduril-progresses-with-sydney-randd-facility.

Kilcullen, David. 'Australian Statecraft: The Challenge of Aligning Policy with Strategic Culture', *Security Challenges* 3, no. 4 (2007): 45–65.

Laird, Robbin. *Joint By Design: The Evolution of Australian Defence Strategy* (Pannsauken: BookBaby, 2021).

Lowy Institute, *Lowy Institute Poll 2023: Security and Defence* (Sydney: Lowy Institute, 2023), https://poll.lowyinstitute.org/themes/security-and-defence (accessed 26 July 2023).

Marles, Richard. 'Address: Center for Strategic and International Studies' (speech, Washington D.C, 12 July 2022), www.minister.defence.gov.au/speeches/2022-07-12/address-center-strategic-and-international-studies-csis.

Newdick, Thomas. 'Australian airbase gets upgrades for American bomber deployments', *The Drive*, 31 October 2022, www.thedrive.com/the-war-zone/australian-airbase-gets-upgrades-for-american-bomber-deployments.

Palazzo, Albert. *The Australian Army: A History of Its Organisation 1901–2001* (Melbourne: Oxford University Press, 2001).

Renouf, Alan. *The Frightened Country* (London: Macmillan, 1979).

Robinson, Dougal. 'A Sustained Tantrum: How the Joint Chiefs of Staff Shaped the ANZUS Treaty', *Australian Journal of International Affairs* 74, no. 5 (2020): 495–510.

Sinnott, D. H. *The Development of the Over-the-Horizon Radar in Australia*, DSTO Bicentennial History Series (Canberra: Commonwealth of Australia, 1988).

Tange, Arthur. *Defence Policy-Making: A Close Up View 1950–1980*, ed. Peter Edwards (Canberra: ANU Press, 2008).

Taylor, Brendan. 'Is Australia's Indo-Pacific Strategy an Illusion?', *International Affairs* 96, no. 1 (2020): 95–109.

Teo, Sarah. 'Middle Powers Amid Sino-U.S. Rivalry: Assessing the "Good Regional Citizenship" of Australia and Indonesia', *The Pacific Review* 35, no. 6 (2022): 1135–1161.

Thompson, Alan. *Defence Down Under: Evolution and Revolution, 1971–88*, Working Paper (London: University of London Sir Robert Menzies Centre for Australian Studies, 1988).

White, Hugh. *Power Shift: Australia's Future between Washington and Beijing*, Quarterly Essay, no. 39 (Collingwood: Blac Inc, 2010).

Young, Thomas-Durell. 'Capabilities-Based Defence Planning: The Australian Experience', *Armed Forces & Society* 21, no. 3 (1995): 349–369.

# 3 The contours of Singapore's defence planning

## Rethinking deterrence, defence diplomacy, and resilience

*Michael Raska*

In the twenty-first century, when geopolitical uncertainties, economic disruptions, and technological advancements collide, Singapore's defence planners and organisations must prepare to meet complex security challenges.[1] The list of potential threats or 'disruptions' is growing and diverse, from countering violent regional and global terrorism, tackling disinformation and hybrid threats in the cyber domain, and responding to ideological and strategic competition between great powers to a range of acute global security challenges such as pandemics, climate change, and mega-disasters.[2] However, the challenge for small powers such as Singapore is not listing the threats that are likely foreseeable or how to prioritise them. The real problem is penetrating the fog of these ever-widening risks and assessing what matters most when three or four elements collide or converge. Traditional defence planning methods based on assessing the environment; identifying specific threats; formulating options; evaluating priorities; and selecting core objectives, strategies, and policies may no longer be sufficient for countries such as Singapore. This is because the converging spectrum of security challenges amplifies complexity that may limit the traditional intelligence bandwidth of small powers—or the capacity of government services to monitor the varying threats 24/7, connect the dots, and provide sound politico-military assessments to decision-makers. Hence, exploring alternatives to static threat assessments and linear response models is no longer an option, but a strategic necessity.[3]

Amid external geostrategic and technological disruptions coupled with internal demographic challenges, Singapore has therefore increasingly turned to emerging technologies such as artificial intelligence (AI) systems, augmented reality, and data-driven decision-making methods, which are redefining traditional defence planning approaches.[4] Singapore's AI systems and machine-learning algorithms are already helping to sort through vast amounts of data across various government and military applications,[5] including predictive maintenance taskings—determining who needs what and when, predictive analysis—identifying indicators to events rather than the events themselves, and in cyber-defences in which sophisticated algorithms detect malicious codes, recognise abnormal system patterns or behaviour, and provide automated threat detection and response.[6] At the same time, however, technology can't solve complex strategic and operational challenges alone. Therefore, Singapore's defence planning focuses on institutional agility through

DOI: 10.4324/9781003398158-3

collaborative security, intelligence, and defence networks, in which traditional organisational boundaries are being erased. Collaborative defence planning relies on diverse networks that can be linked in novel ways—military, cross-agency government collaboration, and increasingly private companies can share data, experiences, and best practices to tackle complex security challenges.[7] Over the past decade, strengthening a nodal resilience between the government, society, and technology has shaped the organisational agility in Singapore's defence planning approaches. The key enabler holding the various horizontal and vertical collaboration networks has depended on ramping up and maintaining internal trust and cohesion between Singapore's government agencies and society, and externally with international partners such as tech companies. The question going forward is whether these changes can effectively sustain Singapore's strategic edge— prevent potential strategic surprises and mitigate security risks and challenges while extending Singapore's deterrence, resilience, and defence.

Seen from this perspective, this chapter provides brief contours of the continuity and change in Singapore's defence planning amid strategic and technological disruptions over the past two decades. It begins by outlining the importance of Singapore's security constants that have shaped the country's defence assumptions since its inception: historical experience of vulnerability, lack of strategic depth, lack of natural resources, and asymmetries in demography. In the twenty-first century, these constants have been augmented by three main disruptions: the return of geopolitics and strategic competitions in the Indo-Pacific, the diffusion of emerging technologies and new domains of military rivalry, and the convergences or complexities of security challenges that combine a mix of 'high-low' warfare, which effectively challenge Singapore's traditional approaches to deterrence, defence, and resilience. The second part then highlights Singapore's ways and means to manage these 'disruptions' through technology, organisational adaptability, and conceptual innovation. In particular, the chapter argues that Singapore's defence planning strategies in the twenty-first century focus on 'niche hedging'—preparing for multiple possible futures by developing a portfolio of capabilities that would provide defensive options across a range of contingencies. The final part of the chapter will then assess Singapore's defence and military innovation trajectories, focusing on the ongoing 'Next-Gen' SAF 2040 transformation. Its military-technological paths include implementing digitisation, robotisation, and sensor revolution in warfare, while its main organisational changes stipulate a shift from adaptation to transformation, particularly with the establishment of the Digital and Intelligence Service (DIS)—a newly minted fourth branch of the SAF. Ultimately, the chapter argues that while three core pillars of Singapore's defence strategy remain unchanged—deterrence, defence diplomacy, and resilience—how these may be achieved in an increasingly dynamic, complex, and uncertain strategic environment is changing and is likely to change further.

## Singapore's defence planning: geostrategic constants and disruptions

Since its independence as a small city-state on August 9, 1965, Singapore has grappled with insecurity and geostrategic uncertainty conditioned by historical legacies, geography—lack of strategic depth and natural resources—and asymmetries in demography. Singapore's baseline geostrategic conditioning factor has been the *historical experience of vulnerability*—nearly an existential 'angst' narrative, originating from the traumatic experiences of the fall of Singapore under the British colonial rule in 1942 and subsequent Japanese occupation and wartime deprivation until 1945, to the experiences in attaining independent international status in 1965, followed by challenges in creating a robust defence capability and socio-economic stability in the immediate post-independence period. In this context, Singapore's independence has been portrayed as accidental, unwanted, and unexpected.[8] S. Rajaratnam, Singapore's first foreign minister, stated that an independent Singapore had a 'near-zero chances of survival—politically, economically, or militarily', while Lee Kuan Yew called the idea of an independent and separate Singapore 'a political, economic and geographical absurdity'.[9]

Apart from its historical sense of vulnerability, Singapore's security has been traditionally defined by its geographical disposition—location, size, lack of natural resources, and physical limitations of Singapore as a small island city-state of 719 square kilometres, the smallest state in Southeast Asia. On the one hand, Singapore's geographic location at the southern end of the Straits of Malacca crosses some of the most important Sea Lines of Communication (SLOCs) in the world, linking the Indian Ocean and the Pacific Ocean, which carry a major portion of the world's trade. The Malacca and Singapore Straits carry more than 40% of the world's commerce, half of the world's oil and 80% of oil bound for China and Japan.[10] Strategically, Singapore's position coupled with its deep natural harbour provides a hub for maritime trade in Southeast Asia, connecting trade routes between Asia, Europe, America, and the Middle East, significantly contributing to Singapore's socio-economic prosperity. On the other hand, however, Singapore's strategic location and absence of natural resources amplify its *extreme dependency* on the outside world. In this context, one could argue that Singapore's national security is not threatened as much by a single country but by a disruption of commerce. According to Huxley, 'Serious disruption of Singapore's physical links with the outside world would threaten not just its economic well-being: its very survival as an independent nation would be at stake'.[11] Singapore's security has been, therefore, inextricably linked to the profound strategic interests, involvement, and influence of major powers in the region with attendant geopolitical and economic risks. S. Rajaratnam, Singapore's first foreign minister, noted: 'We are valuable in war as a strategic asset. We are valuable in peace as a great commercial centre. So, whether it is peace or war, we cannot escape the consequences of whatever happens in the Pacific'.[12]

As a small city-state, Singapore also lacks strategic depth as well as natural resources for sustaining defence against protracted and determined surprise external aggression.[13] In other words, Singapore cannot trade space for time nor lose a single war. As Goh Chok Tong noted, 'the loss of one city would mean the loss of the whole nation'.[14] In times of war, the SAF cannot retreat to safe areas and consolidate before counter-attacking an enemy—an experience learned in February 1942 when the Imperial Japanese Army overwhelmed the island state's British-led defences. The lack of strategic depth may also decrease early warning time. Moreover, Singapore's proximity to potential regional flashpoints such as the

South China Sea and the risks of maritime trade interdiction or transnational crimes at sea, including piracy, play an important role in Singapore's security conceptions. Additional factors related to the lack of strategic depth, such as a high population density, proximity of military and vital installations to residential areas and lack of safe hinterland for the civilian population, have contributed to Singapore's 'strategic constants'—the narrative of critical vulnerabilities—for example, in case of stand-off, precision-strike attacks that would likely cause devastating destruction and casualties.[15]

Parallel to the historical legacies and geostrategic conditions of vulnerability, the narrative of Singapore's traditional security paradigm has pointed towards asymmetries in demographic factors. Facing two regional powers (and potential threat vectors) of Malaysia and Indonesia, predominantly Muslim-Malays with a combined population of around 312 million (2023), Singapore's population of 5.9 million (2023)[16] consisting of 3.61 million Singaporean citizens with a majority ethnic Chinese background (75%), 540,000 permanent residents, and 1.77 million foreign residents had evoked perennial perceptions of 'Chinese island in a Malay sea'.[17] Singapore's internal ethnic and religious complexities have historically reinforced the city-state's perceptions of internal vulnerability, mainly through the linkages between political developments in neighbouring countries and its potential spillovers into communal relations in Singapore.[18] In this context, Singapore's historical narrative often reiterates the traumatic experiences of the 1950 'Maria Hertogh' riots and the 1964 riots.[19] Externally, potential ethnic or cultural conflicts coupled with political internal instabilities in neighbouring countries could develop into armed conflicts that would threaten Singapore's economic assets. At the same time, Singapore's security conceptions have increasingly pointed to its internal demographic and population trajectories reflecting continuous low-level birth rates, an ageing population, and a smaller talent pool. In 2022, Singapore's birth rate reached a record low of 1.05.[20] If these trends continue, the SAF's manpower supply for its National Service (NS) is projected to decrease by one-third by 2030.[21]

Amid these geostrategic and resource constraints, therefore, Singapore's defence planners have been continuously searching for efficient and more productive use of its manpower resources, in terms of not only military expenditures and use of technology but also organisational learning, adaptation, and leadership experience necessary to maintain Singapore's meritocratic model of governance. While Singapore has experienced considerable limitations in balancing its security needs with policies directed at maintaining economic growth, on the other hand, the city-state has gradually developed considerable educational, economic, military-technological resources and human capital that have amplified Singapore's ability to shape its relations with neighbouring countries, sustain a credible defence posture, and strengthen its internal socio-economic stability. In this context, Singapore has devised a robust defence posture based on a comprehensive 'Total Defence Strategy' augmented by three mutually reinforcing pillars: deterrence, diplomacy, and resilience, which have progressively evolved amid shifts in Singapore's security environment and internal socio-economic and political challenges. The key objective of this strategy, however, has been constant—protecting Singapore's sovereignty, territorial integrity, interests, and freedom of action, while ensuring peace and stability in the region. If deterrence and defence diplomacy fail, enable a swift and decisive victory.[22]

Singapore's geostrategic context, historical legacies, and a mix of internal and external security challenges have subsequently shaped the trajectory of its strategic thought, foreign policy and defence planning, and overall force modernisation paths of the SAF.[23] Indeed, one could argue that Singapore's core strategic assumptions have been anchored around a remarkable continuity of its traditional security paradigm—the enduring narrative of geostrategic vulnerability—that has underscored the character of Singapore's evolving defence and foreign policies and continues to resonate to this day, particularly in Singapore's indicative desire to maintain its 'strategic edge' or credible deterrence capacity and freedom of action in the region. As Bilveer Singh noted, 'If lessons can be learnt from history, Singapore made a conscious effort to ensure that history will not repeat itself; that Singapore will not rely entirely on others for its own defence and will not be defenceless in the face of security threats'.[24] In this context, however, the principal challenge for Singapore's defence planners has reflected the same fundamental problem for over six decades: how to translate Singapore's limited resources of a small island nation into an effective defence capability amid the progressive range and complexity of security challenges?

Notwithstanding the seeming continuity of Singapore's security paradigm, its defence planning trajectories in the twenty-first century have been shifting by the confluence of multiple 'disruptive' strategic challenges and uncertainties. There are at least three mutually integrated 'disruptive' vectors: (1) changing geopolitics—planning for attendant consequences of growing Sino-U.S. strategic competition in the Indo-Pacific; (2) regional disparities in addressing unresolved historical legacies and tensions surrounding critically important geopolitical hotspots such as Taiwan, the Korean Peninsula, and the South China Sea; and (3) the introduction of emerging technologies as power projection capabilities in the region and changing forms of warfare, including the rise of cyber-enabled information conflicts and varying forms of 'hybrid warfare' (see Figure 3.1).

First, the resurgence of great power rivalries coupled with intensifying arms competition for advanced military technologies in the Indo-Pacific suggests that while wars and conflicts are not inevitable, neither are they inconceivable.[25] Accordingly, Singapore might be indirectly drawn into a conflict or crisis in the event of significant incidents deriving from a regional environment of 'troubled peace' or in a regional conflict over Taiwan or the South China Sea. While regional economic interdependencies have cushioned some of the effects of strategic rivalries, these have not decelerated a growing arms competition in East Asia or the 'arms dynamic' characterised by a mix of cooperative and competitive pressures, continued purchases of advanced weapon platforms, including the introduction of new types of arms and, therefore, unprecedented military capabilities.[26] These include the introduction of fifth-generation fighter jets, long-range precision strike capabilities, early warning, intelligence, surveillance and reconnaissance assets, and naval

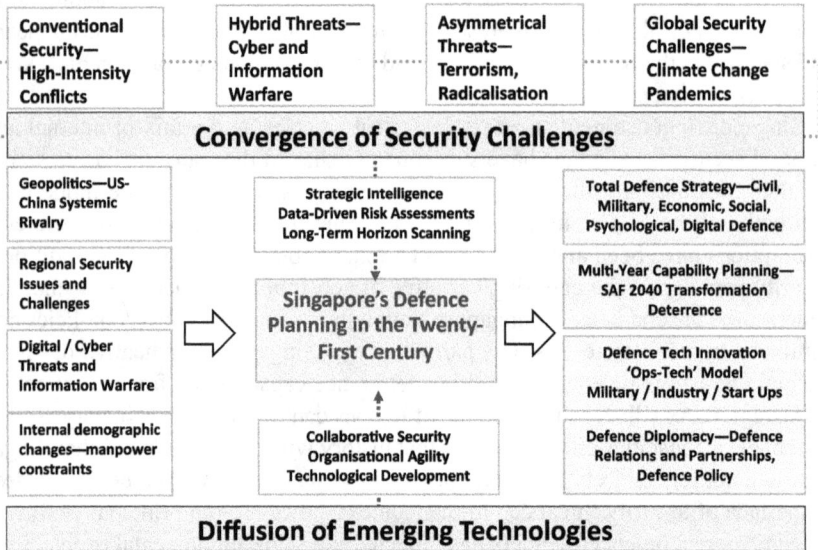

*Figure 3.1* Conceptualising Singapore's defence planning in the twenty-first century.

*Source:* Figure created by author based on this chapter's main argument

assets, including new types of surface combatants, submarines, and unmanned weapon systems.[27] These military technologies have become 'platforms of choice' in the Indo-Pacific as they enable regional powers to project presence, influence, and overcome their 'tyranny of geography'—or the traditional geopolitical entrapment of shared historical path dependencies. From China, Japan, Korea, Taiwan, to select countries in Southeast Asia, regional militaries have also been acquiring military technologies with greater lethality and accuracy at longer ranges.[28]

For Singapore, this means that military-technological gaps in the Indo-Pacific are effectively narrowing, and should a regional conflict occur, Singapore's security could be affected. At the same time, with the increasing convergence of 'old' and 'new' security threats, Singapore faces a competing strategic landscape and relative uncertainty about which types of adversaries and contingencies will be the most consequential. From terrorism to responding to Sino-U.S. strategic competition, the SAF faces competing operational requirements, which are likely to increase further. If so, Singapore's defence planners must answer how the SAF should build a force and doctrine capable of dealing simultaneously with current security threats while anticipating future challenges. This problem can be illustrated in the following examples and relevant questions that effectively challenge Singapore's security conceptions[29]:

- With the ongoing Sino-US strategic rivalries in the Indo-Pacific and potential conflict scenarios in the South China Sea, the Malacca Strait, Taiwan, and the East China Sea, Singapore's security has been increasingly affected by what used to be considered as 'rear areas' or its 'outer circle of defence'. What types of challenges does this present for Singapore and the SAF? How will the SAF operate in contested environments saturated with long-range missiles, submarines, and offensive space and cyberspace assets?

- To sustain its freedom of action and operational access into the South China Sea, the SAF must enhance the security, reliability, and integrity of its mission-critical C4ISR (Command, Control, Communications, Computers, Intelligence, Surveillance and Reconnaissance) systems as well as its combat support and logistics systems that may become increasingly vulnerable, including threats from electromagnetic pulse and high-powered microwave weapons. Disruption of these systems would likely result in cascading effects with ramifications on the entire SAF and its services to carry out operational missions.
- As military-technological capabilities in the Indo-Pacific narrow, the ability to strike with precision is becoming widely available. Given that Singapore lacks strategic depth, this raises prospects for hardening versus dispersion of SAF's assets. How will the SAF make existing platforms more survivable?
- While Singapore is not a member of the US-led regional security alliances, should the SAF invest in technologies that would support emerging US multi-domain operations and military technologies to ensure future interoperability? These may include, for example, developing offensive cyber and electromagnetic spectrum capabilities, AI-enabled air defences, unmanned ISR systems, and in the long term, more robust active defences such as missile defence.
- While certain traditional SAF missions will likely endure—that is, anti-piracy—Singapore is increasingly vulnerable to other emerging threats, particularly cyber-enabled political and hybrid warfare. As conflicts evolve parallel in the cyber, information, and cognitive domains, the value and more importantly, the accuracy and reliability of strategic information relevant to the situational awareness and function of the nation-state as a system will become even more critical. Consequently, how will the SAF prepare for next-generation hybrid conflicts?
- In a similar context, the character of conflicts in East Asia's hotspots, such as the South China Sea, may also likely reflect low-level 'grey-zone' conflicts in 'peripheral campaigns' rather than high-end missions—given the considerable escalatory risks. In the presence of 'little blue men' and fishing militias, how will the SAF redefine its rules of engagement when encountering such threats?
- Singapore is critically dependent upon imports of major weapons platforms and systems, particularly in the maritime and air domains. These must be acquired from foreign suppliers, that is, submarines, fighter aircraft, airborne early warning aircraft, drones, missiles, tanks, and various types of precision weapons. While the problem of sustaining supply chains and logistics is not new, what has changed is the sheer complexity of the many technological, economic, and financial interdependencies in arms production, procurement, and deliveries that create unforeseen risks of disruption. If so, how should the SAF mitigate potential vulnerabilities in defence supply chains?
- Future attacks could entail the use of swarming techniques of extremely large numbers (i.e. in the hundreds or even thousands) of small, relatively cheap but highly accurate air-launched precision-guided weapons. Together with the

proliferation of UAVs, especially mini- and micro-UAVs (many of which are becoming increasingly available on the commercial market), will provide adversaries more opportunities to gather intelligence on the SAF; size and stealth will likely make them more difficult to detect and counter, and if UAVs are deployed in swarms, they will be more difficult to eliminate.

- Can Singapore's conceptions of deterrence provide security guarantees against all escalation levels—from limited, low-intensity subversion of the rule of law to military provocations that may test the SAF's readiness? While there is a robust technological emphasis on developing SAF's future capabilities— these are inherently 'expected' to provide enhanced deterrence, escalation risk management, and decision space—it is unclear how Singapore's MINDEF and the SAF would manage deterrence and escalation in a 'cross-domain' operational environment—particularly in new domains of military rivalries such as space, underwater, cyber, and AI.

### Singapore's defence planning approaches, methods, and strategies

Singapore's defence planners have been responding to the 'disruptive' convergence of contending geopolitics, multidomain conflicts, and advanced military-technological diffusion in the Indo-Pacific by prioritising technological innovation, organisational agility, societal cohesion, and resilience while pursuing external support through strategic partnerships and defence diplomacy. Accordingly, Singapore's defence planning approaches, methods, and strategies must be viewed in a much broader and *comprehensive* framework, starting with the concept of *'Total Defence (TD)*. The strategy was introduced in 1984 to unite all sectors of society— government, business, and people—and to enhance Singapore's resolve, unity, and defence capabilities.[30] Applicable for both peacetime and wartime, TD emerged from Singapore's benchmarking of similar strategies in Sweden and Switzerland, countries with relatively small populations and some form of national military service.[31] Initially consisting of five and now six mutually supportive defence pillars: military, civil, economic, social, psychological, and digital, TD has evolved over the past four decades and essentially become a symbol of Singapore's security identity, resilience and adaptability, including its conscription-based National Service.[32]

In brief, the *Civil Defence* pillar provided an ethnically diverse corps of professionals and national servicemen in the Singapore Civil Defence Force (SCDF) that is responsible for firefighting, rescue operations, and community preparedness through exercises and education programmes, and incorporating smart technologies for disaster response. In parallel, Singapore has continuously developed programmes and invested resources in the twin pillars of *Social and Psychological Defence* to foster racial and religious cohesion, maintaining a strong sense of identity, and societal resilience—against varying and evolving threats such as terrorism, online hate speech and disinformation, as well as mental health challenges. For example, *Social Defence* aims to prevent any exploitation of ethnic unrest by stressing mutual co-existence, cohesion, and harmony based on multicultural consensus and community-building regardless of race, language, and religion. In *Psychological Defence*, the Singapore Government has focused on an array of educational campaigns such as National Education Programmes carried out in schools to strengthen Singapore's citizens' collective will and commitment to defend the country. *Economic Defence* recognises that military power depends on economic strength and vice versa. Without a strong economy, the costs of creating and maintaining an effective military capability would be too high. Hence, Singapore must sustain economic resilience, for example, through supporting diverse trade and supply chains, investment partners, safeguarding critical infrastructure, and promoting economic sustainability. In times of crisis or war, *Economic Defence* also means contingency planning for converting civilian resources, technological skills, and capital investments for the military while ensuring the continuity and function of the economy when these resources are mobilised for war. Ultimately, Singapore's *Military Defence* implies maintaining credible and capable deterrent forces—the SAF—and the development of an indigenous defence industrial base that meets SAF's military-technological requirements.[33] In 2019, Singapore added *Digital Defence* as the sixth pillar to protect and defend Singapore's critical infrastructure and systems against growing cyber and information warfare threats, promoting responsible online behaviour, and developing cybersecurity expertise.[34]

While it would be beyond the scope of this chapter to project the defence planning dynamics for each pillar of the *Total Defence*, including its organisational structures, and strategic, operational, and tactical considerations, it is possible to focus on one of its pillars—*Military Defence*—and ascertain Singapore's approaches and methods to military defence amid continuity and change in its internal institutional and organisational path dependencies and external changes. In particular, Singapore's military defence planning reflects a robust, comprehensive, and adaptable defence system with several distinct features: (1) long-term strategic focus, planning and foresight, backed by consistent defence resource allocation at around 2% of GDP; (2) adaptive capability development through '*Ops-Tech*' defence innovation approach that links defence technology users (military), developers (science and technology base), and producers (defence industry); (3) defence policy, diplomacy, and regional defence cooperation including active engagement in regional and international security dialogues and forums, participating in joint multilateral exercises and disaster relief operations; (4) National Service conscription system coupled with national education and media campaigns to raise awareness about diverse security threats and the importance of national resilience.

To begin with, Singapore's defence planning relies on long-term *strategic foresight*—a combination of long-term horizon scanning risk/net assessment methodologies augmented by strategic intelligence, data collection, and analysis. Singapore's government has dedicated units for scenario planning, horizon scanning, and threat anticipation, integrating diverse functional expertise, sources, and perspectives into the planning processes. The aim is to identify and monitor potential threats or disruptions and deliver potential solutions and policy options in advance of need. Comparative benchmarking, access to varying intelligence sources, scenario planning, and Singapore's global networks augment defence policymakers to detect change and prompt debates on the validity of established approaches, processes, and paradigms. In this context, nearly all strategically significant agencies, including MINDEF with its Strategic and Futures Group, have dedicated foresight teams. For example, the Centre for Strategic Futures—a Strategy Group in the Prime Minister's Office, established in 2009—provides the highest levels of Singapore government with the ability to 'navigate emerging strategic challenges and harness potential opportunities' in multiple ways: building capacity and providing strategic foresight and risk management by training public servants; doing strategic foresight work—gathering insights on emerging trends and identifying signals of change; and by communicating and disseminating insights to decision- and policymakers across all of government.[35] While navigating uncertain futures and forecasting future threats in a hyperconnected world might be increasingly challenging, if not impossible, its value for Singapore's military defence planning process is its long-term strategic orientation and the need to prepare for potential unforeseen contingencies.

Long-term strategic defence planning stipulates the need for consistent defence resource allocation. Over the past decade, Singapore's defence spending has consistently hovered around 2.7% of GDP, which translates to approximately SG$17.9 billion (US$13.4 billion) in 2023.[36] The majority of the expenditures—about 88% is directed towards operating expenditures of the SAF, covering personnel, training and exercises, maintenance, procurement, and acquisitions. The remaining expenditures cover research and development in future technologies such as robotics, AI, and unmanned vehicles for military applications; infrastructure projects such as construction and improvement of military bases, training facilities, and critical infrastructure upgrades, and talent development.[37] While specific allocations within each category may vary depending on the annual budget, strategic and operational priorities, Singapore's defence spending has over the years shown remarkable consistency, particularly in maintaining a balance between the imperatives of developing technologically advanced SAF, the need for sustaining Singapore's socio-economic growth and economic competitiveness, and regional security perceptions. In other words, mitigating potential tensions between military and socio-economic priorities—particularly for a small power with limited defence resources—requires considerable long-term planning and consistent resource allocation.

In this context, Singapore's Ministry of Defence (MINDEF) employs a multi-year capability planning framework (see Figure 3.2). This includes multi-year strategic plans that define the direction and character of the SAF's military modernisation, such as the 'SAF 2040' roadmap and its capability requirements,

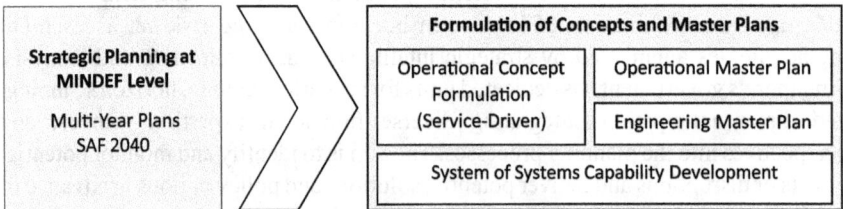

| Strategic Planning at MINDEF Level<br><br>Multi-Year Plans<br>SAF 2040 | Formulation of Concepts and Master Plans | |
|---|---|---|
| | Operational Concept Formulation (Service-Driven) | Operational Master Plan |
| | | Engineering Master Plan |
| | System of Systems Capability Development | |

*Figure 3.2* Singapore MINDEF's multi-year strategic planning process.

*Source:* Adapted from Ministry of Defence (Singapore), 'Defence Science and Technology' website (2023)[41]

technological acquisition priorities, and overall military modernisation trajectories. Parallel, the SAF develops specific operational concepts and master plans, which are classified.[38] According to MINDEF, the design of operational concepts is mostly service-driven, while the varying operational master plans and engineering master plans identify required system architectures for specific capability development. The purpose of the master plans is to ensure that Singapore's defence capabilities not only meet future user requirements for the SAF but can also be sustained in the long term through 'cost-effective' defence management and 'adaptive' systems integration.[39] While specific details remain confidential, the SAF releases fact sheets and provides occasional media briefings that offer a glimpse into its priorities, programmes, and initiatives. Perhaps the most detailed authoritative source of SAF's military modernisation is annual statements from the Singapore Parliament Committee of Supply Debates, which outline SAF's focus areas and resource allocation imperatives. For example, the integration of AI, cyber defence, and unmanned systems in the SAF; continued modernisation of the SAF's Land, Sea, and Air Forces, including the acquisition of F-35B fighter jets and Archer artillery systems; promoting interoperability through enhanced training and exercises to hone operational readiness; and ongoing collaboration with international partners and joint exercises to counter diverse security challenges.[40]

Another core feature of Singapore's defence planning is a focus on indigenous defence innovation and its integration in military applications through a collaborative framework of *Ops-Tech* development approach—integrate operations (Ops) with technology (Tech) across all pillars of national security, including military defence. The idea is to foster a collaborative organisational environment, data-driven decision-making, and interoperability between the varying actors in Singapore's 'defence ecosystem'—*the users* (SAF), *developers* (MINDEF, DSTA, DSO dual-use R&D labs), and *producers* (local defence industries such as ST Engineering).[42] In particular, *Ops-Tech* involves interactions between the Defence Technology Community (DTC) within Singapore's defence establishment, the broader ecosystem that integrates local industries such as the ST Engineering, local research foundations and defence research institutes such as Temasek Labs, the SAF—and Singapore's select foreign strategic partners. The goal of the DTC is a spiral development of technological capabilities and tailored solutions that would integrate these technologies with existing platforms and systems or newly acquired equipment.[43] In other words, Ops-Tech is not only about technology innovation and meeting SAF-user requirements but also about the sustainment of these technologies and capabilities—in creating synergistic applications for defence within the ecosystem.[44] Recent examples of Ops-Tech development include an immersive training system utilising virtual reality (VR) for soldiers to practice combat scenarios in realistic environments; ARTEMIS (Army Tactical Engagement and Information System) battle management system; AI-enabled Command and Control Information System (CCIS)—a high-tech command post that uses AI to process and analyse intelligence data in real time, while providing automatic target detection and classification; and the modification and integration of diverse drones for reconnaissance, surveillance, and target acquisition.[45]

By embracing advanced technologies such as AI, robotics, and unmanned vehicles, the SAF aims to maintain its edge in defence and military innovation. Within the MINDEF HQ, the control tower for the R&D of advanced technologies is *MINDEF Defence Tech Group (DTG)*. Established in 1986, the DTG's key functions include facilitating cutting-edge research on areas such as AI, robotics, unmanned vehicles, cyber security, and advanced materials; collaborating with academia and industry—partnering with universities, research institutes, and private companies, including start-ups to leverage expertise and accelerate technology development; supporting technology acquisition and integration—assisting the SAF in identifying, testing, and deploying new technologies within its operations; and nurturing talent and building expertise—attracting and training skilled personnel in various fields of technology relevant to national defence.[46] Specifically, responsibilities for technological innovation are divided into MINDEF's four main tech departments with distinct technology planning portfolios and policy planning responsibilities. These include *Future Systems & Technology Directorate* (FSTD), responsible for master planning and managing the research and technology requirements of MINDEF and SAF; *Technology Strategy & Policy Office* (TSPO), which develops technology policies and capability development roadmaps, and conducts long-term planning. TSPO also works with the broader civilian/private ecosystem, including start-ups, to optimise defence technology resources and capabilities; *Industry & Resources Policy Office* (IRPO) oversees the local defence industry, land use, logistics, technology security, defence exports, and procurement; *Defence Technology Collaboration Office* (DTCO) is responsible for conceptualising and implementing policies and plans for defence technology-related engagements with local research institutions and international partners.[47]

MINDEF's Defence Tech Group is further supported by the broader layer of Defence Technology Community (DTC) that integrates three technology research arms: (1) *Defence Science and Technology Agency* (DSTA), a statutory board set up under MINDEF to implement defence technology plans, procure defence material and develop defence infrastructure. DSTA builds up a community of engineers and scientists from universities, research institutes, government and industry; (2) *DSO Labs—Defence Science Organisation*, Singapore's largest defence research and development organisation, tasked with both basic and applied defence-technology related research; and (3) *Centre for Strategic Infocomm Technologies* (CSIT)—a technical agency under MINDEF that focuses on cybersecurity, data analytics, software engineering, and cloud infrastructure and services.[48] The varying interactions within the DTC system shape specific military-technological development and acquisition processes, systems development and integration, and capabilities, which are subsequently integrated into the SAF services (see Figure 3.3). In turn, the SAF then provides feedback loops on their performance validation and verification, system safety, reliability and maintainability, logistic support, and other factors.[49] The administrative backbone of the DTC and the entire SAF is the Defence Management Group (DMG) within MINDEF, which is responsible for human resource management, National Service implementation, financial planning and analysis, procurement and contracts, information technology infrastructure

**Multi-Disciplinary Systems Engineering**

- Architecture Design
- Operations Research
- Design for Support
- Risk Assessment
- Life-Cycle Cost Management

Capability
Development
Planning

Capability
Delivery

**Integration to the SAF**

- Analytic Hierarchy Process
- Contracting Strategy
- System Safety
- Quality Management
- Reliability and
  Maintainability
- Logistics Support Analysis
- Performance Validation and
  Verification

**Capability
Sustainment**

- Obsolescence Management
- Integrated Systems
  Monitoring
- Configuration Management

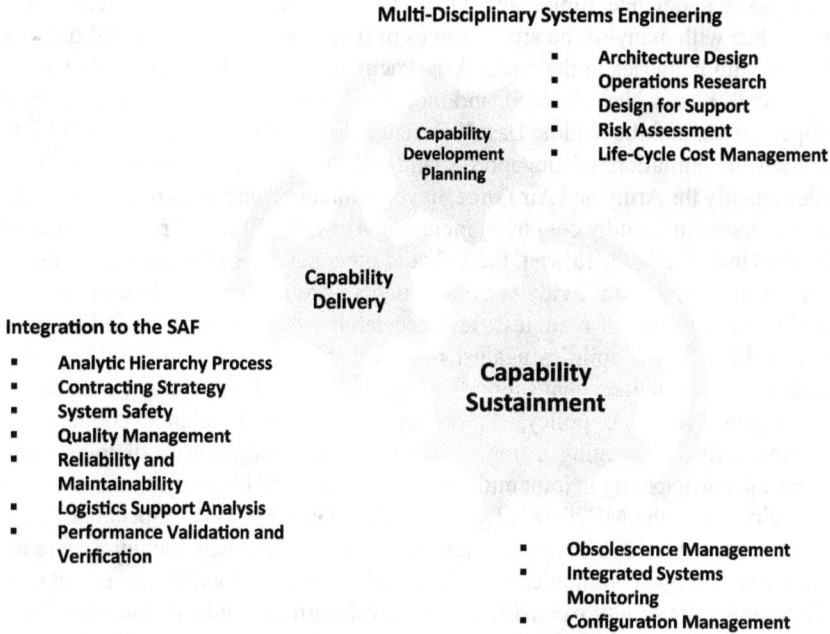

*Figure 3.3* Singapore's approach to defence capability planning.

*Source:* Figure created by author based on Defence Science and Technology Agency, The DSTA Story (2016)[51]

development, and legal and other professional services implementation for MINDEF and the SAF.[50]

Singapore's defence planning, however, is not only primarily about technology and military capability to ensure its strategic edge for deterrence but also about the need to pursue defence diplomacy to maintain cooperative relations with friendly countries, especially its neighbours. In a 2009 lecture on the 'Fundamentals of Singapore's Foreign Policy: Then & Now' at the MFA's Diplomatic Academy, Singapore's founding Prime Minister Lee Kuan Yew explained the rationale for defence diplomacy: 'a small country must seek a maximum number of friends while maintaining the freedom to be itself as a sovereign and independent nation. Both parts of the equation—a maximum number of friends and freedom to be ourselves—are equally important and inter-related'.[52] To achieve this,

> we must make ourselves relevant so that other countries have an interest in our continued survival and prosperity as a sovereign and independent nation; we have to be different from others in our neighbourhood and have a competitive edge. Because we have been able to do so, Singapore has risen over our geographical and resource constraints, and has been accepted as a serious player in regional and international fora.[53]

Singapore's defence diplomacy portfolio is broad and involves close and friendly ties with many of the armed forces of ASEAN countries, bilateral defence relations with countries in the wider Asia-Pacific region, including the US, China, Japan, South Korea, New Zealand, and India, and friendly ties with armed forces in Europe, Africa, and the Middle East.[54] Defence diplomacy has enabled the SAF to overcome the limitations of Singapore's land and airspace. Various units of the SAF, predominantly the Army and Air Force, have maintained long-term training detachments overseas in friendly countries, including Australia, Brunei, France, Germany, New Zealand, Thailand, Taiwan, the US, and other states. SAF's overseas training exercises and presence provide various benefits—from acquiring diverse training experiences, operational readiness, and accelerated technological assimilation to benchmarking SAF's abilities against more capable militaries and building sustained defence diplomacy that supports Singapore's long-term strategic interests.[55]

Singapore's defence policy, diplomacy, and regional defence cooperation, including actively engaging in regional and international security dialogues and forums and participating in joint multilateral exercises and disaster relief operations is coordinated by the MINDEF's Defence Policy Group (DPG).[56] Specifically, the DPG is responsible for Singapore's defence policy and strategy, including defence diplomacy, strategic communications, National Education, Total Defence, and strategic planning. Its structure consists of four key departments with distinct portfolios and responsibilities, including the Defence Policy Office (DPO), MINDEF Communications Organisation (MCO), NEXUS agency, and Information Directorate.[57] While each department supports Singapore's defence planning, the DPO is particularly relevant for developing strategies that shape the direction and character of Singapore's defence diplomacy—bilateral defence relations with select countries while devising policies and positions on specific/thematic issues. In other words, DPO serves as Singapore's defence diplomacy hub, cultivating ties with foreign counterparts, representing Singapore's interests and perspectives on international security issues at various forums and conferences, engaging in policy analysis and research on global and regional security trends and developments, providing strategic intelligence, and engaging in collaborative defence cooperation initiatives. For example, Singapore's position on AI governance and the need for responsible development and use of AI in the military involves various multilateral cooperation initiatives, including representation at the OECD Network of Experts on AI, Global Partnership on AI, AI Partnership for Defence, and REAIM (Responsible AI in Military domain) process. MINDEF's Strategy and Futures Group (SFG)—P7— under the DPO encourages dialogues to discuss AI governance and develops positions on relevant issues such as pursuing an international consensus around shared norms such as responsible, reliable, robust, and safe AI development in the military domain. DPO and SFG formulate these policies and support their implementation with other MINDEF departments and Singapore's government agencies.[58]

### SAF's military innovation and adaptation trajectories

Singapore's defence planning has gradually evolved with the progressive development and capabilities of the SAF: the first-generation or 1G SAF (1960s–70s) aimed at basic capability-development of individual services and the implementation of a purely island-defensive 'poisoned-shrimp' strategy, which envisioned high-intensity urban combat to impose unacceptable human and material costs to potential aggressors. In the 1980s and 90s, the second-generation or 2G SAF, shifted towards combined-arms manoeuvre warfare (1980s–90s), and the 'porcupine' defence strategy that envisioned a limited-power projection in Singapore's near seas and potentially a pre-emptive posture to be able to transfer a conflict into enemy's territory, for example, if Singapore's water supplies would be cut off.[59] At the turn of the twenty-first century, the SAF progressed towards the third-generation or 3G SAF, a transition towards 'network-centric' warfare for the land, air, and sea domains as well as an emphasis on defence diplomacy, operations other than war, counterterrorism, and disaster relief in geographically more distant areas from Singapore.[60]

In retrospect, one could argue that the direction and character of Singapore's defence planning over the past two decades has largely reflected a gradual, evolutionary path—a structured-phased adaptation strategy focusing on developing and sustaining capabilities for force readiness and evolution while preparing for more future-oriented force transformation.[61] In essence, readiness has meant preserving the ability of the SAF to rapidly shift from a peacetime posture to an effective wartime fighting force and gradually developing military-technological capabilities for diverse missions. In doing so, the SAF would first acquire new equipment, introduce progressively more capable systems, and establish new units. Second, the SAF would establish new operational commands to integrate these weapons technologies, systems, and units. In the third phase, the SAF would enhance its leadership and human capital by revising training and education to maintain a steady stream of capable and committed personnel.[62]

Since 2015, however, Singapore's MINDEF and DTC have been gradually unveiling the contours of the Next-Generation SAF or SAF 2040, which arguably signifies much more ambitious military change for the SAF. In particular, the SAF 2040 transformation can be defined along four strategic areas: (1) technological innovation and integration—deployment of advanced technologies such as AI-powered analytics for improved situational awareness and decision-making; (2) organisational changes—restructuring of units and streamlining processes to enhance agility and adaptability; (3) talent development—investing in training programmes and talent acquisition strategies to equip SAF personnel with the necessary skills for future operations; and (4) international collaboration—engaging in joint exercises and technology partnerships with other nations.[63]

While details are scattered in a mosaic of public statements and speeches of MINDEF's senior military leadership, the SAF 2040 envisions that Singapore's military will operate in a strategically uncertain and operationally complex environment that combines high technology with new forms of 'hybrid' warfare, increasingly blurring the lines between peacetime and wartime, civil and military arenas, and new domains such as space, near space, cyberspace, and underwater.[64] The resulting 'high-low' intensity threat spectrum combines diverse and often opposing challenges simultaneously. On one hand, regional extremists or terrorist organisations could evolve by using novel technologies and sheer brute force—for example, by simultaneously using swarms of drones, social media information warfare, cyberattacks, indiscriminate shooting, and use of explosives to attack Singapore's centres of gravity. On the other, with the ongoing China-US strategic rivalries in the Indo-Pacific and potential conflict scenarios in the South China Sea, the Malacca Strait, Taiwan, and the East China Sea, Singapore's security is increasingly affected by the diffusion of advanced military-technological capabilities and strategic competition in the Indo-Pacific.

In the military context, therefore, the SAF aims to preserve its defensive options and offensive choices. In 2023, for example, the acquisition list for the SAF included upgraded early warning systems such as coastal surveillance network and air defence systems; F35s and upgraded F15SG fighter jets, Multi-Role Tanker Transport and G550 Airborne Early Warning aircraft, unmanned aerial vehicles (UAVs) including Orbiter 4 close-range UAVs; new classes of ships that include joint multi-mission ships and multi-role combat vessels (MRCVs); Type 218SG submarines based on the upgraded German-type 214 design, and new types of underwater unmanned vehicles (UUVs); and ultimately, military systems and platforms for more protected and mobile army, such as the indigenous hunter-armoured fighting vehicles, Terrex Infantry Carrier Vehicles, upgraded Leopard tanks, High Mobility Artillery Rocket System (HIMARs) and howitzers, and their supporting systems.[65] Together, these platforms will increase the SAF's freedom of action, make the SAF more survivable due to the increased use of stealth and active defences and improve its capabilities for battlefield knowledge, situational awareness, and command and control.[66]

At the same time, the SAF, together with Singapore's security and intelligence agencies, must grapple with a new era of permanent low-level conflicts in and around the Indo-Pacific that utilise grey-zone strategies—using ambiguity or deniability in everything from disinformation and espionage, hostile influence campaigns, crime and subversion, and cyber means to gain political advantage by projecting instability within countries and a legitimacy crisis on the global stage. While grey zone conflicts do not include the use of violent force, hybrid warfare does. Evolving hybrid warfare strategies often combine timeless 'unconventional methods' with novel technologies to create political, economic, and psychological effects. Like grey zones, hybrid warfare strategies are initially masked in non-military arenas, utilising diplomatic deception and cyber and social media disinformation to influence public opinion. The aim is to maximise non-military forms of influence, political coercion, while seeding chaos and deception to undermine societal resiliency and military resolve. In theory, this creates a psychological advantage for sweeping military actions and undermines the resilience of multicultural societies such as Singapore.[67] Therefore, the SAF aims to revamp its concepts, organisational structures, and technological capabilities to counter 'hybrid' threats in the information and cyber domains; expand counterterrorism capabilities, particularly by strengthening Island Defence and Special Forces; and leverage advanced emerging technologies such as AI, data analytics, and robotics in nearly all aspects of defence planning and military operations.[68]

In particular, the SAF's major organisational change is the formal establishment of the Digital and Intelligence Service (DIS) in October 2022, as the fourth branch of the SAF, on par with the army, navy, and air force. The idea behind the DIS is to consolidate previously compartmentalised Command, Control, Communications, Computers, and Intelligence (C4I), cyber, military intelligence, and supporting units and capabilities. At its onset, the DIS as a Service HQ currently includes four Commands, each responsible for a specific domain: Intelligence, C4 and Cyber, Digital Defence, and Training. It also established the Joint Digital and C4 Department (JDCD) to scale up digitisation within the SAF in collaboration with Singapore's defence innovation ecosystem or Defence Technology Community and the Digital Ops-Tech Centre (DOTC) that aims to serve as the SAF's centre of excellence to integrate AI, data science, and emerging technologies into SAF's operational conduct.[69]

Through the DIS, the SAF's evolving military-technological advances are combined with relevant organisational force structures and increasingly push for more transformative capabilities and conduct of operations (see Figure 3.4). This means that future DIS units will increasingly utilise AI-enabled systems to provide situational awareness, intelligence of the operational environment, and cyber support for SAF's joint operations while defending Singapore's military networks, electronic communications, and information environment.[70] At the high end of warfare, DIS will also enable the so-called sense and strike missions—a follow-up to the evolving 'sensor-to-shooter' concepts that have envisioned integrating diverse automated sensor networks, data sharing, and providing mission taskings to any weapons platforms or units. For example, an uncrewed aerial vehicle or ground robot would be able to spot an enemy tank or ship, share the intelligence and data

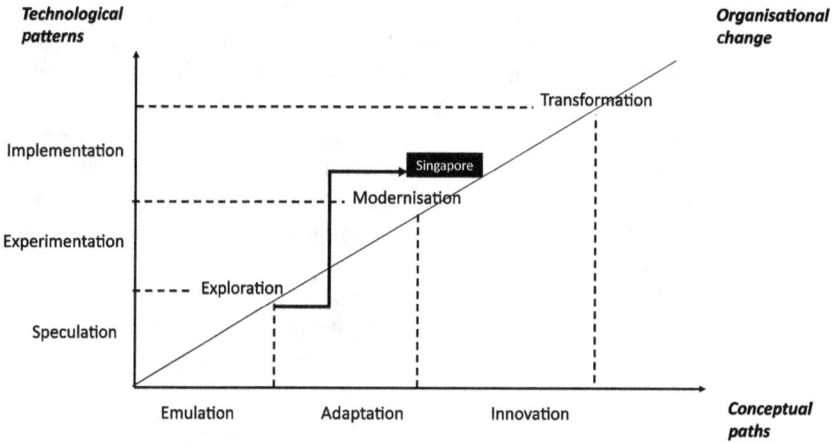

*Figure 3.4* Singapore's military innovation trajectories.

*Source:* Figure created by author based on Michael Raska, Military Innovation of Small States— Creating a Reverse Asymmetry (Routledge, 2016)[72]

in real-time with relevant non-line-of-sight strike systems in the rear, which in turn would be able to use this automated target detection with the human-in-the-loop to launch precision strikes on the target.[71]

In September 2023, the SAF conducted its biennial, large-scale training exercise 'Forging Sabre' at the Mountain Home Air Force Base in Idaho, US. The exercise series, dating back to 2005, has always served as a platform for testing and refining SAF's cutting-edge capabilities, readiness, and proof-of-concepts. The latest edition, however, showed how Singapore's military progressively absorbs defence innovation in military AI technologies, concepts, and organisations in military operations. In particular, the SAF brought its latest version of AI-enabled Command and Control Information System (CCIS)—a high-tech command post that uses AI to process and analyse intelligence data in real-time, while providing automatic target detection and classification, logistics support and predictive maintenance, cyber-defence and overall, human-machine teaming in what the SAF terms 'sense-and-strike' operations. The 2023 Forging Sabre is thus a relevant baseline for understanding the direction and character of Singapore's defence transformation drive under its conceptual umbrella of the SAF 2040. While the Forging Sabre 2023 showcased only early versions of SAF capabilities that are likely to advance considerably between now and 2040, the actual use of AI-enabled systems reflects the pace of innovation and the urgency to incorporate the value of AI and machine learning into SAF military operations, both of which are likely to increase with the acquisition of advanced military weapons platforms, systems, and technologies.[73]

However, implementing the technological, conceptual, and organisational innovation path of the SAF 2040 and its 'sensor-strike' capabilities will be constrained by challenges reminiscent of the 3G SAF 'network-centric' force since the 1990s. The major problem then, as it is now, is that the SAF must be able to effectively (in real time) integrate the various sensor-to-shooter loops, while providing robust, secure, and fast data storage capabilities in terms of terabytes per second connected with diverse inter-service sensor platforms and systems. This means effectively creating a secure, fast, and resilient cloud system that would link the various air force, army, navy, and cyber battle management; command and control, communications and networks; intelligence, surveillance, and reconnaissance; electronic warfare; positioning, navigation, and timing; with precision munitions. Moreover, since many of these systems will aim to leverage AI, SAF operational commanders and National Servicemen must trust these systems. Compounding the technological challenges are increasing organisational barriers and demographic constraints— declining birth rates and an ageing population. Specifically, Singapore's total fertility rate (TFR) dipped to 1.12 in 2022, far below the replacement rate of 2.1 needed to maintain a stable population, while the median age in Singapore is projected to reach 45 by 2030, with the proportion of citizens above 65 exceeding 20%. The ageing population raises challenges for both manpower availability and healthcare costs associated with maintaining a large reservist pool. And herein lies the principal challenge for implementing the automated warfare concepts embedded in the SAF 2040—developing automated SAF operations and embracing the natural complexity of an operational environment and novel technologies requires a new mindset—new ways of thinking and operating at every echelon of the SAF. The question is whether the SAF's continuous adaptation strategy can incorporate disruptive innovation paths, which embrace creativity and innovation and promote the rearrangement of existing rules in Singapore's strategic culture. In the words of Bernard Loo, 'For the SAF to remain strategically relevant, it has had to grapple with the fundamental questions of what its mission is, the security challenges that it is likely to face (at any given moment), and the context (both external security and domestic political) in which the SAF will find itself'.[74] Ultimately, Singapore has no choice but to turn to technology— the SAF 2040 roadmap essentially amplifies technological advantage in digitisation, robotisation, and sensor-revolution across the entire SAF. However, the question is whether a high-tech army equals effective military capability, and therefore provides strong deterrence.

## Conclusion

Taken together, the continuity and change in Singapore's defence planning over the past two decades, as summarised in Figure 3.5, has reflected a seemingly perennial paradox. On the one hand, the SAF—similarly to other military organisations—has been traditionally resistant to major changes, preserving tried and tested strategies and structures to foster continuity amid the prevailing process of institutionalisation and the fact that the cost of error in the face of ubiquitous strategic uncertainty may be exceedingly high. Notwithstanding the continuous military training, increasing technological sophistication, and participation by the SAF in overseas

| SMPS ADJUST THEIR APPROACHES BASED ON SYSTEMIC CHANGE OUT OF THEIR CONTROL | SMPS REACT TO CHANGES IN THE RELEVANT SECURITY COMPLEX(ES) THE SMP IS PART OF | | | | |
|---|---|---|---|---|---|
| *Alliances, Dependencies and National Ambitions* | | | | | |
| SMPS FACE CONSTRAINTS IN SAFEGUARDING NATIONAL SECURITY | FINANCIAL | DEMOGRAPHIC | CULTURAL | GEOGRAPHICAL | OTHER |
| SMPS NEED TO POSITION THEMSELVES VIS-À-VIS GLOBAL AND REGIONAL POWER CONSTELLATIONS | BALANCING | HEDGING | BANDWAGONING | NON-ALIGNING | OTHER |
| SMPS HAVE TO ENGAGE WITH OTHER STATES TO INCREASE THEIR DEFENCE CAPABILITIES | SECURITY GUARANTEES | TRANSACTIONAL | FORCE INTEGRATION | SPECIALISATION | OTHER |
| SMPS HAVE ARMED FORCES FOR DIFFERENT PURPOSES | INSTRUMENT OF DIPLOMACY | PRESTIGE PROJECT | REDUCE STRATEGIC RISK (DEFEND) | DOMESTIC STABILITY | OTHER |
| *Approaches, Processes, Methods and Techniques* | | | | | |
| SMPS TAKE A MORE CONSTRAINED APPROACH TO DEFENCE PLANNING | THREAT-NET ASSESMENT BASED | PORTFOLIO-BASED | TASK-BASED | MOBILISATION | OTHER |
| SMPS HAVE LIMITED RESOURCES TO DESIGN AND IMPLEMENT DEFENCE PLANNING PROCESSES | TAILORED ANALYTICAL PROCESS | OUTSOURCING STRATEGY | ACCEPT DILUTION / FAIR SHARE | EXTERNAL TEMPLATE | OTHER |
| SMP EMPLOY MULTIPLE METHODS AND TECHNIQUES | COPY GREAT POWERS | INFORMAL CONSENSUS | TAILORED METHODS | AD HOC | OTHER |
| SMPS RELY ON DIFFERENT PROCESSES TO DECIDE ON FORCE POSTURE | ANALYTICAL RESULTS | POLITICS DECIDES | MILITARY DECIDES | GREAT POWER DECIDES | OTHER |
| *Military Innovation* | | | | | |
| SMPS FACE CHALLENGES IN KEEPING PACE WITH TECHNOLOGICAL ADVANCES | BUILD DOMESTIC TECH/INDUSTRY | TECH TRANSFER AGREEMENTS | ACQUIRE READY INNOVATIONS | PARTNER WITH GREAT POWER | OTHER |
| SMPS PURSUE VARIOUS MILITARY INNOVATION PATHS AND PATTERNS | EMULATION / SPECULATION | ADAPTATION / EXPERIMENTATION | INNOVATION / IMPLEMENTATION | OTHER / OTHER | OTHER |
| SMPS FIND THEMSELVES AT DIFFERENT STAGES OF ORGANISATIONAL CHANGE | EXPLORATION | MODERNISATION | TRANSFORMATION | OTHER | OTHER |

(Left vertical axis labels: GENERAL COMMONALITIES; VARIATIONS WITHIN THESE COMMONALITIES)

*Figure 3.5* Structured focused comparison framework for Singapore.

*Source:* Figure created by author based on this chapter's main argument

operations other than war—the absence of modern (or any) real combat experiences, the lack of open debates, and general adherence to the established norms and strategic culture based on traditional hierarchy and discipline—has arguably precluded the implementation of a more 'disruptive' or 'transformative' doctrinal and organisational innovation in Singapore's strategic thought and practice. The 3G SAF modernisation has inherently reflected an attempt to balance between preserving tried and tested strategies and structures with adapting elements of modern 'network-centric' warfare in preparation for multi-level conflicts. Yet, despite the inherent risk aversion, the implementation of the 3G SAF has been relatively successful—the SAF has acquired and integrated prerequisite 'network-centric' weapons platforms and technologies while gradually adapting its organisational force structure and operational concepts that have ultimately continued to qualitatively outpace its neighbours in relative terms—today, the SAF remains arguably the best-equipped, organised, and likely the most technologically advanced military in Southeast Asia.

On the other hand, the SAF has recognised that failure to achieve advances in the ways and means of war may result in losing its 'strategic edge' and thus has been motivated to seek both defence and military innovation. As such, Singapore's defence planners have continuously pointed towards the disruptive nature of emerging technologies and digital revolution as both strategic challenge and opportunity—turning to advanced technologies as a 'force multiplier' to strengthen its deterrence, defence, and resilience capabilities, while mitigating the effects of internal demographic changes and vulnerabilities. Therefore, the contours of the SAF 2040 defence planning effectively signify a major military change for the SAF. Integrating multidomain operations requires a new strategy, units, and doctrinal revision to include new missions and career paths, changing the curriculum of SAF's professional military education institutions, and revising training and experimentation. Technology is and will remain SAF's critical enabler in the process, especially as Singapore lacks strategic depth and a limited (and declining) manpower base. It must strive to keep its military as high-tech as possible, including its manpower. At the same time, Singapore must continue to leverage its relatively high level of education and technical training to craft a smaller but more technologically savvy military. The SAF's irreducible priorities are still to protect the Singaporean homeland and to safeguard Singapore's security by working to maintain peace and stability and reduce tensions throughout Southeast Asia. Deterrence, defence, diplomacy, and resilience are still the key watchwords of the SAF, but *how* these might be achieved in the twenty-first century is probably changing and may change further with the diffusion of geostrategic and technological challenges, uncertainties, and disruptions.

## Notes

1 Ministry of Defence (Singapore), "Defence Policy and Diplomacy", last modified April 6, 2021, www.MINDEF.gov.sg/web/portal/MINDEF/defence-matters/defence-topic/defence-topic-detail/defence-policy-and-diplomacy

2 Bernard Fook Weng Loo, "The Management of Military Change: The Case of the Singapore Armed Forces", in *Security, Strategy and Military Change in the 21st Century: Cross-Regional Perspectives*, eds. J.I. Bekkevold, I. Bowers, and M. Raska (London: Routledge, 2015), 70–88.

3 Michael Raska, "How Will SAF Look Like After Its Next Incarnation?", *Today Online*, March 19, 2019, www.todayonline.com/commentary/saf-after-its-next-incarnation

4 Kenneth Cheng, "SAF Looks to Artificial Intelligence to Gain Punch", *Today Online*, October 28, 2016, www.todayonline.com/singapore/saf-looks-artificial-intelligence-gain-punch;

5 Chia Jie Lin, "Singapore's New 'Soldiers': AI, Augmented Reality, and Data Analytics," *GovInsider*, July 18, 2018, https://govinsider.asia/intl-en/article/singapore-defence-dsta-mindef-ai-ar-data-analytics

6 Ministry of Defence (Singapore), "Fact Sheet: Strengthening MINDEF/SAF's Cyber Defence Capabilities", last modified June 30, 2021, www.MINDEF.gov.sg/web/portal/MINDEF/news-and-events/latest-releases/article-detail/2021/June/30jun21_fs7

7 Ministry of Defence (Singapore), "Fact Sheet: Leveraging Technology and Innovation to Drive Digital Transformation", last modified March 1, 2021, www.MINDEF.gov.sg/web/portal/MINDEF/news-and-events/latest-releases/article-detail/2021/March/01mar21_fs4

8 Edwin Lee, *Singapore: The Unexpected Nation* (Singapore: ISEAS Publishing, 2008).

9 Andrew Tan, "Domestic Determinants of Singapore's Security Policy", *APCSS Occasional Paper Series* (2001): 2.

10 Maritime Executive, "Strait of Malacca Key Chokepoint for Oil Trade," last modified August 27, 2018, https://maritime-executive.com/article/strait-of-malacca-key-chokepoint-for-oil-trade

11 Tim Huxley, *Defending the Lion City: The Armed Forces of Singapore* (St Leonards: Allen & Unwin, 2000), 32.

12 Tan, "Domestic Determinant's of Singapore's Security Policy," 1.

13 Collin Koh, "Seeking Balance: Force Projection, Confidence Building, and the Republic of Singapore Navy," *Naval War College Review* 65, no. 1 (2012): 75–92.

14 Kin Wah Chin, "Singapore: Threat Perception and Defence Spending in a City State," in *Defence Spending in Southeast Asia*, ed. Kin Wah Chin (Singapore: ISEAS, 1987), 196.

15 David Boey, "Singapore: A Fragile Nation Toughens Up," *Jane's Intelligence Review* 8, no. 7 (July 1996): 316–320.

16 Xinghui Kok, "Singapore's Population Grows 5% as Foreign Workers Return Post-Pandemic", *Reuters*, September 29, 2023, www.reuters.com/world/asia-pacific/singapores-population-grows-5-foreign-workers-return-post-pandemic-2023-09-29/

17 Lee Kuan Yew, *The Singapore Story: Memories of Lee Kuan Yew* (Singapore: Times Publishing, 1998), 15.

18 Huxley, *Defending the Lion City: The Armed Forces of Singapore*, 33.

19 Pak Sung Ng, "From 'Poisonous Shrimp' to 'Porcupine': An Analysis of Singapore's Defence Posture Change in the Early 1980s," SDSC Working Paper 397 (2005): 14, https://sdsc.bellschool.anu.edu.au/sites/default/files/publications/attachments/2016-03/WP-SDSC-397_0.pdf

20 Abigail Ng, "Singapore's Total Fertility Rate Drops to Historic Low of 1.05," *Channel News Asia*, February 24, 2023, www.channelnewsasia.com/singapore/singapore-total-fertility-rate-population-births-ageing-parents-children-3301846

21 Kor Kian Beng, "SAF Confident of Coping with Tighter Manpower Resources," *The Straits Times*, June 30, 2017, www.straitstimes.com/singapore/saf-confident-of-coping-with-tighter-manpower-resources-thanks-to-increased-automation-and

22 Ministry of Defence (Singapore), "Defence Policy and Diplomacy," last modified April 6, 2021, www.MINDEF.gov.sg/web/portal/MINDEF/defence-matters/defence-topic/defence-topic-detail/defence-policy-and-diplomacy

23 Michael Raska, "A Structured-Phased Evolution: The Third Generation Force Transformation of the Singapore Armed Forces", in *Military Innovation in Small States: Creating a Reverse Asymmetry*, ed. M. Raska (New York: Routledge, 2016), 130–162.
24 Bilveer Singh, "Arming the Singapore Armed Forces: Trends and Implications," *Canberra Papers on Strategy and Defence*, no. 153 (2003): 18, https://sdsc.bellschool. anu.edu.au/experts-publications/publications/3090/arming-singapore-armed-forces-saf-trends-and-implications
25 Chung Min Lee, *Fault Lines in a Rising Asia* (Washington, DC: Carnegie Endowment for International Peace, 2016).
26 Richard Bitzinger, "A New Arms Race? Explaining Recent Southeast Asian Military Acquisitions," *Contemporary Southeast Asia* 32, no. 1 (2010): 50–69.
27 Richard Bitzinger, *Arming Asia: Technonationalism and Its Impact on Local Defence Industries* (New York, NY: Routledge, 2017).
28 Michael Raska, "Strategic Competition and Future Conflicts in the Indo-Pacific Region", *Journal of Indo-Pacific Affairs* 2, no. 2 (2019): 83–97.
29 Author's conceptions.
30 Ministry of Defence (Singapore), *Defence of Singapore 1994–95* (Singapore: Ministry of Defence, 1995), 1.
31 Lee, Hsien Loong, "Security Options for Small States", *The Straits Times*, November 6, 1984.
32 Ron Matthews and Fitriani Bintang Timur, "Singapore's Total Defence Strategy" *Defence and Peace Economics* (2023), https://doi.org/10.1080/10242694.2023.2187924
33 Ron Matthews and Nellie Zhang Yan, "Small Country 'Total Defence': A Case Study of Singapore", *Defence Studies* 7, no. 3 (2007): 376–395.
34 Ministry of Defence (Singapore), "Fact Sheet: Digital Defence", last modified February 15, 2019, www.MINDEF.gov.sg/web/portal/MINDEF/news-and-events/latest-releases/article-detail/2019/February/15feb19_fs
35 Centre for Strategic Futures (Strategy Group, Singapore Prime Minister's Office), "Who we are—Our Approach", last modified October 9, 2023, www.csf.gov.sg/our-work/our-approach/
36 Jon Grevatt and Andrew MacDonald, "Singapore Boost 2023 Defence Budget by Nearly 6%", *Janes*, February 27, 2023, www.janes.com/defence-news/news-detail/singapore-boosts-2023-defence-budget-by-nearly-6
37 *Ibid.*
38 Ministry of Defence (Singapore), "Defence Science and Technology", last modified February 7, 2023, www.mindef.gov.sg/web/portal/mindef/defence-matters/defence-topic/defence-topic-detail/defence-science-and-technology
39 Defence Science and Technology Agency, *The DSTA Story 2000–2015* (Singapore: DSTA, 2016).
40 Ministry of Defence (Singapore), "Committee of Supply Debate 2023", last modified February 27, 2023, www.mindef.gov.sg/web/portal/mindef/news-and-events/latest-releases/article-detail/2023/cos2023
41 Ministry of Defence (Singapore), "Defence Science and Technology", last modified February 7, 2023, www.mindef.gov.sg/web/portal/mindef/defence-matters/defence-topic/defence-topic-detail/defence-science-and-technology
42 Shu Huang Ho, "The Hegemony of an Idea: The Sources of the SAF's Fascination with Technology and the Revolution in Military Affairs", *IRASEC Discussion Paper* (2009), 1–22. www.irasec.com/IMG/UserFiles/Files/04_Publications/Notes/The_Hegemony_of_an_Idea.pdf
43 Ministry of Defence (Singapore), "Defence Science and Technology", last modified February 7, 2023, www.mindef.gov.sg/web/portal/mindef/defence-matters/defence-topic/defence-topic-detail/defence-science-and-technology
44 Defence Science and Technology Agency, *The DSTA Story 2000–2015* (Singapore: DSTA, 2016), 42.

45 Ministry of Defence (Singapore), "Fact Sheet: Enhancing the SAF's Operational-Readiness and Servicemen's NS Experience through Digitalisation and Innovation", last modified June 30, 2022, www.mindef.gov.sg/web/portal/mindef/news-and-events/latest-releases/article-detail/2022/June/30jun22_fs

46 DSO National Laboratories, "Our History", last modified February 2024, www.dso.org.sg/about/history

47 Ministry of Defence (Singapore), "Defence Science and Technology", last modified February 7, 2023, www.mindef.gov.sg/web/portal/mindef/defence-matters/defence-topic/defence-topic-detail/defence-science-and-technology

48 *Ibid.*

49 *Ibid.*

50 Ministry of Defence (Singapore), "Defence Management Group", last modified January 16, 2018, www.mindef.gov.sg/web/portal/mindef/about-us/organisation/organisation-profile/defence-management-group

51 Defence Science and Technology Agency, *The DSTA Story 2000–2015* (Singapore: DSTA, 2016), 42.

52 Kuan Yew Lee, "The Fundamentals of Singapore's Foreign Policy: Then & Now", *S. Rajaratnam Lecture* at the Ministry of Foreign Affairs Diplomatic Academy, April 9, 2009.

53 *Ibid.*

54 Ministry of Defence (Singapore), *Defending Singapore in the 21st Century* (Singapore: MINDEF, 2000), www.mindef.gov.sg/oms/dam/publications/ebooks/more_ebooks/ds21.pdf

55 Ministry of Defence (Singapore), "Fact Sheet: Exercises and Operations", last modified June 13, 2018, www.mindef.gov.sg/web/portal/mindef/defence-matters/exercises-and-operations

56 Ministry of Defence (Singapore), "Defence Policy Group", last modified September 21, 2023, www.mindef.gov.sg/web/portal/mindef/about-us/organisation/organisation-profile/defence-policy-group

57 *Ibid.*

58 In February 2024, for example, MINDEF DPO-SFG in conjunction with the ROK MFA, Dutch MFA, and RSIS Military Transformations programme (led by the author of this chapter) organised Responsible AI in the Military Domain (REAIM) Regional Consultations (Asia) in Singapore.

59 Bernard Fook Weng Loo, "Explaining Changes in Singapore's Military Doctrine: Material and Ideational Perspectives", in *Asia in the New Millennium*, eds. Amitav Acharya and Lee Lai To (Singapore: Marshall Cavendish Academic, 2004), 352–374.

60 *Ibid.*

61 Michael Raska, "A Structured-Phased Evolution: The Third Generation Force Transformation of the Singapore Armed Forces", in *Military Innovation in Small States: Creating a Reverse Asymmetry*, ed. M. Raska (New York: Routledge, 2016), 130–162.

62 *Ibid.*

63 Ministry of Defence (Singapore), "Infographic: Towards SAF 2040", last modified February 24, 2023, www.mindef.gov.sg/web/portal/mindef/news-and-events/latest-releases/article-detail/2023/February/24feb23_infographic

64 Michael Raska, "The SAF After Next Incarnation", *RSIS Commentary*, no. 41, March 8, 2019, https://dr.ntu.edu.sg/bitstream/10356/106416/1/CO19041.pdf

65 Fabian Koh, "Budget Debate: Transformation to Next-Gen SAF on Track Despite Covid-19", *The Straits Times*, March 1, 2021, www.straitstimes.com/singapore/budget-debate-transformation-to-next-gen-saf-on-track-despite-covid-19-says-ng-eng-hen

66 Ministry of Defence (Singapore), "Speech by Minister for Defence, Dr Ng Eng Hen, at The Committee of Supply Debates 2023", last modified September 24, 2023, www.mindef.gov.sg/web/portal/mindef/news-and-events/latest-releases/article-detail/2023/February/24feb23_speech; Richard Bitzinger, "Military-Technological Innovation in

Small States: The Cases of Israel and Singapore", *Journal of Strategic Studies* 44, no. 6 (2021): 873–900.

67  Richard Bitzinger and Michael Raska, "The Contours of Emerging Technologies and Future Warfare", *IDSS Paper*, February 22, 2022, www.rsis.edu.sg/rsis-publication/idss/ip22010-the-contours-of-emerging-technologies-and-future-warfare/

68  Michael Raska, "4G SAF Creating New Advantages", *RSIS Commentary*, no. 102, May 24, 2017, https://dr.ntu.edu.sg/bitstream/10356/83194/1/CO17102.pdf

69  Ministry of Defence (Singapore), "Fact Sheet: Digital and Intelligence Service", last modified October 28, 2022, www.mindef.gov.sg/web/portal/mindef/news-and-events/latest-releases/article-detail/2022/October/28oct22_fs

70  Ministry of Defence (Singapore), "Fact Sheet: Update on the Digital and Intelligence Service (DIS) Capability Development Efforts", last modified February 24, 2023, www.mindef.gov.sg/web/portal/mindef/news-and-events/latest-releases/article-detail/2023/February/24feb23_fs

71  Ministry of Defence (Singapore), "Fact Sheet: Ex Forging Sabre 2023—Multi-Domain Smart Warfighting", last modified September 21, 2023, www.mindef.gov.sg/web/portal/mindef/news-and-events/latest-releases/article-detail/2023/September/21sep23_fs

72  Michael Raska, "A Structured-Phased Evolution: The Third Generation Force Transformation of the Singapore Armed Forces", in *Military Innovation in Small States: Creating a Reverse Asymmetry*, ed. M. Raska (New York: Routledge, 2016), 130–162.

73  Ridzwan Rahmat, "Singapore Validates Enhanced AI-Infused Combat System at US wargames", *Janes*, September 22, 2023, www.janes.com/defence-news/news-detail/singapore-validates-enhanced-ai-infused-combat-system-at-us-wargames; Aqil Hamzah, "Automated Systems, AI at the Forefront of SAF Exercise in the US", *The Straits Times*, September 21, 2023, www.straitstimes.com/singapore/automated-systems-ai-at-the-forefront-of-saf-exercise-in-the-us

74  Bernard Fook Weng Loo, "The Management of Military Change: The Case of the Singapore Armed Forces", in *Security, Strategy and Military Change in the 21st Century: Cross-Regional Perspectives*, eds. J.I Bekkevold, I. Bowers, and M. Raska (London: Routledge, 2015), 70–88.

## Bibliography

Bitzinger, Richard. "A New Arms Race? Explaining Recent Southeast Asian Military Acquisitions." *Contemporary Southeast Asia* 32, no. 1 (2010): 50–69.

———. *Arming Asia: Technonationalism and Its Impact on Local Defence Industries*. New York: Routledge, 2017.

———. "Military-Technological Innovation in Small States: The Cases of Israel and Singapore." *Journal of Strategic Studies* 44, no. 6 (2021): 873–900.

Bitzinger, Richard and Michael Raska. "The Contours of Emerging Technologies and Future Warfare." *IDSS Paper*, February 22, 2022. www.rsis.edu.sg/rsis-publication/idss/ip22010-the-contours-of-emerging-technologies-and-future-warfare/

Boey, David. "Singapore: A Fragile Nation Toughens Up." *Jane's Intelligence Review* 8, no. 7 (July 1996): 316–320.

Centre for Strategic Futures (Strategy Group, Singapore Prime Minister's Office). "Who We Are—Our Approach." Last modified October 9, 2023. www.csf.gov.sg/our-work/our-approach/

Cheng, Kenneth. "SAF Looks to Artificial Intelligence to Gain Punch." *Today Online*, October28,2016.www.todayonline.com/singapore/saf-looks-artificial-intelligence-gain-punch

Chia, Jie Lin. "Singapore's New 'Soldiers': AI, Augmented Reality, and Data Analytics." *GovInsider*, 18 July 2018. https://govinsider.asia/intl-en/article/singapore-defence-dsta-mindef-ai-ar-data-analytics

Chin, Kin Wah. "Singapore: Threat Perception and Defence Spending in a City State." In *Defence Spending in Southeast Asia*, edited by K. W. Chin, Singapore: ISEAS Publishing, 1987.

Defence Science and Technology Agency (Singapore). *The DSTA Story 2000–2015*. Singapore: Defence Science and Technology Agency, 2016.

DSO National Laboratories. "Our History." Last modified February 2024. www.dso.org.sg/about/history

Grevatt, Jon and Andrew MacDonald. "Singapore Boost 2023 Defence Budget by Nearly 6%." *Janes*, February 27, 2023. www.janes.com/defence-news/news-detail/singapore-boosts-2023-defence-budget-by-nearly-6

Hamzah, Aqil. "Automated Systems, AI at the Forefront of SAF Exercise in the US." *The Straits Times*, September 21, 2023. www.straitstimes.com/singapore/automated-systems-ai-at-the-forefront-of-saf-exercise-in-the-us

Ho, Shu Huang. *The Hegemony of an Idea: The Sources of the SAF's Fascination with Technology and the Revolution in Military Affairs*. IRASEC Discussion Paper 5. Bangkok: Research Institute on Contemporary Southeast Asia, 2009. www.irasec.com/IMG/UserFiles/Files/04_Publications/Notes/The_Hegemony_of_an_Idea.pdf

Huxley, Tim. *Defending the Lion City: The Armed Forces of Singapore*. St Leonards: Allen & Unwin, 2000.

Koh, Collin. "Seeking Balance: Force Projection, Confidence Building, and the Republic of Singapore Navy." *Naval War College Review* 65, no. 1 (2012): 75–92.

Koh, Fabian. "Budget Debate: Transformation to Next-Gen SAF on Track Despite Covid-19." *The Straits Times*, March 1, 2021. www.straitstimes.com/singapore/budget-debate-transformation-to-next-gen-saf-on-track-despite-covid-19-says-ng-eng-hen

Kok, Xinghui. "Singapore's Population Grows 5% as Foreign Workers Return Post-Pandemic." *Reuters*, September 29, 2023. www.reuters.com/world/asia-pacific/singapores-population-grows-5-foreign-workers-return-post-pandemic-2023-09-29/

Kor, Kian Beng. "SAF Confident of Coping with Tighter Manpower Resources." *The Straits Times*, June 30, 2017. www.straitstimes.com/singapore/saf-confident-of-coping-with-tighter-manpower-resources-thanks-to-increased-automation-and

Lee, Chung Min. *Fault Lines in a Rising Asia*. Washington, DC: Carnegie Endowment for International Peace, 2016.

Lee, Edwin. *Singapore: The Unexpected Nation*. Singapore: ISEAS Publishing, 2008.

Lee, Hsien Loong. "Security Options for Small States." *The Straits Times*, November 6, 1984.

Lee, Kuan Yew. *The Singapore Story: Memories of Lee Kuan Yew*. Singapore: Times Publishing, 1998.

———. "The Fundamentals of Singapore's Foreign Policy: Then & Now." S. Rajaratnam Lecture, Ministry of Foreign Affairs Diplomatic Academy (Singapore), April 9, 2009.

Loo, Bernard Fook Weng. "Explaining Changes in Singapore's Military Doctrine: Material and Ideational Perspectives." In *Asia in the New Millennium*, edited by Amitav Acharya and Lee Lai To, 352–374. Singapore: Marshall Cavendish Academic, 2004.

———. "The Management of Military Change: The Case of the Singapore Armed Forces." In *Security, Strategy and Military Change in the 21st Century: Cross-Regional Perspectives*, edited by J. I. Bekkevold, I. Bowers, and M. Raska, 70–88. London: Routledge, 2015.

Maritime Executive. "Strait of Malacca Key Chokepoint for Oil Trade." August 27, 2018. https://maritime-executive.com/article/strait-of-malacca-key-chokepoint-for-oil-trade

Matthews, Ron and Fitriani Bintang Timur. "Singapore's Total Defence Strategy." *Defence and Peace Economics* (2023), https://doi.org/10.1080/10242694.2023.2187924

Matthews, Ron and Nellie Zhang Yan. "Small Country 'Total Defence': A Case Study of Singapore." *Defence Studies* 7, no. 3 (2007): 376–395.

70    *Michael Raska*

Ministry of Defence (Singapore). *Defence of Singapore 1994–95*. Singapore: Ministry of Defence, 1995.

———. "Committee of Supply Debate 2023." Last modified February 27, 2023. www.mindef.gov.sg/web/portal/mindef/news-and-events/latest-releases/article-detail/2023/cos2023

———. "Defence Management Group." Last modified January 16, 2018. www.mindef.gov.sg/web/portal/mindef/about-us/organisation/organisation-profile/defence-management-group

———. "Defence Policy and Diplomacy." Last modified April 6, 2021. www.MINDEF.gov.sg/web/portal/MINDEF/defence-matters/defence-topic/defence-topic-detail/defence-policy-and-diplomacy

———. "Defence Science and Technology." Last modified February 7, 2023. www.mindef.gov.sg/web/portal/mindef/defence-matters/defence-topic/defence-topic-detail/defence-science-and-technology

———. *Defending Singapore in the 21st Century*. Singapore: Ministry of Defence, 2000. www.mindef.gov.sg/oms/dam/publications/ebooks/more_ebooks/ds21.pdf

———. "Fact Sheet: Defence Policy Group." Last modified September 21, 2023. www.mindef.gov.sg/web/portal/mindef/about-us/organisation/organisation-profile/defence-policy-group

———. "Fact Sheet: Digital and Intelligence Service." Last modified October 28, 2022. www.mindef.gov.sg/web/portal/mindef/news-and-events/latest-releases/article-detail/2022/October/28oct22_fs

———. "Fact Sheet: Digital Defence." Last modified February 15, 2019. www.MINDEF.gov.sg/web/portal/MINDEF/news-and-events/latest-releases/article-detail/2019/February/15feb19_fs

———. "Fact Sheet: Enhancing the SAF's Operational-Readiness and Servicemen's NS Experience through Digitalisation and Innovation." Last modified June 30, 2022. www.mindef.gov.sg/web/portal/mindef/news-and-events/latest-releases/article-detail/2022/June/30jun22_fs

———. "Fact Sheet: Ex Forging Sabre 2023—Multi-Domain Smart Warfighting." Last modified September 21, 2023. www.mindef.gov.sg/web/portal/mindef/news-and-events/latest-releases/article-detail/2023/September/21sep23_fs

———. "Fact Sheet: Exercises and Operations." Last modified June 13, 2018. www.mindef.gov.sg/web/portal/mindef/defence-matters/exercises-and-operations

———. "Fact Sheet: Leveraging Technology and Innovation to Drive Digital Transformation." Last modified March 1, 2021. www.MINDEF.gov.sg/web/portal/MINDEF/news-and-events/latest-releases/article-detail/2021/March/01mar21_fs4

———. "Fact Sheet: Strengthening MINDEF/SAF's Cyber Defence Capabilities." Last modified June 30, 2021. www.MINDEF.gov.sg/web/portal/MINDEF/news-and-events/latest-releases/article-detail/2021/June/30jun21_fs7

———. "Fact Sheet: Update on the Digital and Intelligence Service (DIS) Capability Development Efforts." Last modified February 24, 2023. www.mindef.gov.sg/web/portal/mindef/news-and-events/latest-releases/article-detail/2023/February/24feb23_fs

———. "Infographic: Towards SAF 2040." Last modified February 24, 2023. www.mindef.gov.sg/web/portal/mindef/news-and-events/latest-releases/article-detail/2023/February/24feb23_infographic

———. "Speech by Minister for Defence, Dr Ng Eng Hen, at The Committee of Supply Debates 2023." Last modified September 24, 2023. www.mindef.gov.sg/web/portal/mindef/news-and-events/latest-releases/article-detail/2023/February/24feb23_speech

Ng, Abigail. "Singapore's Total Fertility Rate Drops to Historic Low of 1.05." *Channel News Asia*, February 24, 2023. www.channelnewsasia.com/singapore/singapore-total-fertility-rate-population-births-ageing-parents-children-3301846

Ng, Pak Sung. *From 'Poisonous Shrimp' to 'Porcupine': An Analysis of Singapore's Defence Posture Change in the Early 1980s*. SDSC Working Paper 397. Canberra: Strategic and

Defence Studies Centre, 2005. https://sdsc.bellschool.anu.edu.au/sites/default/files/publications/attachments/2016-03/WP-SDSC-397_0.pdf

Rahmat, Ridzwan. "Singapore Validates Enhanced AI-Infused Combat System at US Wargames." *Janes*, September 22, 2023. www.janes.com/defence-news/news-detail/singapore-validates-enhanced-ai-infused-combat-system-at-us-wargames

Raska, Michael. "A Structured-Phased Evolution: The Third Generation Force Transformation of the Singapore Armed Forces." In *Military Innovation in Small States: Creating a Reverse Asymmetry*, edited by M. Raska, 130–162. New York: Routledge, 2016.

———. "4G SAF Creating New Advantages." *RSIS Commentary*, no. 102, May 24, 2017. https://dr.ntu.edu.sg/bitstream/10356/83194/1/CO17102.pdf

———. "Strategic Competition and Future Conflicts in the Indo-Pacific Region." *Journal of Indo-Pacific Affairs* 2, no. 2 (2019a): 83–97.

———. "The SAF After Next Incarnation." *RSIS Commentary*, no. 41, March 8, 2019b. https://dr.ntu.edu.sg/bitstream/10356/106416/1/CO19041.pdf

———. "How Will SAF Look Like After Its Next Incarnation?" *Today Online*, March 19, 2019c. www.todayonline.com/commentary/saf-after-its-next-incarnation

Singh, Bilveer. *Arming the Singapore Armed Forces: Trends and Implications*. Canberra Papers on Strategy and Defence 153. Canberra: Strategic and Defence Studies Centre, 2003. https://sdsc.bellschool.anu.edu.au/experts-publications/publications/3090/arming-singapore-armed-forces-saf-trends-and-implications

Tan, Andrew. *Domestic Determinants of Singapore's Security Policy*. APCSS Occasional Paper Series. Honolulu, HI: Asia-Pacific Centre for Security Studies, 2001.

# 4 Israel's innovation as a main pillar of defence planning

*Eitan Shamir*

## Introduction

Since its foundation, Israel has accounted for a disproportionate share of the world's military innovation compared to many larger countries. Israeli innovation culture emphasises practical solutions that assure a battlefield advantage, delivered quickly and at low cost. In a volatile political and strategic environment, long-term plans become obsolete and the ability to forecast is limited. The Israel Defense Forces (IDF) therefore rely on their innovation capacity to retain a military advantage over diverse and ever-changing adversaries and threats.

The first part of this chapter explores the sources of Israel's military innovation, its evolution and the drivers that propel it. The chapter applies Michael Raska's comprehensive model of military change and measures innovation across three dimensions: technological patterns (speculation, experimentation, implementation); conceptual paths (emulation, adaptation and innovation) and magnitude (exploration, modernisation, transformation).[1] It explains how Israel's 'customised' emulation and improvisation has evolved into adaptation and innovation.

The second part explores the sources of Israel's innovation culture, which can be traced to the unique and long history of the Jewish nation and the circumstances surrounding the state's foundation in 1948. Scarcity, a sense of 'no choice' and human talent combined to encourage local improvisation, adaptation and innovation. From the mid-1960s, the US became increasingly generous, providing the IDF with some of the best weapon systems available. This should have dissuaded the IDF from homegrown innovation, but the contrary occurred. Other contributing factors included elements of the Israeli system such as conscription, special programmes, the reserves, the cultivation of human talent and the special relationship between the IDF and Israel's military and research and development (R&D) industry.

The third part presents the IDF's recent transformation effort through its five-year force build-up plan, Momentum (*Tnufa*). The plan shows the IDF's strength in applying innovation to state-of-the-art technologies and using designated operational units as experimental labs to generate new tactics and techniques. However, it also demonstrates its limitations in the form of budgetary constraints and a constant need to divert attention to immediate security concerns. These constraints

DOI: 10.4324/9781003398158-4

became evident during the Hamas attack on Israel on October 7, 2023, prompting a potential shift in emphasis within Israel's defence planning. As of the current composition, the conflict in Gaza and the ongoing regional crisis persist. However, a substantial realignment of national priorities in Israel is anticipated in the aftermath of the surprise attack on October 7, 2023, leading to increased allocations for security and military endeavours. Israeli leadership has articulated intentions to augment investments in military industries, extend the conscription period of service, and reinstate focus on enhancing its reserve forces.

The chapter concludes with a discussion of the connections between the three themes of the book: national ambitions and alliances, innovation and the methods and processes of deciding defence planning and the ways in which they have affected force development.[2]

## Explaining Israel's innovation and adaptation in context

Major changes in the structure, doctrine and conduct of operations of a given military can stem from three venues: innovation, adaptation or emulation.[3] Innovation emphasises the new, while adaptation emphasises adjusting existing structures. The third path for change, emulation—studying and copying others' successful ideas and practices—has been regarded as the quickest way to improve a state's competitive advantage and combat power versus other means like homegrown innovation or slow adaptation.[4]

Michael Raska's comprehensive model of innovation includes two dimensions describing technological patterns (speculation, experimentation, implementation) and conceptual paths (emulation, adaptation, innovation), which determine the magnitude of innovation, exploration, modernisation and transformation (the highest).[5] Raska concludes that Israel is high on innovation towards transformation.[6] The path to transformation and innovation was evolutionary. Israel's starting point was selective emulation, adaptation and exploration, gradually moving towards innovation and transformation, sometimes going through modernisation and adaptation, at other times skipping these stages. It would probably be more accurate to say that these stages existed in parallel but the proportion between them changed.

### *Multi-source influence: learning from others and improving*

The State of Israel was founded on 14 May 1948. Weeks later, on 26 May 1948, Israel's armed forces, the IDF, were established. The country was in a *de facto* state of war as of the partition decision by the UN on 30 November 1947. What would become Israel's War of Independence lasted until 29 June 1949.

The first stage of the war was characterised by small operations as the enemy was local Palestinian armed groups. Following Israel's declaration of independence in May 1948, five Arab armies invaded Israel. The newborn IDF had to quickly adapt to fight regular war in large formations such as brigades. Under fire it also developed its then infant air force, armour and navy. Unlike many European ex-colonies that simply copied the organisation, structure and tactics of their former coloniser, the IDF was an original creation. The IDF had a few men with experience commanding large formations and leading them in regular warfare, but most of its officers who had served in the British military had reached no higher than the rank of major. The Israelis therefore had to teach themselves everything from scratch and rely on experimentation and improvisation. The weapon embargo imposed by the UN also forced the Israelis to utilise, improve on and make good use of old weapon systems sometimes deemed unfit for battle.

At the time of its establishment, the IDF was primarily influenced by two sources: the British Army and the illegal Jewish insurgency movements, chiefly the *Hagana* and its special military force, the *Palmach*. David Ben-Gurion, Israel's founding father and first prime minister, regarded the British army as a model for a professional army in a democratic state. Consequently, he promoted people who had served in the British Army and asked them to set up the IDF so it would resemble the British Army as much as possible.[7] Former *Palmach* members remained predominant in the IDF, however, and represented a more eclectic, creative kind of thinking. The *Palmach* merged different military traditions—including that of the Red Army—and modified them to match the special circumstances of Palestine at that time. Therefore, in reality, the IDF in its early years reflected a society of immigrants with various experiences and languages and was as a consequence influenced by a variety of military sources. Ideas were borrowed and implemented after they had been adjusted to fit the specific conditions of the IDF. The process was thus not a straightforward emulation but rather trial and error, through experimentation and debate. The Israelis taught themselves, so the result—while inspired by many sources—is an original creation.[8]

Other than the influence of the British Army through many serving veterans, other sources of learning were the German Army and the Soviet Red Army as well as the Polish, Austrian, American and other armies.[9] For example, Israel's Air Force (IAF) was headed by pilots who had served in and were very much influenced by the British Royal Air Force (RAF).[10] Accordingly, they demanded an independent status for the air force, as is the case with the RAF. The solution finally chosen was not the British one but a compromise custom-tailored to the IDF: the air force was granted a special status and additional powers but remained subservient to the general staff. A similar solution was adopted for the Israeli Navy.[11] The IDF reserve system is a mix of the Swiss model with the historic Prussian mobilisation model in which the main force of the army is based on fully equipped and manned reserve units that can become operational at short notice.[12] During the 1950s, when the IDF started to develop armoured combat, its interest in the German Army increased, focusing on the Germans' armoured campaigns in the Second World War and, more specifically, their tactics in the Western Desert, which highly resembles the Negev and Sinai deserts' terrain and climate.[13]

Despite a growing American supply of military technology from the 1970s, Israel's need for niche solutions coupled with a wide spectrum of ever-changing threats led to increased emphasis on homegrown tactical and technological solutions. The growth of an Israeli military-industrial complex and its constant friction and testing in battle resulted in a spiral movement towards reliance on local defence innovation.

**Israel's primary and secondary drivers of military innovation**

*Primary driver 1: 'Ein Breyra'—no choice*

The old cliché that necessity is the mother of innovation is true in Israel's case. Since its foundation in 1948, it has experienced seven major wars.[14] Between the wars, Israel endured constant attacks by militant and terrorist organisations and conducted many military operations short of war. Not a year has passed since the country's founding without violent political incidents.

This reality coupled with a strong sense of scarcity was a major driver of innovation. The country is much smaller in territory and population, and lacks natural resources, compared to its many adversaries. Sources and countries willing to provide weapons to Israel were scarce, as were the means available to the poor country for defence acquisition and maintenance of equipment. Even if some weapon systems were available, as was the case in the 1970s when the US became the major weapon provider to Israel, the available equipment did not always fit Israel's special operational needs. Therefore, the Israelis had to come up with their own solutions to their unique operational and tactical challenges. As a result of Israel's resource inferiority, the strategic imperative called for reliance on maximising human talent, specifically its scientific advantage.[15]

*Primary driver 2: grand strategy and strategic culture*

Because it is a small state, Israel's grand strategy inherently differs from that of a great power. A small state such as Israel has a very limited ability to influence global security arrangements or lead global economic trends or policies. Israel has no choice but to cope with global events through adaptation.

Israel represents a unique case of a small state that has adopted the stance of a regional power over the past two decades. It has done so by leveraging its exceptional advances in the technology and innovation sector, strong defence industry and military experience, newfound energy deposits and strong ideational appeals to the US. Employing these means, it has exploited the dramatic shifts that have occurred in the Middle East region's alliance structure since the beginning of the twenty-first century to evolve its grand strategy and actively shape that region.

While Israel maintains a special relationship with the US as a great power and relies on it for a wide umbrella of diplomatic and economic benefits, it does not represent a classic 'bandwagon' or balancing approach. Israel is not part of a formal military alliance with the US and largely conducts itself as an independent military power on the local and regional levels. Moreover, Israel occasionally acts in defiance of US interests and diplomatic pressure when it comes to military operations. However, due to the variety and magnitude of threats it faces, Israel must rely on superpower support, and it has no substitute for the US as its main patron for the foreseeable future.

Israel is a liberal democracy by most measures. However, its defence planning process and military budget are closed to the public and are decided solely by the Ministry of Defense and the Ministry of Finance. The only oversight is done by the Foreign Policy and Security Committee of the Knesset, Israel's parliament.

The committee's members have high security clearances, but they can only raise questions; the committee has no power to overrule decisions made by the Ministry of Defense.

As a result of this and other factors, the Ministry of Defense is considered the most powerful ministry in Israel. In terms of force development, every four to five years the IDF formulates a force development plan against high-probability, high-impact threat scenarios. The plan is then approved and receives an annual budget. The plan and budget are thus constantly adjusted against unforeseen events. The last IDF five-year plan, Momentum, was launched in 2020.

As mentioned, the small state of Israel has had to cope with existential threats ever since its inception. Necessity alone is not, however, sufficient to allow innovation. The second major driver for Israeli innovation is Israel's national and strategic culture. Israel's culture is the creation of many Jewish immigrant groups, largely from Eastern Europe and later from Middle Eastern Islamic countries, as well as other groups from around the world. The various groups created a unique Israeli culture united by Jewish culture and religion and the national ethos of the Zionist movement.[16] Despite a strong early influence by Soviet Russian and later American culture, Israel's society is neither conformist nor 'individualistic American' in its emphasis on self-reliance as part of the Zionist idea of the 'new Jew' who lives off the land.

Despite its shift in recent decades towards a more liberal western and individualistic society, Israeli society is still more committed to the pursuit of its national goals and collective objectives than most other western countries.[17] Israeli attitudes towards rules and regulations are more relaxed and are characterised by informality and little attention to hierarchy. Israel is defined by sociologists as a 'small power distance society',[18] and these cultural traits are also characteristic of the defence establishment and the IDF. The IDF has a relatively lean bureaucracy and a simple structure and is marked by informal and egalitarian relations.

These characteristics foster much innovation through the bottom-up flow of ideas within the informal web of the organisation. Israeli culture, both in general and in the military, is uncomfortable with formality and ceremony.[19] This often leads to little respect for traditions and constant testing and retesting of long-held established practice, in an atmosphere that allows everything to be questioned. The absence of a military tradition allowed the IDF to experiment with new ideas and methods as 'military problems were approached in [an] intellectual and open-minded manner'.[20] The IDF exercises mission command as its command approach and its commanders are expected to lead by initiative, to take action and to provide on-the-spot solutions.[21]

This practical, hands-on problem-solving approach has led to a strong inclination towards improvisation and flexibility. Improvisation is necessary for creativity.[22] Adding to these traits, the Israeli fascination with technological solutions and development completes the necessary ingredients for innovation.[23]

Along with the two primary drivers of necessity and national culture are secondary drivers that combine to create a system that facilitates innovation.

### Secondary driver 1: conscription

The IDF has a three-tier force structure comprised of conscripts, regulars and reserves. Universal conscription for all men and women was always the foundation of the IDF. Nearly all regulars and reservists (enlisted troops, NCOs and officers) served as conscripts, and all members of Israel's Internal Security Agency, the Mossad and the police have undergone compulsory military service. Officially, conscription has not changed since the establishment of the state and the IDF in 1948. Practically, however, the recruiting model has gone through many changes and adjustments in responding to changes in the threats facing the nation and in Israeli society.[24] These changes to recruiting practices are designed to support innovation, specifically targeting the technology sector. Simply put, by using conscription, the IDF controls the country's talent pool. The main source of motivation in the country's first decades was, to a large degree, the patriotic desire to serve the country.

Although the motivation to serve has not changed dramatically, a gradual shift in Israel's society has changed the sources of motivation in more recent years. Young Israelis look to realise dual purposes during their military service. When recruits speak of 'meaningful service' (*sherut mashmauti*), a relatively recent Israeli term, they mean that they wish to serve in military postings that supply valuable service both to the nation and to their own personal development.[25] In other words, they want to feel they are accomplishing something that furthers them personally, but also serves the greater collective good. The IDF is skilfully channelling the large and motivated talent pool into many technology and intelligence programmes. A large number of these programmes effectively lengthen the time served within the IDF and hence make troops undergoing them available for deployment beyond formally stipulated periods. These programmes are designed to provide training in a plethora of specialised occupations and roles that the modern-day IDF needs, including *Shchakim*,[26] the *Havatzalot* programmes,[27] and the famed crème de la crème *Talpiot* programme.[28] These programmes take place in either military or civilian settings and train recruits in such subjects as Arabic and Farsi or computer and telecommunications skills. During participation in civilian institutions (paid for by the IDF), potential recruits are not yet soldiers but do receive a small salary from the Ministry of Defense. The IDF also runs a number of technical high schools and is involved in programmes (such as computing) in regular civilian high schools that train future specialists.

To join most of these programmes, recruits must agree to lengthen their term of service by between a few months and a few years. The result is that the additional periods of training are organically related to the jobs conscripts take on. In addition, about a thousand individuals per year are allowed to study for an undergraduate degree (all universities in Israel are civilian), mainly in the sciences, medicine, engineering and social sciences. The students' summers are devoted to basic training and NCO and officer courses. Individuals enrolled in these programmes must commit, beyond their obligatory conscription term, to a further three years of service as short-term regulars. Obtaining university degrees usually guarantees that they will be placed in corresponding military roles.

Participation in these programmes can secure a good career and salary in the civilian market, so they are quite competitive. The same applies to elite combat units. Draftees must volunteer and compete to be accepted by one of the elite reconnaissance units and special forces. Some of the special forces, such as *Sayeret Matkal*, and similar top-tier Special Operations Forces (SOF) units, necessitate a commitment on the part of every soldier for a further three fully salaried years. However, unlike in many other countries where units of this type are very insular, many of those who serve in Israel's special forces graduate from an officer course and are sent as junior officers to command regular combat units. SOF units in the IDF as in other militaries are often used as the forefront and as laboratories of new methods and technologies. Thus, when embedded in regular units, the young officers bring some of their SOF open-mindedness and willingness to experiment.

Many graduates of elite IDF programmes are posted in units like the renowned 8200 and 81. Unit 8200's core function is to collect and provide signal intelligence (SIGINT). In many aspects it is the equivalent of the US National Security Agency (NSA). According to some sources, it is the unit responsible for cyber-offensive operations.[29] The essential ideology of 8200 is that nothing is impossible. Its general atmosphere and working culture resemble those of a very high-energy version of a high-tech startup company more than a regular army unit. 8200 commanders are well aware of this and do what they can to encourage unrestrained ambition.[30] Moreover, 8,200 officers employ an 'Open Door' policy that extends to the commanding officer. Anyone in the unit, including a new recruit convinced that a matter needs higher-level attention, is free to present him- or herself to a more senior officer, up to and including the commanding officers, without having to respect the official chain of command. The unit organises lectures, seminars and enrichment programmes in many fields, with fun days to break the routine.[31] An example of such an activity is the TED Project, where soldiers get to choose a field of interest, whether or not it is linked to their military work and present a TED talk about it to their colleagues.[32]

### *Second driver 2: the role of the reserves*

The basic asymmetry between Israel and its opponents meant the IDF would have to rely on a large reserve force that could mobilise on a few hours' notice in case of a situation of high-intensity regular war. Such was the case in the Six-Day War of 1967 and the 1973 Yom Kippur War. From the day of its foundation following a decree by Prime Minister David Ben-Gurion on May 26, 1948, 12 days after the Declaration of Independence,[33] the IDF included the crucial innovation of a *reserve-centred* structure, whereby the training of personnel to man reserve formations with stored weapons and equipment is a primary purpose of active-duty ground forces. With multiple Arab armies already invading its borders from the south, east and north, it was obvious that the newborn state could not possibly field an army large enough to protect a total Jewish population of 650,000, as it was on May 15, 1948.[34]

It was quickly discovered that the reserve component compensates for young conscripts who lack experience and expertise, the universal downside of the conscription model. IDF reservists also outnumber career officers. In contrast to the former, the latter have only one objective: to get the job done with as few distractions and as little interference as possible. Reservists serve in all parts of the IDF and their military expertise covers every branch. Their civilian expertise ranges from astrophysics to zoology with all forms of engineering in between, along with management and more. They produce a constant stream of expertise and know-how and introduce it into the IDF.

While the two other major recruiting alternatives of a part-time militia or full-time regular army were considered, the answer had to be an original invention: an army largely composed of fully equipped reserve units manned, when necessary, by former conscripts recalled to active duty. Until then, those reservists could lead ordinary and productive civilian lives. That way, the country could field a disproportionately large army in wartime while otherwise keeping in uniform only the current crop of conscripts, as well as a relatively small cadre of professional officers and assorted specialists.

There was nothing original about placing conscripts on the rolls of reserve units after their term of national service, a practice well established in European armies in the nineteenth century, to add mass to regular standing forces upon mobilisation for war. There is also the famous Swiss model of a full citizen army. Yet there was, and is, a critical difference: the ongoing wars and operations short of war the IDF has fought since its establishment. Israeli reservists have to be called up to provide extra strength even between wars, to face unending security threats large and small in between major outbreaks of war. Those serving in combat and support formations are also recalled on an annual basis to keep up their combat readiness with refresher training and can be recalled yet again to learn to operate and maintain new equipment issued to their unit since the last annual recall. When it comes to ground forces, in particular, reserves greatly outnumber active-duty formations.

To ensure they maintain the required skills, reservists are first trained thoroughly when they are full-time conscripts and are then recalled for annual refresher training stints lasting as long as a month. These reserve stints must be efficient and productive because civilians taken away from their families, jobs and businesses have very little patience with useless drills and poorly run exercises. The IDF's high ratio of reservists to active-duty personnel connects army and society at every level and to a unique extent, especially now that universal military service has been abandoned in almost every other country.[35]

Reservists have been important innovators in the IDF. The innovation connection can be very direct: IDF reservists, and not only scientist-reservists or engineer-reservists, who are unhappy about whatever is issued to them by way of weapons or equipment, or the lack of some item of equipment that perhaps exists only in their minds, frequently initiate proposals within the IDF and beyond, perhaps contacting research centres or one of the country's aerospace and defence firms. They can always find a fellow reservist on the inside, or at least the friend of a fellow reservist. They then press from the outside and keep at it until the relevant IDF command either takes up the evaluation or turns it down—and even then, they might keep trying. Persistence is in the Israeli national culture and in most cases, doors are always open for reservists at the command headquarters of their own unit. From there it is rarely difficult to reach anyone, including the IDF's most senior generals.

### Secondary driver 3: a unique military-industrial complex

When it comes to the development and production of new weapons, platforms and systems, the IDF benefits from a uniquely close relationship with the country's aerospace and military industries because of the dominance of its reservists in their management, research departments and labour force.[36] Regardless of the ownership of the different firms, some entirely private, others entirely state-owned and some in-between, employees can never forget that what they design, develop and produce could be used by themselves and their kin when the next violent event erupts on Israel's borders.

Aside from the emotional aspect, there is direct communication between the IDF customer and military suppliers staffed in great part by IDF reservists. All over the world, such communications are carefully calculated and guarded because huge amounts of money are involved and any military acquisition can easily become politically controversial. There are a great many bureaucratic rules to guard against improprieties and ensure strict impartiality.

Almost everywhere in the world, military purchasing begins with some type of RFP (request for proposals) fuelled by intense efforts by potential vendors to understand what the military customer actually wants. This is crucial because not infrequently it happens that the military purchaser is not allowed to freely specify what it wants, either because the defence ministry would object for political reasons or because the finance or industry ministry would object for industrial reasons. The military purchaser must therefore compromise its own preferences to accommodate broader defence, financial or industrial priorities. Often, obscure details in an RFP are designed to favour a specific contractor—usually the traditional national domestic supplier.

The weapon development process should be very fluid to accommodate changes quickly—otherwise, by the time the weapons system is produced and delivered to the armed forces, it may no longer fit changed military requirements, or even be obsolescent. But fluidity is drastically restricted by contractual obligations and specifications that cannot be simply changed as needed without formal and highly detailed 'change orders' that require elaborate renegotiations, conducted by exchanging legally vetted correspondence. This means lawyers participate at each step, not just engineers and cost analysts. The result is often a rigid and not very change-responsive system.

In Israel, by contrast, the active-duty IDF people doing the buying and the mostly reserve IDF people doing the selling—or rather the research, development, fabrication, testing, evaluation, modification and retesting—talk to each other all the time, before and after the actual contract is signed, with no lawyers involved except when all the work has been done and final contracts can be signed. There is no waiting for contracts to be renegotiated to pursue design changes—which are essential to keep everything up-to-date as components change—and there is no waiting for periodic 'progress reviews'.

There are instead informal channels of coordination between the IDF unit doing the buying and the industrial teams working to develop or produce equipment for them, with a single coordinating institution in the middle: the Administration for the Development of Weapons and Technological Infrastructure (known by its Hebrew acronym MaFat). MaFat was jointly formed by the Ministry of Defense and the IDF to coordinate with all state-owned entities engaged in R&D and pro-duction of IDF equipment: Israel Military Industries, Israel Aerospace Industries, *Rafael* Advanced Defense Systems, the Institute for Biological Research and the Space Agency.

The MaFat director, a full member of the General Staff, is a brigadier general, but his civilian suit indicates his hybrid position. MaFat's brief is to preserve the IDF's qualitative edge in weapons and infrastructures by directing domestic R&D projects and joint projects with foreign partners, and by nurturing exceptional man-power for them, not least through the Talpiot programme.

Its internal structure reflects the diversity of the disciplines MaFat is supposed to coordinate. Applied science is the province of its Technological Infrastructure and Research Unit, which supplies useful applications for the R&D projects it has prioritised; the Space Administration, which is tasked with the R&D, manufac-ture, launch, orbit placement and subsequent operation of all satellites; the missile-defence directorate, which overviews all anti-missile R&D projects in cooperation with its US counterpart, the Department of Defense's (DoD) Missile Defense Agency (MDA); the drone directorate, which advances unmanned aerial vehicle (UAV) capabilities and technologies and a variety of other units in charge of budg-eting, one-off projects and liaison with foreign partners.[37]

MaFat was the product of a 1971 debate between the IDF and the Ministry of Defense over the fire-control system needed for the new battle tank—Israel's first such effort—whose development had just started and would eventually produce the *Merkava*. The question was whether it would have to be imported, as with the diesel engine, whose local design and production were unimaginable, or whether Israel's fledging electronic industry was up to the task.[38]

The unexpected outcome of this debate, which featured the Office of the Chief Scientist of the Ministry of Defense on one side and the IDF's weapon develop-ment department on the other, was a decision to merge the two into a joint civilian-military R&D unit. The innovation in MaFat's structure lies in its hybrid nature—its head attends both IDF General Staff meetings as a general and the Defense Minis-try's staff meetings of department chiefs as an administrator. Indeed, he is both a general in the reserves and a civilian administrator.[39]

Another part of the defence establishment, *Rafael*, the Armament Development Authority mentioned earlier, also experienced a revolutionary metamorphosis, in fact several. It was formed early in 1948 as the 'Science Corps' to gather individual scientists to try to invent things that could help the desperately overtaxed fighting units in some way or other—but no miracles are recorded. In 1952, with a bit of funding, the Science Corps became the Research and Design Directorate, with both a research element and a weapons development unit. It was reorganised in 1958 as *Rafael*, after the Hebrew acronym for 'Authority for the Development of Armaments', and later further renamed as the current *Rafael* Advanced Defense Systems Ltd, incorporated as a limited company in 2002.

Though still entirely state-owned, as a self-standing company *Rafael* is a fair competitor for the country's private companies and is responsible for a string of innovations: the *Python* series of air-to-air missiles; the *Spike* family of fire-and-forget surface-to-surface missiles; the *Popeye* very long-range air-to-ground missile, which is believed to be the basis of a nuclear-armed, submarine-launched cruise missile; the *Iron Dome* system for the low-cost interception of cheaper rockets, but expensive missiles as well; *Trophy*, the first effective active defence system for armoured vehicles, the world's first operational unmanned surface vehicle; and *David's Sling*, an anti-missile system of much longer range than *Iron Dome*.

*Rafael's* company culture still echoes the ethos of Major General Moshe 'Musa' Peled, a retired major-general and armour officer who played a large role in the Golan Heights battles in October 1973 as commander of an armoured division of reservists.[40] He was appointed president of *Rafael* in 1987. In his new post, Peled mounted another persistent offensive, this time against bureaucratic proclivities and tendencies to play it safe.[41] He did not think it worthwhile to pursue incremental innovation—Peled demanded real breakthroughs or nothing. He reportedly told his engineers: 'If every project results in a success it means you are not daring enough. I would expect an overall failure rate of 50%.'[42] Peled's struggle against mediocrity came at a time when he feared that *Rafael* might lose its edge, and *Rafael* became more of a risk taker than ever before.

### Current and future directions for Israel's innovation

Following the relatively poor performance of the IDF land forces in the 2006 Lebanon war, Israel's defence establishment and the IDF were embroiled in internal soul searching and debates about the best way forward in terms of force development. It was clear that many years of counterterrorism and policing duties during low-intensity battles against unsophisticated enemies had led to a neglect of combined arms fire and movement, which had once been the hallmark of the IDF.[43] When the IDF faced a more formidable opponent it realised it needed to make major adjustments to maintain effectiveness.

In the period of about 15 years until Lt. General Aviv Kochavi came into office as IDF Chief of Staff (COS) in 2019, three major schools of thought surfaced.[44] One, led mainly by Lt. General COS Gabi Ashkenazi (2007–2011), was the 'back to basics school'. The remedy, according to this approach, was to reclaim basic tactical skills that had been lost. The second approach, represented primarily by COS Lt. General Benny Gantz (2011–2015), suggested that following the changes in the character of war and the nature of the threats to Israel, the IDF should focus on areas of relative strength: airpower, intelligence dominance and SOF. This school advanced the argument that future conflicts with Hezbollah and Hamas no longer required large manoeuvring formations, as was demonstrated in Lebanon in 2006 and in Gaza in 2008, 2012 and 2014.

Kochavi represented a third approach. His multi-year force build-up plan, *Tenufa* (Momentum), is an attempt to find a middle ground. The main idea is the recognition that large manoeuvres will be required in the future, but just as the Blitzkrieg in 1940 represented a new approach to warfare, the IDF needs to reinvent manoeuvres for the twenty-first century.

The purpose of the Momentum programme, according to COS Kochavi, is to build a force adapted to the challenge posed by 'decentralised terrorist armies' equipped with thousands of rockets aimed at Israel's population centres.[45] Changes to the IDF force structure should allow the IDF to become significantly more lethal and achieve unequivocal decisions. New capabilities would enable the IDF to expose and attack targets at a much higher rate and accuracy than in the past.

In addition to counterfire and manoeuvring, the plan includes defence of the rear, which will be a central area in any future campaign. The conceptual foundation of the plan traces back to the technological developments of the twenty-first century, which constitute, according to its proponents, the primary driver for the change. These developments were described by the chairman of the Davos Forum, Klaus Schwab, as the 'fourth industrial revolution'.[46] The essence of this revolution is a combination of 'new technologies, including artificial intelligence (AI) machine learning, natural language coding, robotics, sensors, cloud computing, nanotechnology, 3D printing and the Internet'.

The Momentum plan aims to harness the vision of the fourth industrial revolution to encourage a military revolution. Brigadier General (BG) Eran Ortal, commander of the Dado research centre that led the conceptual work, laid out the main ideas in a series of articles:

> The technological innovations that have recently become available allow the IDF to implement a much more aggressive Offense-Defense Integration approach. It includes a link between various weapons and sensing systems, and a combination of unified launchers for the purposes of intercepting and attacking sources of fire on the battlefield.[47]

According to the plan's advocates, the era of artificial intelligence, autonomous machines and advanced information processing will enable the establishment of sensing combinations, information processing and rapid attack through reconnaissance units that will operate on the battlefield as part of the manoeuvring force. The tactical information-gathering elements will be based on UAVs, drones and radars linked in a data network and will receive and decode the emission of signatures and their locations. All this is in order to create 'defence-attack' combinations.[48] The intention then is to build a 'multidimensional combat capability, both through better integration between the arms and through imparting multidimensional capabilities to the tactical ranks themselves, all through more advanced combat methods and proper exploitation of the current technological revolution'.

The five-year plan's force build-up has the following main themes: considerably increasing the lethality of forces in scope, quality and precision; strengthening tactical engaging forces by providing high-quality information in real time; rapidly exposing low-signature enemy formations in all the spaces where the enemy operates, such as urban, subterranean and mountains; using multidimensionality as a basis for action, meaning the synchronised coordination of the IDF's different strengths into actions that increase operational achievement; using connectivity for multidimensional and multi-armed synergy by harnessing artificial intelligence to increase quality and shorten intelligence and operational processes and using robotics and autonomous machines to expand the scope of action on the battlefield while reducing risk to IDF forces.[49]

The meaning of the transformation for ground forces is explained by another officer from the IDF doctrine department.[50] As far as land forces are concerned, the challenge is not necessarily to create networked connectivity between one unit and others, but to sustain a unit that has a host of capabilities, such as UAVs and drones, unmanned ground vehicles for reconnaissance and attack, cyber capabilities, diverse defence capabilities including in the electronic medium, and diverse attack capabilities like missiles of different ranges, precision mortars and the like.[51]

COS Kochavi has established an experimental unit in the IDF, the 'Ghost' unit, which is about the size of a battalion. It is intended to operate independently and has at its disposal a large arsenal of capabilities that until now existed at the division level and higher. The Ghost unit essentially calls for a multitude of small infantry teams assisted by 'swarms' of small remotely piloted aircraft to help uncover enemy positions and communicate them by network technology to aircraft and ground launchers, which would then destroy the targets within minutes or even seconds by standoff fire from afar.

The Ghost unit is being used as a laboratory to conduct experiments and exercises to validate the concept and carry out continuous learning. As such, the unit is a kind of 'playground' and is given a lot of freedom for trial and error. Its ultimate mission is to serve as the vanguard for the entire IDF.[52] During the May 2021 escalation of fighting with Gaza, the unit was successfully deployed on the Gaza border and conducted operations to detect and destroy Palestinian rocket launchers and guided anti-tank missile launchers firing into Israel.[53]

The Momentum five-year plan has not been implemented at the pace and scale envisioned by Gen. Kochavi. The setbacks are typical of the barriers in Israel's system to wide organisational transformation as opposed to specific innovation in a weapon system or a unit. The plan was stalled due to the ever-changing need to respond to strategic developments such as events on Israel's border and the COVID-19 pandemic. The other barrier is Israel's political instability, which resulted in many months without budget approval.[54]

On October 7, 2023, Israel faced an unexpected, all-out attack by Hamas. Approximately 3,000 armed Hamas militants infiltrated Israeli territory, simultaneously targeting military bases and villages along the Gaza border. The attackers utilised drones, anti-tank missiles (ATGM), and explosives to breach Israel's sophisticated barrier, equipped with remote-controlled cameras, sensors, and weapons.

After more than 24 hours of intense fighting, Israel managed to repel the attack, but not without significant losses. The casualties included 1,112 civilians and security forces killed, with thousands more injured. Additionally, 252 individuals were abducted into Gaza, and around 220,000 residents were evacuated from their homes.[55] The events of October 7 marked Israel's most devastating single-day catastrophe since its independence.

In response, Israel declared war and initiated a full mobilisation of its reserves. The counter-offensive began with extensive air raids, followed by a ground operation involving several IDF divisions. Escalations also occurred on Israel's northern border with Hezbollah, and the Yemeni Houthis launched cruise missiles and ballistic missiles toward Israel's southern port city of Eilat on the Red Sea. The Houthis also targeted commercial shipping in the Red Sea associated with Israeli ports.

As of the current writing, the war is ongoing, and its final outcome remains uncertain. Nonetheless, some observations can be made regarding force planning in the aftermath of this strategic crisis. It is evident that previous Israeli retaliation operations against Hamas (in 2008, 2012, 2014 and 2021) did not achieve their intended objectives, as Hamas was not deterred and continued to strengthen its forces in preparation for such an attack. Israel recognises the need for large ground forces to conduct extended and substantial operations to effectively neutralise Hamas or Hezbollah capabilities. Consequently, substantial investments in ground formations and reserve forces can be expected.

The conflict underscores the necessity of ground formations and the need for high readiness of the reserve forces. Another lesson is the need for ample stockpiles of munitions in prolonged attrition wars, leading Israel's industry to prioritise the IDF over exports in the near future.[56] Moreover, the advantage of a networked military and the importance of jointness between air and ground forces, fire and manoeuvre, and other capabilities developed through the momentum plan have been evident. It is likely that the IDF will continue to invest in enhancing collaboration and integration among these different elements while also investing in more traditional elements such as armour, infantry and artillery units. Just as in the case of Lebanon in 2006 and the Ukraine war, it has been confirmed again that technology alone does not substitute the skill, training, and professionalism of soldiers and commanders.

## Michael Raska's analytical framework—discussion

Utilising Raska's three-dimensional matrix (Figure 4.1), Israel has achieved high scores on both the *organisational change* (*x*-axis) and the *technological patterns* (*y*-axis). Time and again, Israel has demonstrated its ability to translate innovative ideas into practical and highly effective systems, proving their prowess on the battlefield. Noteworthy examples include the pioneering use of the first UAVs in combat, the introduction of the first missile boats in the western world and more recent advancements like the Iron Dome and Trophy systems. Moreover, Israel's innovation extends beyond technology, encompassing numerous tactical, personnel practices and doctrinal advancements. Based on these considerations, when combining the scores to establish Israel's position on the *organisational change* graph, I contend that Israel falls above *modernisation* and is well towards the *transformation* point; however, it is short of scoring a perfect 3.3 for the reasons mentioned at the end of the previous section, limited resources and frequent security and domestic distractions.

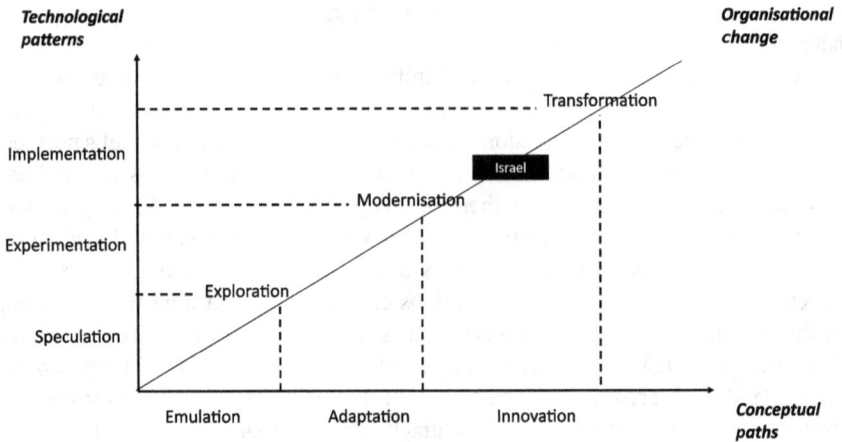

*Figure 4.1* Israel's military innovation trajectories.

*Source:* Figure created by the author based on Michael Raska, Military Innovation of Small States—Creating a Reverse Asymmetry (Routledge, 2016)[57]

## Conclusion

In the context of structured comparative analysis, Israel, despite its small size, has transformed into a significant regional power across several critical domains, notably in its military and technological prowess (see Figure 4.2). In its first decades of existence, Israel's ambitions were limited to survival in a hostile region and its options in terms of alliances were very limited. Its national strategy doctrine dictated that it relied on a major power to provide necessary diplomatic material support while owning its internal resources and ingenuity. Continuous trial and error through numerous wars and operations and the necessity for rapid response led to force development based on adaptation and innovation. Gradually and more rapidly since the revolution of military affairs (RMA) in the 1990s, Israel became a technology and high-tech leader using its military technology sector to feed its civilian sector and vice versa. The export of high technology, both military and civilian, boosted Israel's economy to the level of top-tier economies.[58]

Despite its considerable capabilities, Israel's security doctrine dictates a continued reliance on a major power for the diplomatic and material assistance it necessitates, particularly during times of conflict. This reliance has been notably evident since the 1970s, with the United States extending steadfast support to Israel. This enduring partnership involves essential military aid, encompassing acquisitions such as cutting-edge fighter jets and substantial munitions supplies. However, it is crucial to note that Israel's doctrine emphasises the intrinsic responsibility of its own armed forces to conduct military operations.

In the twenty-first century, Israel's geostrategic landscape has completely changed, though the country remains under constant serious threats that force it to keep its edge. The rift between the Shia Iranian camp—which also represents

| | SMPS ADJUST THEIR APPROACHES BASED ON SYSTEMIC CHANGE OUT OF THEIR CONTROL | SMPS REACT TO CHANGES IN THE RELEVANT SECURITY COMPLEX(ES) THE SMP IS PART OF | | | | |
|---|---|---|---|---|---|---|
| | *Alliances, Dependencies and National Ambitions* | | | | | |
| | SMPS FACE CONSTRAINTS IN SAFEGUARDING NATIONAL SECURITY | FINANCIAL | DEMOGRAPHIC | CULTURAL | GEOGRAPHICAL | OTHER |
| | SMPS NEED TO POSITION THEMSELVES VIS-À-VIS GLOBAL AND REGIONAL POWER CONSTELLATIONS | BALANCING | HEDGING | BANDWAGONING | NON-ALIGNING | OTHER |
| | SMPS HAVE TO ENGAGE WITH OTHER STATES TO INCREASE THEIR DEFENCE CAPABILITIES | SECURITY GUARANTEES | TRANSACTIONAL | FORCE INTEGRATION | SPECIALISATION | OTHER |
| | SMPS HAVE ARMED FORCES FOR DIFFERENT PURPOSES | INSTRUMENT OF DIPLOMACY | PRESTIGE PROJECT | REDUCE STRATEGIC RISK (DEFEND) | DOMESTIC STABILITY | OTHER |
| | *Approaches, Processes, Methods and Techniques* | | | | | |
| | SMPS TAKE A MORE CONSTRAINED APPROACH TO DEFENCE PLANNING | THREAT-NET ASSESSMENT BASED | PORTFOLIO-BASED | TASK-BASED | MOBILISATION | OTHER |
| | SMPS HAVE LIMITED RESOURCES TO DESIGN AND IMPLEMENT DEFENCE PLANNING PROCESSES | TAILORED ANALYTICAL PROCESS | OUTSOURCING STRATEGY | ACCEPT DILUTION / FAIR SHARE | EXTERNAL TEMPLATE | OTHER |
| | SMP EMPLOY MULTIPLE METHODS AND TECHNIQUES | COPY GREAT POWERS | INFORMAL CONSENSUS | TAILORED METHODS | AD HOC | OTHER |
| | SMPS RELY ON DIFFERENT PROCESSES TO DECIDE ON FORCE POSTURE | ANALYTICAL RESULTS | POLITICS DECIDES | MILITARY DECIDES | GREAT POWER DECIDES | OTHER |
| | *Military Innovation* | | | | | |
| | SMPS FACE CHALLENGES IN KEEPING PACE WITH TECHNOLOGICAL ADVANCES | BUILD DOMESTIC TECH/INDUSTRY | TECH TRANSFER AGREEMENTS | ACQUIRE READY INNOVATIONS | PARTNER WITH GREAT POWER | OTHER |
| | SMPS PURSUE VARIOUS MILITARY INNOVATION PATHS AND PATTERNS | EMULATION / SPECULATION | ADAPTATION / EXPERIMENTATION | INNOVATION / IMPLEMENTATION | OTHER / OTHER | OTHER |
| | SMPS FIND THEMSELVES AT DIFFERENT STAGES OF ORGANISATIONAL CHANGE | EXPLORATION | MODERNISATION | TRANSFORMATION | OTHER | OTHER |

*(Left margin labels: GENERAL COMMONALITIES; VARIATIONS WITHIN THESE COMMONALITIES)*

*Figure 4.2* Structured focused comparison framework for Israel.

*Source:* Author

the main threat to Israel—and the moderate Sunni states, coupled with the ongoing disengagement of the US from the region, has positioned Israel as a regional leader in opposition to Iran's nuclear aspirations and regional expansion efforts.[59] In a highly significant development, Israel is now viewed by the Arab Gulf States as a strategic ally. The region's countries are now interested in obtaining Israeli military technology and know-how in areas such as phone surveillance, missile defence, drone operations and counterterrorism tactics. Thus, Israel's military innovation has become its foremost means of leveraging its diplomatic status and influence. In contrast to its earlier isolation, Israel is now operating in a large arc from Azerbaijan to Greece, Cyprus and the Gulf countries, strengthening its relations through military and defence collaboration, commerce, technology and science. Its forces participate in exercises with many other militaries. Its relocation under US CENTCOM command is another strong indication of these countries' desire to work closely with Israel and benefit from what it has to offer.[60]

From the first day of its independence, Israel was at war; therefore, its defence planning had to be driven by a threat-based analysis. Israel also faced unique challenges and problems and therefore had to tailor its unique solutions. This situation led to innovation that primarily focuses on addressing immediate problems and niche, low-cost solutions. This approach has both its advantages and limitations. The primary impetus for Israel's innovation arises from its geopolitical vulnerability, resource scarcity and constant encounters with numerous threats and wars. Consequently, a culture of 'no choice' and improvisation has taken centre stage in driving innovation forward. Over time, Israel has developed an entire ecosystem to support its military innovation. This ecosystem includes young conscripts who participate in special programmes and reservists who often hold dual roles in their civilian lives, working in high-tech companies and military industries while serving as reserve officers. Additionally, the unique structure of Israel's military-industrial complex is tightly intertwined with the armed forces.

Currently, Israel is vigorously pursuing transformation efforts to leverage the vision of the fourth industrial revolution, propelling its military revolution further ahead. Units like the Ghost unit serve as laboratories to conduct experiments and exercises, validating concepts and facilitating continuous learning. However, despite its ambitious five-year plan, Israel's limitations as a small country are evident in the form of budgetary constraints and the constant need to divert attention to immediate security concerns. These latter factors significantly influence the context in which innovation thrives.

The surprise attack by Hamas on October 7, 2023, is poised to prompt a shift in Israel's priorities. Anticipated changes include an augmentation of the defence budget, with a particular focus on enhancing both the size and quality of the army. This shift implies increased investments in the ground forces and the increase of large armour formations. Importantly, this emphasis on conventional forces is not intended to replace but rather complement the ongoing efforts in advancing military technology.

It is therefore safe to predict that as long as Israel continues to experience security pressures from its belligerents, innovation will remain its key approach. Innovation provides not only a much-needed battlefield advantage but also high economic dividends and diplomatic leverage.

## Notes

1 Michael Raska, *Military Innovation in Small States, Creating a Reverse Asymmetry* (Abingdon: Routledge, 2016), 168.
2 Yossi Yehoshua, 'The Lessons of The War: The IDF Will Grow, the Service in It Will Be Extended, and a Solution to the Permanent Crisis Will Be Found', *Ynet*, 7 November 2023, www.ynet.co.il/news/article/yokra13666011
3 Theo Farrell and Terry Terriff, *The Sources of Military Change: Culture, Politics, Technology* (New York, NY: Lynne Rienner Publishing, 2002), 6.
4 Farrell and Terriff, *The Sources of Military Change*, 6.
5 Raska, *Military Innovation*, 168–169.
6 Raska, *Military Innovation,* 192.
7 Edward Luttwak and Dan Horowitz, *The Israeli Army* (London: Allen Lane, 1976), 74.
8 Luttwak and Horowitz, *The Israeli Army,* xi.
9 Yigal Shefi, *The Platoon Leader's Clip, Military Thinking in Officers Courses in the Hagana* (Tel Aviv, Israel: Ministry of Defense, 1991), 57. (Hebrew)
10 Interview with Major General Dan Tolkowsky, Israel Air Force Commander between 1953–1958, Tel Aviv, Israel, 27 July 2015.
11 Zahava Ostfeld, *An Army Is Born* (Tel Aviv, Israel: The Ministry of Defence, 1994), 560. (Hebrew)
12 Gunther Rothenberg, *The Anatomy of the Israeli Army* (London: Redwood Burn, 1979), 82.
13 Aryeh J.S. Nusacher, 'Sweet Irony: The German Origins of Israel Maneuver Warfare Doctrine' (MA thesis, Royal Military College, Canada, 1996).

14 1948 Israel's Independence War, 1956 Sinai Campaign, 1967 Six-Day War, 1969–1970 War of Attrition, 1973 Yom Kippur War, 1982 First Lebanon War, 2006 Second Lebanon War.
15 Dima Adamsky, 'The Israeli Approach to Defense Innovation', *SITC Research Briefs* 10, no. 7 (2018): 3.
16 Motti Regev, 'To Have a Culture of Our Own', *Ethnic and Racial Studies* 23, no. 2 (2000): 223–247.
17 Dmitry Adamsky, *The Culture of Military Innovation: The Impact of Cultural Factors on the Revolution in Military Affairs in Russia, the US, and Israel* (Stanford, CA: Stanford University Press, 2010), 110.
18 Adamsky, *The Culture*, 110.
19 *Ibid.*, 110–111.
20 Eitan Shamir, *Transforming Command: The Pursuit of Mission Command in the US, British, and Israeli Armies* (Stanford, CA: Stanford University Press, 2011), 85.
21 Shamir, *Transforming Command*, 198; Adamsky, *The Culture*, 117.
22 Adamsky, *The Culture*, 117.
23 Meir Finkel, 'Worshiping Technology in the IDF: Recovering the Balance for Ground Forces Development', *Ma'arachot* 407 (2006): 40–45 (Hebrew).
24 See: Eyal Ben-Ari, Elisheva Rosman, and Eitan Shamir, 'Neither a conscript army nor an all-volunteer force: Emerging recruiting models', *Armed Forces & Society* 49, no. 1 (2023): 138–159.
25 Reuven Gal, 'Motivation Levels for IDF Enlistment Over the Years', in *Military Service in Israel: Challenges and Ramifications*, eds. Elran Meir and Sheffer Gabi (Tel Aviv: Institute for National Security Studies, September 2016, Memorandum 159), 49–60 (Hebrew); Roni Tiargan, 'Different Reflections on the Motivation to Serve in the IDF', in *Military Service in Israel*, 61–76 (Hebrew).
26 Shchakim is a programme preparing potential conscripts for telecommunication roles in intelligence postings. Those found suitable for the programme are invited to participate in a pre-service programme that does not count towards the time they must serve. After completing the course, they must sign up for additional service, ranging from six extra months to two extra years. Women must also sign on for 32 months of service, equal to men's compulsory service.
27 Havatzalot is the flagship programme of the Intelligence Corps. Those selected for the programme complete a B.A. programme in Middle Eastern Studies and Political Science together with telecommunication skills. They must sign on for a minimum of six years of additional service and attend officers' course.
28 Talpiot is the premier IDF programme and the most prestigious training program for recruits who have demonstrated outstanding academic ability in the exact sciences and have shown leadership potential. Graduates pursue university degrees while they go through long military training that exposes them to a variety of IDF units. After graduation they are posted in various research and development units in technological leadership positions and are expected to lead the IDF's technological breakthroughs.
29 Josh Halliday, 'Stuxnet worm is the "work of a national government agency"', *The Guardian*, 24 September 2010, www.theguardian.com/technology/2010/sep/24/stuxnet-worm-national-agency.
30 From interviews with serving officers and NCOs in Unit 8200.
31 Lieutenant Colonel Uri, 'Chisel Water from Rock—Change and Changeability in the IDF's R & D Mechanisms', *Between the Poles: Power Building—Part 2* 7 (2016): 41–59. (Hebrew)
32 Interviews with serving Unit 8200 officers.
33 See 'David Ben Gurion and the IDF', *IDF & Defense Establishment Archives*, 12 September 2017, www.archives.mod.gov.il/pages/exhibitions/bengurion/bigImages/hakamat_tzahal.jp.

34  Dov Tamari, *The Armed Nation: The Rise and Decline of the Israel Reserve System* (Moshav Ben Shemen, Israel: Modan Publishing House and Ma'arachot, 2012), 59–214. (Hebrew).

35  Even after recent cuts. Only the Finnish armed forces are comparable: as of 2016 they consisted of some 24,000 on active duty and 230,000 in equipped reserve formations.

36  The state-owned Israel Aerospace Industries, or IAI; *Rafael*, the Hebrew acronym for 'Authority for the Development of Armaments', officially Advanced Defense Systems Ltd, and the private Elbit Systems, whose proprietary technology includes the F-35's chief advance, its helmet-mounted display system.

37  See information on MaFat on the MOD website: www.mod.gov.il/Departments/Pages/ Research_and_Development_Agency_Mafaat.aspx. (Hebrew)

38  Uzi Eilam, *Eilam's Bow* (Tel-Aviv: Miskal, Yedioth Aharonoth, Chemed Books, 2009), 153–162 (Hebrew). The book was also published in English as *Eilam's Arc: How Israel Became a Military Technology Powerhouse* (Brighton, UK: Sussex Academic Press, 2011).

39  Eilam, *Eilam's Bow*, 354–370.

40  In the Patton Museum of Kentucky's Fort Knox, the high temple of the armour fraternity, Peled's photo is displayed alongside those of Patton, Erwin Rommel, Creighton Abrams, Marshal Georgy Zhukov and Israel Tal.

41  Eilam, *Eilam's Bow*, 374–378.

42  Eilam, *Eilam's Bow*, 378.

43  Itay Brun, 'Where Has the Maneuver Disappeared?', *Ma'arachot Journal* 420/421 (2008): 4–14. (Hebrew)

44  Eran Ortal, 'Turn on the Light, Put Out the Fire', *Between the Poles—Land Forces Part A*, 31/32 (2021): 53–69. (Hebrew)

45  Aviv Kochavi, 'COS Forward', *Between the Poles—Military Superiority and the Momentum Plan,* 28–30 (2020): 7–8. (Hebrew)

46  Klaus Schwab, *The Fourth Industrial Revolution* (New York, NY: City Crown Publishing House, 2016).

47  Eran Ortal, 'Resuming the Offensive: A Theoretical Framework for the Momentum Plan', *Between the Poles: Military Superiority and the Momentum Plan* 28–30 (2020): 38.

48  Ortal, 'Resuming the Offensive', 39–40.

49  Ortal, 'Turn on The Light', 66.

50  Yoel Strik and Moshe Tsach, 'A Tale of Champions', *Between the Poles: Military Superiority and the Momentum Plan* 28–30 (2020): 213–227. (Hebrew)

51  Gal Ritz, 'Establishing an Independent Land Air Capability: Revolution in Land Warfare', *Between the Poles* 37 (2022). (Hebrew)

52  Shir Sachar, 'Where Is the Ghost Unit Going?', *Ma'arachot Journal*, 2 December 2021, www. maarachot.idf.il/2021/%D7%9E%D7%90%D7%9E%D7%A8%D7%99%D7%9D/% D7%99%D7%97%D7%99%D7%93%D7%AA-%D7%A8%D7%A4%D7%90%D7% 99%D7%9D-%D7%9C%D7%90%D7%9F/.

53  Amir Bokhbot, 'Flocks of IDF Quadcopters Participated in the Fighting in Gaza and the Rules of the Game Are Expected to Change', *Walla News*, 5 June 2021, https://news. walla.co.il/item/3439695. (Hebrew)

54  Noam Amir, 'The Security Establishment Is Afraid: The Dissolution of the Knesset Will Eliminate the Momentum Plan: Chief of Staff Kochavi's Multi-Year Plan Continues to Be Postponed Due to Non-Transfer of the Budget', *Makor Rishon*, 15 October 2020, at: www.makorrishon.co.il/news/288009/; Shmuel Aven, 'The *"Tnufa"* Multi-Year Plan for the IDF: Where are the Cabinet Approval and the Budgets?', *INSS Insight*, no. 1,357, 5 August 2020, at: www.inss.org.il/publication/tnufa-where-is-the-cabinet.

55  "Israel At War—Data Updated in Real Time", INSS, Israel, www.inss.org.il/he/ publication/war-data/

56  Yehoshua, 'The Lessons of the War'.

57 Michael Raska, 'A Structured-Phased Evolution: The Third Generation Force Transformation of the Singapore Armed Forces', in *Military Innovation in Small States: Creating a Reverse Asymmetry*, ed. M. Raska (New York: Routledge, 2016), 130–162.

58 Evidence of this development is Israel's joining the Organization for Economic Cooperation and Development (OECD) in 2010. Israel's ranking in 2022 in terms of GDP per capita is 14, see: 'List of countries by GDP (nominal) per capita', *Wikipedia*, accessed 5 January 2023, https://en.wikipedia.org/wiki/List_of_countries_by_GDP_(nominal)_per_capita.

59 Eitan Shamir, 'From Isolation to Desired Partner: The Success of Israel's Grand Strategy in the 21st Century', *BESA Center Perspectives Paper*, no. 2,175, 25 January 2023, https://besacenter.org/from-isolation-to-desired-partner-the-success-of-israels-grand-strategy-in-the-21st-century/.

60 David Levy and Shay Shabtai, 'Israel's Move to USCENTCOM is Transformational', *BESA Center Perspectives Paper*, no. 2,184, 27 February 2023, https://besacenter.org/israels-move-to-uscentcom-is-transformational/.

## Bibliography

Adamsky, Dima. 'The Israeli Approach to Defense Innovation'. *SITC Research Briefs* 10, no. 7 (2018): 1–4.

Adamsky, Dmitry. *The Culture of Military Innovation: The Impact of Cultural Factors on the Revolution in Military Affairs in Russia, the US, and Israel*. Stanford, CA: Stanford University Press, 2010.

Amir, Noam. 'The Security Establishment Is Afraid the Dissolution of the Knesset Will Eliminate the Momentum Plan: Chief of Staff Kochavi's Multi-Year Plan Continues to Be Postponed Due to Non-Transfer of the Budget'. *Makor Rishon*, 15 October 2020. www.makorrishon.co.il/news/288009/.

Aven, Shmuel. 'The *"Tnufa"* Multi-Year Plan for the IDF: Where Are the Cabinet Approval and the Budgets?'. *INSS Insight*, No. 1357, 5 August 2020. www.inss.org.il/publication/tnufa-where-is-the-cabinet.

Ben-Ari, Eyal, Rosman, Elisheva, and Shamir, Eitan. 'Neither a Conscript Army Nor an All-Volunteer Force: Emerging Recruiting Models'. *Armed Forces & Society* 49, no. 1 (2023): 138–159.

Bokhbot, Amir. 'Flocks of IDF Quadcopters Participated in the Fighting in Gaza and the Rules of the Game Are Expected to Change'. *Walla News*, 5 June 2021. https://news.walla.co.il/item/3439695. (Hebrew)

Brun, Itay. 'Where Has the Maneuver Disappeared?'. *Ma'arachot Journal* 420/421 (2008): 4–15. (Hebrew)

Eilam, Uzi. *Eilam's Bow*. Tel-Aviv: Miskal, Yedioth Aharonoth, Chemed Books, 2009. (Hebrew). Published in English as Eilam, Uzi. *Eilam's Arc: How Israel Became a Military Technology Powerhouse*. Brighton: Sussex Academic Press, 2011.

Farrell, Theo and Terriff, Terry. *The Sources of Military Change: Culture, Politics, Technology*. New York: Lynne Rienner Publishing, 2002.

Finkel, Meir. 'Worshiping Technology in the IDF: Recovering the Balance for Ground Forces Development'. *Ma'arachot* 407 (2006): 40–45. (Hebrew)

Gal, Reuven. 'Motivation Levels for IDF Enlistment over the Years'. In *Military Service in Israel: Challenges and Ramifications*, edited by Elran Meir and Sheffer Gabi, 49–60. Tel Aviv: Institute for National Security Studies, September 2016, Memorandum 159. (Hebrew)

Halliday, Josh. 'Stuxnet Worm Is the "Work of a National Government Agency"'. *The Guardian*, 24 September 2010. www.theguardian.com/technology/2010/sep/24/stuxnet-worm-national-agency.

Kochavi, Aviv. 'COS Forward'. *Between the Poles—Military Superiority and the Momentum Plan* 28–30 (2020): 7–10. (Hebrew)

Levy, David and Shabtai, Shay. 'Israel's Move to USCENTCOM Is Transformational'. *BESA Center Perspectives Paper*, No. 2,184, 27 February 2023. https://besacenter.org/israels-move-to-uscentcom-is-transformational/.

Luttwak, Edward and Horowitz, Dan. *The Israeli Army*. London: Allen Lane, 1976.

Nusacher, Aryeh J. S. 'Sweet Irony: The German Origins of Israel Maneuver Warfare Doctrine'. MA thesis, Royal Military College, Canada, 1996.

Ortal, Eran. 'Resuming the Offensive: A Theoretical Framework for the Momentum Plan'. *Between the Poles—Military Superiority and the Momentum Plan* 28–30 (2020): 35–49.

———. 'Turn on the Light, Put Out the Fire'. *Between the Poles—Land Forces Part A* 31/32 (2021): 53–69. (Hebrew)

Ostfeld, Zahava. *An Army Is Born*. Tel Aviv: The Ministry of Defense, 1994. (Hebrew)

Raska, Michael. *Military Innovation in Small States, Creating a Reverse Asymmetry*. Abingdon: Routledge, 2016.

Regev, Motti. 'To Have a Culture of Our Own'. *Ethnic and Racial Studies* 23, no. 2 (2000): 223–247.

Ritz, Gal. 'Establishing an Independent Land Air Capability: Revolution in Land Warfare'. *Between the Poles* 37 (2022). (Hebrew)

Rothenberg, Gunther. *The Anatomy of the Israeli Army*. London: Redwood Burn, 1979.

Sachar, Shir. 'Where Is the Ghost Unit Going?'. *Ma'arachot Journal*, 2 December 2021. www.maarachot.idf.il/2021/%D7%9E%D7%90%D7%9E%D7%A8%D7%99%D7%9D/%D-7%99%D7%97%D7%99%D7%93%D7%AA-%D7%A8%D7%A4%D7%90%D7%99%D7%9D-%D7%9C%D7%90%D7%9F/.

Schwab, Karl. *The Fourth Industrial Revolution*. New York: City Crown Publishing House, 2016.

Shamir, Eitan. *Transforming Command: The Pursuit of Mission Command in the US, British, and Israeli Armies*. Stanford, CA: Stanford University Press, 2011.

———. 'From Isolation to Desired Partner: The Success of Israel's Grand Strategy in the 21st Century'. *BESA Center Perspectives Paper*, No. 2,175, 25 January 2023. https://besacenter.org/from-isolation-to-desired-partner-the-success-of-israels-grand-strategy-in-the-21st-century/.

Shefi, Yigal. *The Platoon Leader's Clip, Military Thinking in Officers Courses in the Hagana*. Tel Aviv: Ministry of Defense, 1991. (Hebrew)

Strik, Yoel and Tsach, Mosche. 'A Tale of Champions'. *Between the Poles: Military Superiority and the Momentum Plan* 28–30 (2020): 213–227.

Tamari, Dov. *The Armed Nation, the Rise and Decline of the Israel Reserve System*. Moshav Ben Shemen: Modan Publishing House and Ma'arachot, 2012. (Hebrew)

Tiargan, Roni. 'Different Reflections of the Motivation to Serve in the IDF'. In *Military Service in Israel: Challenges and Ramifications*, edited by Elran Meir and Sheffer Gabi, 61–76. Tel Aviv: Institute for National Security Studies, September 2016, Memorandum 159. (Hebrew)

Uri, Lieutenant Colonel. 'Chisel Water from Rock—Change and Changeability in the IDF's R & D Mechanisms'. *Between the Poles: Power Building—Part 2* 7 (2016): 41–59. (Hebrew)

Yehoshua, Yossi. 'The Lessons of the War: The IDF Will Grow, The Service in It Will Be Extended, and a Solution to the Permanent Crisis Will Be Found'. *Ynet*, 7 November 2023. www.ynet.co.il/news/article/yokra13666011.

# 5 Finland's defence planning in times of geopolitical disruption

'Never again alone'

*Olli Pekka Suorsa*

## Introduction

Finland's geostrategic maxims have changed little since the country declared its independence from the Russian Empire on 6 December 1917. First, the Soviet Union and then, modern-day Russia have manifested as Helsinki's primary security challenge. Finland's many wars with its larger neighbour, including during the Second World War—the Winter War (November 1939—March 1940) and the Continuation War (June 1941–September 1944)—remain painful reminders of the sheer power asymmetry between the two countries. This history, geographic adjacency and power asymmetry forced Finland to adopt a distinctively threat-based defence planning.

To cope with these inescapable circumstances, Finland has developed comprehensive security and territorial defence models as the basis of Helsinki's national defence and crisis preparedness. The Finnish Defence Forces' (FDF's) defence planning and force development are based on careful studies and long-term planning and government oversight. However, limited resources have placed significant constraints on Finland's military modernisation and defence planning. In the early 2010s, the FDF was forced to make painful cuts in expenses, manpower and bases. These cuts also led to substantial transformations within the FDF's organisation to deal with, not just the smaller defence outlays but also to cope with the reduced manpower. Budgetary vows have limited the pace of capability modernisation.

In the new era of geopolitical disruption, which began with Russia's annexation of Crimea, in 2014, and full-scale invasion of Ukraine on 24 February 2022, Finland sought to maximise its security through internal and external balancing. Finland accelerated its military modernisation and abandoned its long-held military non-alignment policy by joining the North Atlantic Treaty Organisation (NATO). The war in Ukraine has demonstrated that the work the FDF has done over the past decade to increase the military's preparedness and readiness to address the full spectrum of conflicts—ranging from 'hybrid' threats to high-intensity conflict—has been largely correct and timely.

This chapter takes a closer look at Finland's defence planning in times of geopolitical disruptions that range from the end of the Cold War to Russia's invasion of Ukraine. It highlights the fundamental continuities in Finnish defence planning

DOI: 10.4324/9781003398158-5

as well as adaptation to changes in the security environment and technological disruption. The chapter is divided into three parts, which look at Finland's alignment choices, military innovation and adaptation in the face of geopolitical and technological disruption, and defence planning based on careful analysis and long-term planning.

### Part 1: Finland's alignment choices

Finland's historical experience has been dominated by the acute feeling of geographic isolation and fear of facing military aggression alone. This insecurity became prominent in Finland's defence thinking, with a strong feeling of self-reliance and an attempt to ensure that Finland would never again need to face its larger neighbour (to the East) alone.[1] During the Cold War, Helsinki was forced to compromise on its strategic autonomy to safeguard Finland's sovereignty. Total defence and military non-alignment became synonymous with Finland's defence planning during this time.[2] The end of the Cold War and the dissolution of the Soviet Union enabled Helsinki to seek security in the West by aligning itself fully with the Euro-Atlantic institutions, including the European Union, in 1995, and by joining NATO's Partnership for Peace (PfP) programme, in 1994, and participation in European and NATO-led peace-keeping and peace-enforcement operations, most prominently in the Balkans and Afghanistan.[3] Despite clear political alignment with the West, Helsinki, however, held fast to its military neutrality. Nevertheless, during the past decade or so, Finland's military neutrality became more a political mantra than a reflection of reality.[4]

During the past decade, Finland has constructed a wide-ranging defence cooperation with a growing number of partners in the Nordics, Europe and beyond. Since the 2010s, Helsinki has signed various bilateral, minilateral and multilateral military agreements with Sweden, Norway, Denmark, Estonia, Poland, Germany, France, the United Kingdom and the United States, among others. The focus of Finland's defence network build-up was to add value to Finland's own security and defence in the High North and in the Baltic Sea areas.

Finland's historically close defence partnership with Sweden, which includes operational planning, frequent joint-training and joint-operative units, has deepened further with no 'predetermined limitations'.[5] The close partnership between Finland and Sweden is Helsinki's most important bilateral relationship. Finland's close historical association with Sweden witnessed a renewed effort to increase bilateral security and defence cooperation in the face of the rapidly deteriorating security situation in northern Europe throughout the past decade. Today, Finland-Sweden defence ties are characterised by operational units that can be used 'in all situations'.[6] These units include a joint Finnish-Swedish naval unit and frequent cross-border training and exercises.

In the Nordic context, the NORDEFCO (Nordic Defense Cooperation) arrangement, which comprises Finland, Sweden, Norway, Denmark and Iceland, intends to improve each nation's defence capacity through materiel acquisition and defence technology collaboration, planning and joint training and improving overall interoperability amongst the five countries, among other initiatives.[7] Hindering closer cooperation has been the participating states' different alignment choices, with Denmark, Norway and Iceland being members of NATO, whereas Finland and Sweden have maintained a 'military neutrality'. With Finland's accession to NATO in April 2023 and Sweden's impending accession into the alliance, prospects for deeper Nordic defence cooperation become a possibility.

In addition, bilateral defence cooperation between Finland and the US has grown steadily since the end of the Cold War. The Finnish Air Force's selection of the F-18C/D Hornet in 1992 represented an important milestone and a political signal of Helsinki's strong western alignment. The increasing depth and breadth of the bilateral military cooperation has been clearly visible in Finland's defence acquisitions ever since, including the JASSM (Joint Air-to-Surface Standoff Missile), the ESSM (Evolved SeaSparrow Missile), GMLRS-ER (Extended-Range Guided Multiple Launch Rocket System) and the F-35A Block IV Lightning II.[8] Besides military sales, joint training and exercise cooperation has increased substantially with, for example, Finnish Air Force participation in the Red Flag Alaska exercise in 2018 and increasingly frequent joint training in Finland since 2015.[9] From Helsinki's perspective, the joint exercises are helpful in developing strategic understanding and meaningful capacities to integrate with one another in ways that benefit Finland's security and defence. In addition, to further deepen military cooperation between Finland, Sweden and the US, a trilateral statement of intent was signed in 2018.[10] Finally, Finland and the US signed the Defence Cooperation Agreement in December 2023, which allowed for the prepositioning of supplies in Finland, access to military bases and other infrastructure, and eased training and exercise rotation of American forces in Finland, further reinforcing Finland's security and enabling fulfilment of NATO's obligations.[11]

Besides Finland's two most important bilateral military partners, Sweden and the US, Helsinki has sought to diversify its military partnerships to include several important European powers. Finland participates in the UK-led Joint Expeditionary Force (JEF) and the German initiative, the Framework Nation Concept (FNC).[12] The build-up of a comprehensive network of defence partnerships helps address various insecurities, such as the security of supply, and boost deterrence against aggression. In addition, the FDF engages in 60 to 90 international training and exercise events, which also help to increase readiness and preparedness.[13] Due to the influx of large numbers of defence engagement invitations, Helsinki tries to pick ones that benefit Finland's own national defence capability and planning.

Finally, EU membership has been perceived by the Finnish leadership as a source of increased security. President Sauli Niinistö has been active in pushing for deeper security cooperation within the EU based on the Lisbon Treaty Articles 222 and 42 on solidarity and assistance during emergencies.[14] Helsinki placed a lot of hope in the development of the European Common Security and Defence Policy and concepts closer to a collective defence, outside of NATO. Before Russia's invasion of Ukraine, a common European defence would arguably have been Helsinki's preference.[15] Despite these efforts, a majority of European NATO members wished to maintain the North-Atlantic alliance as the foundation of Europe's defence.[16] Helsinki's realisation of this fact became clear following Russia's invasion of Ukraine with Finland and Sweden's application to join NATO. Finland's decision to apply for NATO membership was therefore likely a deadly blow to Helsinki's insistence on European common defence as a parallel security structure in Europe.

### Implications of NATO accession for Finland's defence planning

The geopolitical disruption in Europe, which began with Moscow's annexation of Crimea in 2014 and culminated in Russia's full-scale invasion of Ukraine in 2022, changed the Finnish public's perceptions of Russia's intentions and Finland's security environment in fundamental ways.[17] On 17 May 2022, Helsinki and Stockholm moved firmly to seek NATO membership, which ended both Nordic countries' long-maintained 'military neutrality', with significant implications for each country and the alliance in the High North and the Baltic Sea region.

Aside from the attention on Finland during the NATO accession process, Finland's collaboration with the organisation is not new. Finland joined NATO's PfP framework in 1994 and has incrementally deepened cooperation with the alliance ever since.[18] Finland actively took part in NATO's peace-enforcement operations in the Balkans, in the 1990s, and in Afghanistan and Iraq, in the 2000s.[19] In 2004, Finland was invited to become NATO's Enhanced Opportunities Partner (EOP), a framework which includes only six countries, and signed the Memorandum of Understanding for the Provision of NATO's Host Nation Support in 2014.[20] Helsinki has also enjoyed NATO's support in defence technology R&D, establishing the Center of Excellence for Countering Hybrid Threats (Hybrid CoE) in Helsinki, and cooperated closely in cyber defence.[21] Russia's invasion of Ukraine altered the Finnish public's perceptions of the surrounding security environment in fundamental ways, leading up to Finland's to becoming a NATO member in May 2022.[22]

Finland's NATO accession is unlikely to change the basic pillars of Finland's defence: comprehensive security and wide-based conscription and broad reserves. Helsinki will join the alliance's command and control, nuclear planning, intelligence and common situational awareness networks.[23] Even in NATO, the FDF's priority would remain the defence of Finland's own territory and, hence, NATO's northeastern border—the longest alliance border with Russia. The Finnish Army, one of the largest in Europe, adds a significant force to the alliance's structure with a large wartime reserve and one of Europe's largest artillery forces.[24] The Finnish Air Force could contribute to NATO's Baltic Air Policing (BAP) and Iceland Air Policing (IAP) missions. Finland's support of the BAP mission would come naturally due to the country's geographic proximity to the three Baltic states.[25] The Finnish Navy, on the other hand, could contribute to the alliance's Standing NATO Mine Countermeasures Group (SNMCG) and Standing NATO Maritime Groups (SNMG) missions in the Baltic Sea with significant expertise to offer.[26] The FDF would, however, have to make some structural changes to enable Finland's participation in NATO's Very High Readiness Joint Task Force (VHRJTF).

Moreover, Helsinki would need to make legislation changes to enable the use of conscripts or reserve personnel in any overseas deployment. Participation in the VHRJTF would also likely pose new demands on FDF's logistical and airlift capabilities.

It is also critical to make note of the strategic consequences of Finland's NATO accession to Russia's security, especially in the Baltic Sea and the high North regions. As both Finland and Sweden enter NATO, the Baltic Sea will be turned into 'NATO's lake', which will make the movement of Russian forces and supplies from St Petersburg to the Kaliningrad exclave next to impossible in times of conflict.[27] This reality, however, also addresses one of Helsinki's long-lasting security vulnerabilities: protection of Finland's supply chains across the Baltic Sea during wartime. Furthermore, Finland's accession to NATO will bring the military organisation's border next to the Kola Peninsula, home to Russia's Northern Fleet, Arctic Command and a significant portion of Russia's nuclear weapons. Finland's sparsely populated Lapland is sandwiched between the northern regions of Sweden and Norway and the Russian Kola Peninsula. The growing competition in the Arctic and NATO's expansion to the North will likely result in Russian countermoves to strengthen its military capabilities in the Kola Peninsula. Finland (and NATO) will have to increase military capabilities of their own in the North in response. Finland will struggle to divert limited resources to two different primary directions, the Southeast (towards the Karelian Isthmus) and the High North.

### Part 2: threat-based military modernisation

Finland has adopted a pragmatic and careful long-term strategy for defence procurement and capability development. Its military modernisation is driven by Helsinki's threat perception and historical experience associated with its larger neighbour, Russia. Limited defence budgets have restricted the pace of military modernisation, demanding Helsinki to adopt a ubiquitous defence acquisition strategy. Finland makes balanced acquisitions between select high-tech weapon systems ('spearhead' capabilities) and second-hand ('good enough') arms. In addition, Finland's own defence industrial base manufactures small arms, mortar systems, armoured vehicles and ammunition for Finland's own needs and to ensure security of supply.[28]

During the Cold War, Finland was forced to maintain a careful balance between the East and the West, both politically and militarily. As part of this balance, Helsinki sought to please Moscow by buying military hardware from the Soviet Union and the Eastern Bloc countries while seeking access to western technologies, especially through Sweden.[29] The end of the Cold War ended these political constraints and opened the door for Helsinki to the West and, in particular, to the US.[30] As a proof of the new alignment and opportunities, Finland bought the McDonnel Douglas (now Boeing) F-18C/D Hornet from the US to replace the air force's ageing fleet of Soviet-built MiG-21Bis and Swedish-made J-35CS/S Draken.[31]

### Second-hand military procurement

Finland has an established tradition of seeking out second-hand hardware at a low cost. For example, following the end of the Cold War and German reunification, in the early 1990s, the East German Army (NVA) disposed of vast amounts of its equipment. Helsinki signed agreements for the importation of large arms and other materiel packages in 1991 and 1994, which included, for instance, Soviet-made T-72 main-battle tanks (MBT), armoured-fighting vehicles and armoured personnel carriers, self-propelled artillery and howitzers, anti-tank missiles, man-portable air-defence systems (MANPADS) and large stocks of spares and ammunition.[32] These deals had significant importance for Finland's defence planning. They upgraded the Finnish Army's firepower in terms of both quality and quantity, provided massive war reserves and enabled a more active training of conscripts and reserve forces.[33]

In addition, between 2002 and 2009, Finland acquired 139 used Leopard 2A4 MBTs, again from Germany.[34] Furthermore, significant defence cuts in the Netherlands' military force structure in the 2000s put large numbers of its MBTs and MLRS on sale. Finland was quick to seize the opportunity by acquiring 100 little-used Leopard 2A6 MBTs for €200 million in 2014.[35] Similarly, between 2006 and 2014, Finland bought a total of 40 M270 MLRS systems from the Netherlands, the US and the Danish Army's stocks to satisfy the Army's requirement for long-range fires.[36] In addition, in 2022, Finland procured Guided MLRS (GMLRS) rockets from the US Army's reserve stocks to rapidly deepen the 'magazine' of the M270 MLRS system amid Russia's invasion of Ukraine in the same year.[37]

These acquisitions were based on an internal impact study, which was carried out in the early 2000s, which concluded that heavy MLRS were deemed the most effective system for Finland's defence system.[38] Moreover, as a result of the Finnish Army's Operational Artillery programme (launched in 2014) to replace obsolete equipment and support the operational forces, Finland purchased 48 K9 self-propelled howitzers, worth €146 million, directly from the Republic of Korea Army surplus stocks in 2017.[39] In addition, Finland expanded the capability by exercising the option to acquire 48 additional K-9s from the Republic of Korea Army in two batches between 2021 and 2023, worth €134 million.[40]

In the early 2000s, the Finnish Air Force conducted a study for the life extension and modernisation of its 30-year-old BAe Hawk Mk51/51A advanced training jet fleet. As a result, Finland bought 18 BAe Hawk Mk66 advanced training aircraft from the Swiss Air Force, which had been stored due to the air force's changed training requirements.[41] This made the well-maintained and low-flight-hour Hawks available for sale. Finland acquired the Hawks for a mere €41 million, an amount worth two new-build aircraft and extending the service life of the Finnish Air Force's Hawk fleet well into the 2030s.[42]

These examples demonstrate the acquisitions' favourable capability-cost return. Helsinki acquired second-hand but modern and little-used hardware at a fraction of the cost of new equipment. Furthermore, the equipment came with significant spare parts stocks to ensure future viability. Finland has since upgraded many of the acquired capabilities to better integrate them into the Finnish defence system and to bring them to the latest hardware and software standards. These acquisitions helped add new capabilities to the Finnish Defence Force, replace large stocks of outdated Warsaw Pact-origin equipment with NATO-standard hardware and help maintain mass at low cost. Additionally, instead of acquiring hardware and effectors from the marketplace, Finland dealt directly with national militaries to cut manufacturing lead-time and queueing times from the materiel delivery schedule, thus providing a faster fielding of the required capability. This acquisition strategy worked well for Helsinki during the post-Cold War period as many countries left a lot of the 'heavy iron' available for international sale. However, the window for such opportunities is rapidly closing as the West is awakening to the Russian threat and begins addressing its own military deficiencies.

### Selective acquisition of high-end capabilities

Finland has selectively acquired 'spearhead' capabilities as part of its long-term defence planning. Over the past decade, the focus has been on the Finnish Air Force and Navy's modernisation. Resulting from the FDF's long-range fires and impact study, Finland acquired 70 AGM-158A Joint-Surface to Surface Missiles in 2011.[43] The procurement made Finland only the third customer for the weapon system after the US and Australia. Further acquisitions were directed at improving Finland's air defences, with the purchase of the NASAMS-II (Norwegian Advanced Surface-to-Air Missile System) (2009) and the Evolved Sea-Sparrow Missile (2019). In its latest acquisition, Finland filled a long-lasting capability gap by selecting the Israeli David's Sling air defence system to provide high-altitude and long-range air defence for the capital and other major population centres.[44]

On 10 December 2021, Helsinki selected the F-35A Block IV Lightning II, the only fifth-generation combat aircraft on the market, to replace the country's fleet of F/A-18C/D Hornets,[45] and is preparing for the construction of the first of four new *Pohjanmaa* class multirole corvettes, the largest surface combatants in the Finnish Navy.[46] These strategic programmes are part of Finland's long-term capability replacement plans and ensure the entire defence system's viability for the future. In terms of the *Pohjanmaa* class, Finland is not just replacing vessels one by one but expanding the Navy's capabilities significantly, including in underwater, surface attack and anti-air domains. The acquisition will also significantly increase the Navy's ability to protect the nation's maritime supply chains, a known vulnerability, in times of conflict.

In February 2022, just weeks before Russia's invasion of Ukraine, Finland's Ministry of Defence announced the acquisition of the latest version of the GMLRS rockets, the extended range ER GMLRS, worth €70 million, with a range exceeding 150 km.[47] Furthermore, in response to the war in Ukraine, Helsinki announced additional funds for emergency defence materiel purchases, which have included a range of munitions and effectors for artillery and air defences as well as anti-tank missiles.[48] These acquisitions have helped replenish FDF's war materiel reserves and replace older hardware donated to Ukraine.

Together, the second-hand and select high-end capability acquisition pro-grammes are a critical element of Finland's long-term force development. The combination of the two foreign hardware acquisition strategies allows the maximi-sation of the value reaped from the limited budgetary resources available.

### Finland's defence industrial base

Finland's defence industrial base supports the FDF in several important ways. Fin-land's defence industry is historically focused on providing hardware, small arms and ammunition and other materiel for the Finnish military, which remains its pri-mary customer. Moreover, the FDF relies on the local industry for equipment over-haul, upgrades and training services. The Patria Oy, the largest defence industry organisation in Finland, provides the Finnish Air Force and the Army's fixed- and rotary-wing aircraft life cycle and maintenance support. Patria has also executed planned mid-life updates to each aircraft type. Patria also provides MRO (mainte-nance, repair and overhaul) services for the Navy's gas-turbine engines. The Finn-ish Defence Force is the main customer for Patria's family of armoured personnel carriers, armoured fighting vehicles and turreted mortar systems. Together with its international support base, NAMMO Oy produces ammunition and munitions for the FDF. Helsinki has long relied on the local industry for ammunition and effec-tors, which has become increasingly difficult with the growing sophistication of the effectors and munitions with precision guidance.

### Tactical and technological innovation and adaptation

Historically, Finland is known to have developed ubiquitous tactics and operational concepts fit for the country's own geographic, climatic and technological context and circumstances. Finland has had to innovate and adapt—both in terms of tech-nology and in operational concepts and tactics—to cope with the large asymmetry in power between itself and Russia. Thus, the concept of 'hybridity' (i.e. the com-bination of conventional and irregular warfare concepts) is built into the Finnish operational art, which is seen as bridging eastern (Russian) and western (NATO) concepts of warfare.[49]

In its past wars, Finland has proven resourceful at employing innovative con-cepts of operation, tactics and making innovative use of the limited resources avail-able. This was famously demonstrated during the Winter War in the Finnish '*motti*' (or envelopment) tactics.[50] The *motti* tactic can be seen as a variation of manoeu-vre warfare, utilising small mobile units to encircle road-bound enemy forces and destroy them in separate smaller engagements.[51] Similarly, the Finns' resource-fulness was shown in technological solutions like, for example, the invention of the 'Molotov Cocktail' and use of logwood, in the absence of dedicated anti-tank weapons, to first stop and, then, burn Soviet tanks.[52] This resourcefulness and adap-tation still characterises the Finnish armed forces today.

The FDF has developed ubiquitous operational concepts, such as the Army Doctrine 2015, which utilises small unit tactics, dispersion and devolved decision-making to achieve surprise against a numerically superior adversary. The Finnish operational art resembles what can be described as 'high-tech guerrilla warfare'. These ideas are drawn from Finland's unique position and ability to bridge lessons in warfare from both East and West. Critically, this demonstrates how Finland does not simply follow either the western (e.g. NATO) or eastern (e.g. Russia) concepts of warfare but is able to combine them in unique ways.[53] Furthermore, the same doctrinal innovation is visible in all services. The FDF has proven adept in using and flexibly combining forms of conventional and irregular warfare into its own operational art.[54]

In Michael Raska's framework for assessing military innovation through conceptual paths and technological patterns, Finland can be considered as undergoing constant adaptation through its analytical long-term defence planning. Instead of simply emulating great powers, Finland has followed its own distinct conceptual path. As part of the careful analysis of both geopolitical and technological disruption and their immediate and long-term impacts on Finland's defence planning, a speculative pattern of technology adoption prevails. This, again, is closely linked to careful analysis and adaptation rather than emulation of great powers and trends in technology adoption. Nevertheless, to better understand trends in emerging technologies, Finland's defence industry is well placed to collaborate on research and development and experiment with key emerging technologies in close partnership with the FDF and allies and partners. Military technology cooperation remains a key rationale for Finland's choice of defence partners. As demonstrated in Figure 5.1, Finland is firmly on the path of military modernisation with selective adoption or speculation of the latest technologies.

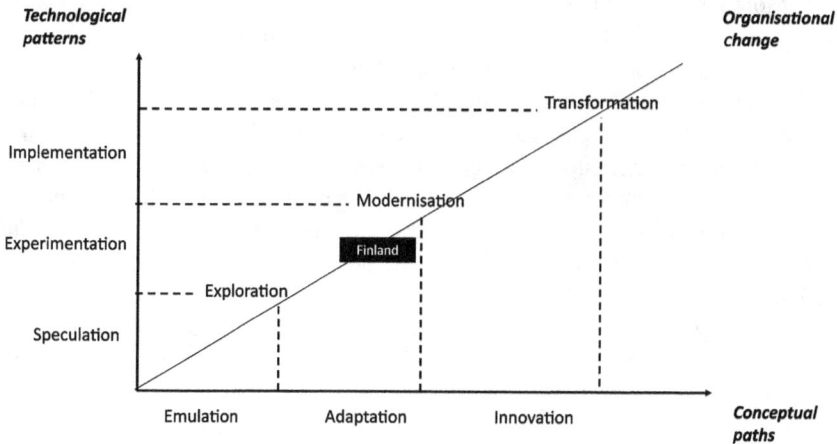

*Figure 5.1* Finland's military innovation trajectories.

*Source:* Figure created by author based on Michael Raska, Military Innovation of Small States— Creating a Reverse Asymmetry (Routledge, 2016)[55]

## Part 3: long-term defence planning approach

Defence planning in Finland is conducted under civilian leadership, where the Finnish Defence Forces act as military advisors for the process. The President is the Commander in Chief of the FDF and works together with the Ministerial Committee on Foreign and Security Policy (*Tasavallan Presidentti-Ulko- ja Turvallisuuspoliittinen ministerivaliokunta*, TP-UTVA), which is chaired by the Prime Minister and includes among others the President of Finland, Ministers of Defence, Foreign Affairs and Interior and adresses issues related to Finland's national security and defence. Depending on the issue, decisions taken in the TP-UTVA may also go through further parliamentary processes for broad-based approval. Decisions are then directed to the Ministry of Defence's planning committee for implementation. Finland's NATO accession was a good example of this process demonstrating the Government and Parliament's flexibility and responsiveness in the face of major geopolitical disruption.

### *Finland's defence choice: a reserve-based military and territorial defence*

The end of the Cold War in Europe and the demise of the Soviet Union altered Europe's geopolitics in fundamental ways—the *raison d'être* of NATO's existence was no more, and the former Warsaw Pact countries of eastern Europe began joining the western politico-economic and security institutions, the European Union and NATO, one by one. These changes led to what is often described as the 'peace dividend', characterised by massive downsizing of armed forces, removing conscription service as unnecessary and broad-based cuts in defence spending. New focus was on international stability, peace-keeping and peace-enforcement operations—instead of territorial defence and resistance against large-scale invasion. Moreover, since the early 2000s, participation in the US-led Global War on Terrorism, following the 9/11 attack, became a new focus. A peer or near-peer conflict was considered unlikely, if not impossible at the time.

Finland became an outlier by not following the popular trend. Instead, Finland retained its well-developed concepts of comprehensive security and territorial defence to guard her territorial integrity and sovereignty. Finland's comprehensive security concept has a long tradition, wherein the country's leadership, businesses, non-governmental organisations and citizens take responsibility for the state's vital functions.[56] The concept has its roots in the post–World War II doctrine of Total Defence, which called for the entire society to be mobilised for a military defence of the country.[57] Moreover, the Finnish concept has served as a model for other countries too.[58] Comprehensive security helps build resilience against emergencies ranging from natural catastrophes and pandemics to interstate armed conflict. It includes preparations that cover both military and civilian personnel, equipment and infrastructure and calls for training and exercises between the different entities.[59]

Throughout the 1990s and 2000s, Finland maintained all-male population conscription and broad reserves as the basis of the country's defence. In addition, since 1997, women have been permitted to join military service alongside men.[60] In 2022,

1,040 women finished voluntary military service, which brought the total number of women in the reserve to 12,000.[61] However, due to economic and demographic pressures, Helsinki was forced to cut defence spending to 1.22–1.28 per cent of GDP in 2014–2018.[62] In 2015, as part of the FDF's modernisation, Finland's war-time strength was reduced to 230,000.[63] This was corrected in 2018 to reach the current force strength of 280,000.[64] Moreover, the number of retraining days has been steadily increased since Russia's annexation of Crimea in 2014. In 2022, the number of reservists called up for retraining increased to 29,000, which was over 10,000 more than in the previous year.[65] Moreover, legislation is being drafted for the call-up of an entire age bracket in 2025.[66]

Importantly, a big part of the deterrence effect of the vast reserves is formed by the Finns' interest in defence matters and willingness to defend their country, which is well documented in annual surveys measuring European nations' willingness to defend their homeland with arms. Finland frequently appears at the top of the list. In 2015, amid Russia's annexation of Crimea, 74 per cent of Finns registered their willingness to wield arms in the country's defence.[67] Following Russia's invasion of Ukraine in February 2022, the percentage of Finns registering their willingness to defend the country in case of an armed aggression rose to a record 83 per cent.[68] Similarly, a record number of Finns, including women, sought voluntary and additional military training amid growing concern over Moscow's intentions. Therefore, Finland's decision-makers can count on the reservist force's motivation to defend the country.

The concept of strict area defence based on holding territory had proven obsolete by the early 2000s due to, not least, advances in intelligence, surveillance and reconnaissance technologies, especially in the space domain, making large, massed units vulnerable to intelligence. Thus, Finland adapted its territorial defence model to move away from holding fixed positions to a more flexible and mobile approach, which utilises small unit tactics, manoeuvre, strong indirect fire support, correct use of geography, deception, cover and surprise to overcome adversary force.[69] The heavy long-range fires remained under the command of FDF HQ. The new Army doctrine called for adding more firepower to each unit, including detachment-level light mortars, machine guns, anti-tank weapons and man-portable air defence systems.[70] In addition, to enable more devolved decision-making, each unit was equipped with robust communications systems for command and coordination. These basic doctrinal tenets are visible throughout the entire FDF. The Finnish Air Force had adopted the dispersed operations concept already in the late 1960s, which relied on the use of civilian airfields and national highways for dispersed operations.[71] Similarly, the Finnish Navy, which operates a small fleet of guided-missile armed fast attack craft and mine layers that are adept at operating within the Finnish coastal archipelago,[72] has adopted an operational concept, which can be best characterised as 'guerrilla warfare' at sea.

Since the 2014 Russian annexation of Crimea, Finland began boosting its military readiness and preparedness in earnest against grey zone aggression and 'hybrid' threats. The changes implemented included the establishment of high-readiness units in all services, consisting of professional soldiers, conscripts and select reservists.[73] To further improve the scalability of response, very-high readiness forces were added, which consisted solely of professional soldiers. Together, these units can employ land, sea and air capabilities. Moreover, legislation changes for the use of force were made to prevent any 'little green men' situations from developing in Finland.[74] Further changes in legislation have been introduced to accelerate flexible mobilisation, which remains the Achilles' heel of the Finnish defence system.[75] Nevertheless, Helsinki's clear-eyed understanding of the changes in Finland's security environment has resulted in rapid adaptation to enable a faster and scalable armed response against any aggression if necessary.

## Conclusion

Finns follow military developments in Russia carefully and have proven to be able to make sound strategic, operational and technological decisions best suited to Finland's own circumstances and limited resources as the current war in Ukraine has demonstrated. Russia's invasion of Ukraine became a watershed moment in Finland's foreign and security policy with the Nordic country making a firm move from military non-alignment to balancing against the perceived threat by joining NATO for the alliance's security guarantees. The accession process has proven seamless as the FDF was already well integrated into the alliance structure and operations and its capabilities were well known amongst its members. In addition to NATO membership, Finland continues to cultivate close bilateral defence relations with, especially, Sweden and the US and minilateral cooperation with other important regional partners.

The core tenets of Finland's defence planning have remained remarkably stable throughout the decades following the Second World War, driven by the government's acute perception of the threat to the East. This is demonstrated in the Structured Focused Comparison Framework in Figure 5.2. Due to its historical experience and feeling of geographic isolation, Helsinki's defence planning has been twofold. On the one hand, Finland had no choice but to balance internally against the perceived threat by relying on a popular well-defined conscription service and large well-trained reserves as the core of Finland's defence. Moreover, due to the perceived 'smallness', concepts of comprehensive security and whole-of-society approaches were adopted. On the other hand, the fear of entanglement into major power conflicts or having to face a military aggression alone forced Helsinki to adopt a non-aligned position—balancing between the East and the West during the Cold War. However, the end of the Cold War and the collapse of the Soviet Union did not result in moves to rapidly align itself militarily to the West to follow a similar change in Finland's foreign policy. It was only the abrupt geopolitical disruptions in Finland's security environment that started in 2014 with the Russian annexation of Crimea and, finally, the invasion of Ukraine on 24 February 2022, which triggered a change in Helsinki's position. Finland, along with Sweden, moved rapidly to balance against the perceived Russian threat by jointly acceding to NATO in May 2022 for security guarantees, ending decades of military non-alignment policy. Moreover, Helsinki's build-up of a network of defence partnerships has helped integrate the Finnish Defence Forces with its European and North American allies and partners for mutual benefit and to reduce the strategic risk associated with negative geopolitical developments linked to Russia.

As shown in the framework (Figure 5.2), defence planning in Finland is carried out in relatively steady long-term planning. In force planning and design, financial constraints have called for a careful analysis. The FDF bases its force design on long-term planning with necessary adjustments depending on the fiscal climate and allocation of the defence budget. Capability reviews and disruptive technology studies provide important analytical structures for military modernisation and future force design and analytical outcomes. Because of the limited resources available, compromises need to be made in capability acquisitions. In this environment, Finland has adopted a unique acquisition strategy, which comprises the active search for viable second-hand options at low cost and selective procurement of high-end, or 'spearhead', capabilities. Finland developed a local defence industry early on to ensure security of supply and MRO services through technology transfers. Despite its small size, Finland's defence industry produces some successful niche products from turreted mortar systems and wheeled armoured vehicles to modern sea mines. Finland has had to source its 'spearhead', or strategic, capabilities from foreign OEMs (original equipment manufacturers). To address technological disruption brought about by emerging technologies, the Finnish industry and FDF collaborates with allies and partners in joint research and development programmes. Finland's high-tech information and communications technology sector (e.g. 6G networks and quantum computing) is well suited to partner with international programmes for mutual benefit. In Raska's framework (Figure 5.1), Finland can be found constantly adapting to the magnitude of the threat to its East and rapidly evolving technologies. Much of the adaptation is achieved through speculation on future technology requirements and ubiquitous operational concepts.

| SMPS ADJUST THEIR APPROACHES BASED ON SYSTEMIC CHANGE OUT OF THEIR CONTROL | SMPS REACT TO CHANGES IN THE RELEVANT SECURITY COMPLEX(ES) THE SMP IS PART OF | | | | |
|---|---|---|---|---|---|
| *Alliances, Dependencies and National Ambitions* | | | | | |
| SMPS FACE CONSTRAINTS IN SAFEGUARDING NATIONAL SECURITY | FINANCIAL | DEMOGRAPHIC | CULTURAL | GEOGRAPHICAL | OTHER |
| SMPS NEED TO POSITION THEMSELVES VIS-À-VIS GLOBAL AND REGIONAL POWER CONSTELLATIONS | BALANCING | HEDGING | BANDWAGONING | NON-ALIGNING | OTHER |
| SMPS HAVE TO ENGAGE WITH OTHER STATES TO INCREASE THEIR DEFENCE CAPABILITIES | SECURITY GUARANTEES | TRANSACTIONAL | FORCE INTEGRATION | SPECIALISATION | OTHER |
| SMPS HAVE ARMED FORCES FOR DIFFERENT PURPOSES | INSTRUMENT OF DIPLOMACY | PRESTIGE PROJECT | REDUCE STRATEGIC RISK (DEFEND) | DOMESTIC STABILITY | OTHER |
| *Approaches, Processes, Methods and Techniques* | | | | | |
| SMPS TAKE A MORE CONSTRAINED APPROACH TO DEFENCE PLANNING | THREAT-NET ASSESSMENT BASED | PORTFOLIO-BASED | TASK-BASED | MOBILISATION | OTHER |
| SMPS HAVE LIMITED RESOURCES TO DESIGN AND IMPLEMENT DEFENCE PLANNING PROCESSES | TAILORED ANALYTICAL PROCESS | OUTSOURCING STRATEGY | ACCEPT DILUTION / FAIR SHARE | EXTERNAL TEMPLATE | OTHER |
| SMP EMPLOY MULTIPLE METHODS AND TECHNIQUES | COPY GREAT POWERS | INFORMAL CONSENSUS | TAILORED METHODS | AD HOC | OTHER |
| SMPS RELY ON DIFFERENT PROCESSES TO DECIDE ON FORCE POSTURE | ANALYTICAL RESULTS | POLITICS DECIDES | MILITARY DECIDES | GREAT POWER DECIDES | OTHER |
| *Military Innovation* | | | | | |
| SMPS FACE CHALLENGES IN KEEPING PACE WITH TECHNOLOGICAL ADVANCES | BUILD DOMESTIC TECH/INDUSTRY | TECH TRANSFER AGREEMENTS | ACQUIRE READY INNOVATIONS | PARTNER WITH GREAT POWER | OTHER |
| SMPS PURSUE VARIOUS MILITARY INNOVATION PATHS AND PATTERNS | EMULATION | ADAPTATION | INNOVATION | OTHER | OTHER |
| | SPECULATION | EXPERIMENTATION | IMPLEMENTATION | OTHER | |
| SMPS FIND THEMSELVES AT DIFFERENT STAGES OF ORGANISATIONAL CHANGE | EXPLORATION | MODERNISATION | TRANSFORMATION | OTHER | OTHER |

*Figure 5.2* Structured focused comparison framework for Finland.

*Source:* Author

## Notes

1　See, for example, Risto E., J. Penttilä, *Finland's Security in a Changing Europe: A Historical Perspective* (Helsinki: National Defence College, 1994).
2　Mikko Karjalainen, Vesa Tynkkynen and Pentti Airio, eds., *Suomen Puolustusvoimat 100 Vuotta* [Finnish Defence Force 100 Years] (Helsinki: National Defense University, 2018).
3　Penttilä, *Finland's Security in a Changing Europe*.
4　See, Matti Pesu and Tuomas Iso-Markku, *Finland as a NATO Ally: First Insights into Finnish Alliance Policy* (Helsinki: Finnish Institute of Internal Affairs (FIIA), December 2022), www.fiia.fi/en/publication/finland-as-a-nato-ally (Accessed 24 March 2023).
5　Charly Salonius-Pasternak and Henri Vanhanen, *Finnish-Swedish Defence Cooperation: What History Suggests about Future Scenarios* (Helsinki: Finnish Institute of International Affairs (FIIA), June 2020), 5–6, www.fiia.fi/julkaisu/finnish-swedish-defence-cooperation?read (Accessed 22 March 2023).
6　Salonius-Pasternak and Vanhanen, *Finnish-Swedish Defence Cooperation*.
7　See 'Nordic Defence Cooperation (NORDEFCO)', *NORDEFCO*, www.nordefco.org/the-basics-about-nordefco (Accessed 24 March 2023).
8　Reuters, 'Finland Buys More Arms from United States as It Boosts Defence', *Reuters,* 29 November 2022, www.reuters.com/world/europe/finland-buys-more-arms-united-states-it-boosts-defences-2022-11-29/ (Accessed 24 March 2023).
9　Ilmavoimat [Finnish Air Force], 'Finnish Air Force to Participate in RED FLAG-Alaska 19–1 in October', *Air force,* 15 August 2018, https://ilmavoimat.fi/en/-/ilmavoimat-red-flag-harjoitukseen-alaskaan-lokakuussa?__cf_chl_tk=1NcGcBcluZu.uZTCKX4S5Jv4IlAyCWBMhz9eHnts_1g-1679643539–0-gaNycGzNC7s (Accessed 24 March 2023).
10　Ministry of Defence (Finland). The Trilateral Statement of Intent agreement was based on earlier bilateral agreements reached between Finland and US and between Sweden and US in 2014 and 2016. For the Trilateral Statement of Intent, see 'Trilateral Statement of Intent Among the Department of Defence of the United States of America and the Ministry of Defence of the Republic of Finland and the Ministry of Defence of the Kingdom of Sweden', 2018, www.defmin.fi/files/4231/Trilateral_Statement_of_ Intent.pdf (Accessed 3 March 2023).
11　Ministry of Foreign Affairs (Finland). 'Defence Cooperation Agreement with the United States (DCA)'. Undated. https://um.fi/defence-cooperation-agreement-with-the-united-states-dca- (Accessed 16 January 2024).
12　Niclas von Bonsdorf, 'Suomen puolustusyhteistyö Saksan kanssa [Finland's Defence Cooperation with Germany]', *Maanpuolustus Lehti*, 7 September 2018, www. maanpuolustus-lehti.fi/suomen-puolustusyhteistyo-saksan-kanssa/ (Accessed 5 January 2023).
13　Puolustusvoimat [Finnish Defence Force], 'Kansainväliset Harjoitukset ja Koulutukset' [International Training and Exercises], undated, https://puolustusvoimat.fi/kansainvaliset-harjoitukset (Accessed 24 March 2023).
14　Tapio Pajunen, 'Sauli Niinistö: Selvitämme Saako Suomi Sotilaallista Apua EU:sta' [Sauli Niinisto: Can Finland Get Military Assistance from EU?], *YLE News*, 25 January 2018, https://yle.fi/a/3-10040887 (Accessed 24 March 2023).
15　Henri Vanhanen, 'Artikla 42.7:llä On Vielä Matkaa EU:n Todelliseksi Turvatakuuksi' [A Long Way Ahead for the Article 42.7 to Be Considered as EU's Security Guarantee], *The Ulkopolitist*, 28 November 2018, https://ulkopolitist.fi/2018/11/28/artikla-42-7lla-on-viela-matkaa-eun-todelliseksi-turvatakuuksi/ (Accessed 24 March 2023).
16　Vahanen, 'Artikla 42.7:llä On Vielä Matkaa EU:n Todelliseksi Turvatakuuksi'.
17　Pesu and Iso-Markku, *Finland as a NATO Ally*.
18　NATO, 'Relations with Finland', 5 July 2022, www.nato.int/cps/en/natohq/topics_49594.htm (Accessed 3 March 2023).
19　NATO, 'Relations with Finland'.

20 Ministry of Defence (Finland), 'Memorandum of Understanding (MOU) between the Government of the Republic of Finland and Headquarters, Supreme Allied Commander Transformation as well as Supreme Headquarters Allied Powers Europe Regarding the Provision of Host Nation Support for the Execution of Operations/Exercises/ Similar Military Activity', 2014, www.defmin.fi/files/2898/HNS_MOU_FINLAND.pdf (Accessed 03 March 2023).
21 NATO, 'Relations with Finland'.
22 Ministry of Foreign Affairs (Finland), 'Finland's Application for NATO Membership', *Ministry for Foreign Affairs of Finland*, https://um.fi/ finland-is-applying-for-nato-membership (Accessed 22 March 2023).
23 See, Pesu and Iso-Markku, *Finland as a NATO Ally*.
24 Heljä Ossa and Tommi Koivula, 'What Would Finland Bring to the Table for NATO?', *War on the Rocks*, 9 May 2022, https://warontherocks.com/2022/05/what-would-finland-bring-to-the-table-for-nato/ (Accessed 24 March 2023).
25 Pesu and Iso-Markku, *Finland as a NATO Ally*, 30.
26 See, for example, Robin Häggblom, 'Finnish Navy Kicks Off Freezing Winds 22 Annual Exercise', *Naval News*, 25 November 2022, www.navalnews.com/naval-news/2022/11/finnish-navy-kicks-off-freezing-winds-22-annual-exercise/ (Accessed 24 March 2023).
27 See, for example, Zachary Selden, 'Will Finland and Sweden Joining NATO Deepen the Alliance's Problems?', *War on the Rocks,* 31 May 2022, https://warontherocks.com/2022/05/will-finland-and-sweden-joining-nato-deepen-the-alliances-problems/ (Accessed 24 March 2023).
28 Finland's defence industry base's primary customer continues to be the Finnish Defence Forces. However, the defence industrial base has also undergone a transition towards international cooperation and joint production, especially in Europe. Today, Finnish defence industry companies, such as Patria Oy, actively seek export opportunities globally.
29 Karjalainen, Tynkkynen and Airio, eds., *Suomen Puolustusvoimat 100 Vuotta* [Finnish Defence Force 100 Years], 297–299.
30 Ibid., 320–325.
31 During the Cold War, Helsinki had to carefully balance its military hardware acquisitions from the East and the West. The Finnish Air Force transitioned to Mach 2- and air-to-air missile-era with the political acquisition of the then brand-new Soviet MiG-21F-13 and K-13 missile in 1963. In the early 1970s, Helsinki procured the Swedish Saab J-35 Draken to expand the air force's fighter capacity. Sweden appeared as Helsinki's window to the West. Since the 1970s, Finland made repeat orders of both the Soviet MiG-21 and the Swedish J-35 Draken to maintain balance between the East and the West. The collapse of the Soviet Union and the end of the Cold War saw Helsinki gaining access to American the fighter as well as an opportunity to consolidate the air force's fleet composed entirely of Western combat aircraft.
32 Karjalainen, Tynkkynen and Airio, eds., *Suomen Puolustusvoimat 100 Vuotta* [Finnish Defence Force 100 Years], 313–319.
33 Ibid., 319.
34 Anne-Pauliina Rytkönen, 'Suomi Hankkii Käytettyjä Panssarivaunuja Hollannista' [Finland Acquires Used Main-Battle Tanks from the Netherlands], *YLE,* 16 January 2014, https://yle.fi/a/3-7034964 (Accessed 22 March 2023).
35 Jarmo Huhtanen, 'Suomi Ostaa Hollannista "Maailman Parhaita Panssarivaunuja", Viimeiset Vaunut Saapuivat Perjantaina' [Finland Acquires 'the World's Best MBTs' from the Netherlands], *Helsingin Sanomat*, 25 October 2019, www.hs.fi/kotimaa/art-2000006285346. html (Accessed 22 March 2023).
36 Aleksei Kettunen, 'Suomesta GMLRS-Ohjussuurvalta?' [Finland Acquires GMLRS Rockets], *Suomen Sotilas*, 7 November 2022, https://suomensotilas.fi/2022/11/07/suomesta-gmlrs-ohjussuurvalta/ (Accessed 22 March 2023).
37 Kettunen, 'Suomesta GMLRS-Ohjussuurvalta?'.

38  Karjalainen, Tynkkynen and Airio, eds., *Suomen Puolustusvoimat 100 Vuotta* [Finnish Defence Force 100 Years], 217–219.

39  Maavoimat [Finnish Army], 'Self-Propelled Howitzer K9 Thunder—From Research to Procurement Programme', undated, https://maavoimat.fi/en/self-propelled-howitzer-k9-thunder (Accessed 22 March 2023).

40  Jukka Harju, 'Suomi Jatkaa Aseostoja Venäjän Hyökkäyssodan Varjossa: Nyt Ostoslistalla On Liki 40 "Moukaria"' [Finland Buys Weapons Amid Russia's Offensive in Ukraine], *Helsingin Sanomat*, 18 November 2022, www.hs.fi/kotimaa/art-2000009211493.html (Accessed 22 March 2023).

41  Jyrki Laukkanen, *BAe Hawk Suomen Ilmavoimissa* [BAe Hawk in Finnish Air Force] (Tampere: Koala Kustannus, 2015).

42  The sum includes the aircraft and associated equipment and tools, as well as the aircraft's overhaul. Laukkanen, *BAe Hawk*.

43  Ministry of Defence (Finland), 'Finland Procures a Long-Range Air-to-Surface Missile System', 1 March 2012, www.defmin.fi/en/topical/press_releases_and_news/press_releases_archive/2012/ finland_procures_a_long-range_air-tosurface_missile_system.4907.news (Accessed 22 March 2023).

44  Shephard Media. 'Finland Purchases David's Sling Long-Range Air Defence System from Israel', 13 November 2023. www.shephardmedia.com/news/landwarfareintl/finland-signs-for-davids-sling-long-range-air-defence-system/. Accessed 16 January 2024.

45  Valtioneuvosto [Finnish Government], 'The Lockheed Martin F-35A Lightning II Is Finland's Next Multi-Role Fighter', 10 December 2023, https://valtioneuvosto.fi/-/lockheed-martin-f-35a-lightning-ii-on-suomen-seuraava-monitoimihavittaja?languageId=en_US (Accessed 21 March 2023).

46  Puolustusvoimat [Finnish Defense Force], 'Pohjanmaa-Luokka Vahvistaa Suomen Puolustuskykyä' [Pohjanmaa-Class Corvettes Will Strengthen Finland's Defence], undated, https://puolus-tusvoimat.fi/web/kansainvalinen-kriisinhallinta/-//1951215/pohjanmaa-luokka-vahvistaa-suomen-puolustuskykya (Accessed 21 March 2023).

47  Nicholas Fiorenza, 'Finnish Defence Forces to Procure Extended-Range GMLRS Munition', *Janes*, 16 February 2022, www.janes.com/defence-news/news-detail/finnish-defence-forces-to-procure-extended-range-gmlrs-munition (Accessed 22 March 2023).

48  See, for example, Ministry of Defence (Finland), 'Ministry of Defence Decided on Procurement of Short and Long-Range Missiles', 9 December 2022, https://valtioneuvosto.fi/en/-/ministry-of-defence-decided-on-procurement-of-short-and-long-range-missiles (Accessed 22 March 2023); and 'Finnish Military to Procure Surveillance Drones Valued at €14M', *YLE News*, 11 April 2022, https://yle.fi/a/3-12400277 (Accessed 22 March 2023).

49  Col.Lt., Dr. Juha Mälkki, 'Suomalainen sotataito' [Finnish Art of War], interview, *Puolustusvoimat* [Finnish Defence Force] *Radio Kipinä Podcast* (in Finnish), 24 March 2016, 43:02, https://puolustusvoimat.fi/asiointi/ kanavat/radio-kipina/ podcastit (Accessed 17 March 2023).

50  Franz-Stefan Gady, 'Breaking the Mannerheim Line: Soviet Strategic and Tactical Adaptation in the Finnish-Soviet Winter War', *War on the Rocks*, 8 February 2023, https://warontherocks.com/2023/02/breaking-the-mannerheim-line-soviet-strategic-and-tactical-adaptation-in-the-finnish-soviet-winter-war/ (Accessed 17 February 2023).

51  Pasi Tuunainen, 'Motti Tactics in Finnish Military Historiography since World War II', *International Bibliography of Military History* 33, no. 2 (2013): 121–147, doi: https://doi.org/10.1163/22115757-03302003.

52  See, for example, Olli Vehviläinen, *Finland in the Second World War: Between Germany and Russia* (Hampshire: Palgrave, 2002).

53  Lt.Col. Mälkki, 'Suomalainen sotataito' [Finnish Art of War].

54  Ibid.

55 Michael Raska, "A Structured-Phased Evolution: The Third Generation Force Transformation of the Singapore Armed Forces", in *Military Innovation in Small States: Creating a Reverse Asymmetry*, M. Raska (New York: Routledge, 2016), 130–162.
56 The Security Committee (Finland), 'Concept of Comprehensive Security—Building National Resilience in Finland', undated, https://turvallisuuskomitea.fi/concept-of-comprehensive-securi-ty-building-national-resilience-in-finland/ (Accessed 2 March 2023).
57 The Security Committee (Finland), 'Concept of Comprehensive Security'.
58 See, for example, Ron Matthews and Fitriani Bintang Timur, 'Singapore's "Total Defence" Strategy', *Defence and Peace Economics* (2023): 1–21.
59 See, The Security Committee (Government of the Republic of Finland), 'Government Resolution: Security Strategy for Society', 2017, https://turvallisuuskomitea.fi/en/security-strategy-for-society/ (Accessed 17 March 2023).
60 Emilia Kemppi and Sirpa Jegorow, 'Naiset Armeijaan' [Women Join Military], *YLE News*, 16 September 2009, https://yle.fi/aihe/artikkeli/2010/09/16/naiset-armeijaan (Accessed 17 March 2023).
61 Puolustusvoimat [Finnnish Defence Force], 'Ennätysmäärä vapaaehtoisia naisia suoritti varusmiespalveluksen—kotiutuneita yli 1000' [A Record Number of Women Concluded Military Service—More than 1,000 Joining Reserve], 30 December 2022, https://puolustusvoimat.fi/-/ennatysmaara-vapaaehtoisia-naisia-suoritti-varusmiespalveluksen-kotiutuneita-yli-1000 (Accessed 17 March 2023).
62 Ministry of Defence, Republic of Finland, 'Puolustusmenojen Jakautuminen' [Defence Budget Allocation], undated, www.defmin.fi/ministerio/toiminta_ja_talous/puolustusmenot#5645 ab81 (Accessed 17 March 2023).
63 Valtioneuvoston Kanslia [Finnish Government], 'Suomen Turvallisuus- ja Puolustuspolitiikka 2012' [Finland's Security and Defence Policy], May 2012, https://vnk.fi/documents/10616/622970/ J0512_Suomen+turvallisuus-+ja+puolustuspolitiikka+2012.pdf/b534174a-13bc-4684-beb0a093 be30ce2a? version=1.0 (Accessed 17 March 2023).
64 Valtioneuvoston Kanslia [Finnish Government], 'Valtioneuvoston Puolustusselonteko 2021' [Government's Defence Report 2021], 9 September 2021, https://julkaisut.valtioneuvosto.fi/handle/10024/163407 (Accessed 17 March 2023).
65 Tuomas Hyytinen, 'Koko Ikäluokka Voi Saada Käskyn Kutsuntoihin Vuonna 2025' [An Entire Age Bracket May Be Called Up in 2025], *YLE News*, 23 January 2023, https://yle.fi/a/74-20014190 (Accessed 17 March 2023).
66 Hyytinen, 'Koko Ikäluokka Voi Saada Käskyn Kutsuntoihin Vuonna 2025'.
67 Teri Schultz, 'In Defense, Finland Prepares for Everything', *DW*, 10 April 2017, www.dw.com/en/finland-wins-admirers-with-all-inclusive-approach-to-defense/a-40806163 (Accessed 17 March 2023).
68 'Poll: Citizens' Willingness to Defend Finland, Support for NATO Hit All Time High', *YLE News*, 1 December 2022, https://yle.fi/a/74-20006876 (Accessed 17 March 2023).
69 See, Puolustusvoimat [Finnish Defence Force], 'Uudistettu Taistelutapa—Perusteet' [Army Doctrine 2015—Basics], 18 February 2013, 7:29, www.youtube.com/watch?v=2crAx8kibis (Accessed 17 March 2023).
70 Puolustusvoimat, 'Uudistettu Taistelutapa'.
71 Jyrki Laukkanen, *Finnish air Force 100 Years* (Helsinki: Koala-Kustannus, 2017).
72 Jouko Pirhonen, *Laivastotoiminta Saaristotaistelussa Suomen Olosuhteissa* (Tampere: Juvenes Print, 2015).
73 International Institute for Strategic Studies, 'Europe: Finland', in *The Military Balance* (London: Routledge, 2021), 77–82.
74 International Institute, 'Europe: Finland', 78.
75 'Jussi Niinistö: "Liikekannallepanon hitaus on puolustusjärjestelmämme Akilleen kantapää"' [Jussi Niinistö: Inertia as the Achille's Heel of Finland's Defence System], *Uusi-Suomi*, 12 August 2017, www.uusisuomi.fi/uutiset/jussi-niinisto-liikekannallepanon-hitaus-on-puolustusjar-jestelmamme-akilleen-kantapaa/eef0efca-e087-320f-8117-a9fa8b61cf88 (Accessed 17 March 2023).

## Bibliography

Bitzinger, Richard, Michael Raska, Collin Koh Swee Lean and Kelvin Wong Ka Weng, 'Locating China's Place in the Global Defense Economy'. In *Forging China's Military Might: A New Framework for Assessing Innovation*, ed. Tai Ming Cheung (Baltimore, MD: Johns Hopkins University Press, 2013).

Häggblom, Robin. 'Finnish Navy Kicks Off Freezing Winds 22 Annual Exercise'. *Naval News*, 25 November 2022. www.navalnews.com/naval-news/2022/11/finnish-navy-kicks-off-freezing-winds-22-annual-exercise/. Accessed 16 January 2024.

Harju, Jukka. 'Suomi Jatkaa Aseostoja Venäjän Hyökkäyssodan Varjossa: Nyt Ostoslistalla On Liki 40" Moukaria"' [Finland Buys Weapons Amid Russia's Offensive in Ukraine]. *Helsingin Sanomat*, 18 November 2022. www.hs.fi/kotimaa/art-2000009211493.html. Accessed 16 January 2024.

Huhtanen, Jarmo. 'Suomi Ostaa Hollannista "Maailman Parhaita Panssarivaunuja", Viimeiset Vaunut Saapuivat Perjantaina' [Finland Acquires 'the World's Best MBTs' from the Netherlands]. *Helsingin Sanomat*, 25 October 2019. www.hs.fi/kotimaa/art-2000006285346.html. Accessed 16 January 2024.

Hyytinen, Tuomas. 'Koko Ikäluokka Voi Saada Käskyn Kutsuntoihin Vuonna 2025' [An Entire Age Bracket May Be Called Up in 2025]. *YLE News*, 23 January 2023. https://yle.fi/a/74-20014190. Accessed 16 January 2024.

Ilmavoimat [Finnish Air Force]. 'Finnish Air Force to Participate in RED FLAG-Alaska 19–1 in October'. 15 August 2018. https://ilmavoimat.fi/en/-/ilmavoimat-red-flag-harjoitukseen-alaskaan-lokakuussa?__cf_chl_tk=1NcGcBcluZu.uZTCKX4S5Jv4I1AyCWBMhz9eHnts_1g-1679643539-0-gaNycGzNC7s. Accessed 16 January 2024.

International Institute for Strategic Studies (IISS). 'Europe: Finland'. In *The Military Balance 2021*, 77–82. London: Routledge, 2021.

Karjalainen, Mikko, Vesa Tynkkynen and Pentti Airio, editors. *Suomen Puolustusvoimat 100 Vuotta* [Finnish Defence Force 100 Years]. Helsinki: National Defense University, 2018.

Kemppi, Emilia and Sirpa Jegorow. 'Naiset Armeijaan' [Women Join Military]. *YLE News*, 16 September 2009. https://yle.fi/aihe/artikkeli/2010/09/16/naiset-armeijaan. Accessed 16 January 2024.

Kettunen, Aleksei. 'Suomesta GMLRS-Ohjussuurvalta?' [Finland Acquires GMLRS Rockets]. *Suomen Sotilas*, 7 November 2022. https://suomensotilas.fi/2022/11/07/suomesta-gmlrs-ohjussuurvalta/. Accessed 16 January 2024.

Laukkanen, Jyrki. *Finnish Air Force 100 Years*. Helsinki: Koala-Kustannus, 2017a.

———. *BAe Hawk Suomen Ilmavoimissa* [BAe Hawk in Finnish Air Force]. Tampere: Koala Kustannus, 2017b.

Maavoimat [Finnish Army]. 'Self-Propelled Howitzer K9 Thunder—From Research to Procurement Programme'. Undated. https://maavoimat.fi/en/self-propelled-howitzer-k9-thunder. Accessed 22 March 2023.

Mälkki, Lt. Col., Dr. Juha. 'Suomalainen Sotataito'. Interview, *Puolustusvoimat* [Finnish Defence Force] *Radio Kipinä Podcast* (in Finnish), 24 March 2016, 43:02. https://puolustusvoimat.fi/asiointi/kanavat/radio-kipina/podcastit. Accessed 16 January 2024.

Matthews, Ron and Fitriani Bintang Timur. 'Singapore's "Total Defence' Strategy"'. *Defence and Peace Economics* (2023): 1–21.

Ministry of Defence (Finland) 'Finland Procures a Long-Range Air-to-Surface Missile System'. 1 March 2012. www.defmin.fi/en/topical/press_releases_and_news/press_releases_archive/2012/finland_procures_a_long-range_air-tosurface_missile_system.4907.news. Accessed 16 January 2024.

———. 'Memorandum of Understanding (MOU) between the Government of the Republic of Finland and Headquarters, Supreme Allied Commander Transformation as Well as Supreme Headquarters Allied Powers Europe Regarding the Provision of Host Nation Support for the Execution of Operations/Exercises/Similar Military Activity'. 2014. www.defmin.fi/files/2898/HNS_MOU_FINLAND.pdf. Accessed 3 March 2023.

———. 'Trilateral Statement of Intent Among the Department of Defence of the United States of America and the Ministry of Defence of the Republic of Finland and the Ministry of Defence of the Kingdom of Sweden'. 2018. www.defmin.fi/files/4231/Trila-teral_Statement_of_Intent.pdf. Accessed 3 March 2023.

———. 'Ministry of Defence Decided on Procurement of Short and Long-Range Missiles'. 9 December 2022. https://valtioneuvosto.fi/en/-/ministry-of-defence-decided-on-procurement-of-short-and-long-range-missiles. Accessed 16 January 2024.

———. 'Puolustusmenojen Jakautuminen' [Defence Budget Allocation]. Undated. www.defmin.fi/ministerio/toiminta_ja_talous/puolustusmenot#5645 ab81. Accessed 17 March 2023.

Ministry of Foreign Affairs (Finland). 'Finland's Application for NATO Membership'. Undated. https://um.fi/finland-is-applying-for-nato-membership. Accessed 22 March 2023.

———. 'Defence Cooperation Agreement with the United States (DCA)'. Undated. https://um.fi/defence-cooperation-agreement-with-the-united-states-dca-. Accessed 16 January 2024.

NATO. 'Relations with Finland'. 5 July 2022. www.nato.int/cps/en/natohq/topics_49594.htm. Accessed 16 January 2024.

Nicholas Fiorenza. 'Finnish Defence Forces to Procure Extended-Range GMLRS Munition'. *Janes*, 16 February 2022. www.janes.com/defence-news/news-detail/finnish-defence-forces-to-procure-extended-range-gmlrs-munition. Accessed 16 January 2024.

Niclas von Bonsdorf. 'Suomen Puolustusyhteistyö Saksan Kanssa [Finland's Defence Cooperation with Germany]'. *Maanpuolustus Lehti*, 7 September 2018. www.maanpuolustus-lehti.fi/suomen-puolustusyhteistyo-saksan-kanssa/. Accessed 16 January 2024.

'Nordic Defence Cooperation (NORDEFCO)'. *NORDEFCO*. www.nordefco.org/the-basics-about-nordefco. Accessed 24 March 2023.

Ossa, Heljä and Tommi Koivula. 'What Would Finland Bring to the Table for NATO?'. *War on the Rocks*, 9 May 2022. https://warontherocks.com/2022/05/what-would-finland-bring-to-the-table-for-nato/. Accessed 16 January 2024.

Pajunen, Tapio. 'Sauli Niinistö: Selvitämme Saako Suomi Sotilaallista Apua EU:sta' [Sauli Niinisto: Can Finland Get Military Assistance from EU?]. *YLE News*, 25 January 2018. https://yle.fi/a/3-10040887. Accessed 16 January 2024.

Penttilä, Risto E. J. *Finland's Security in a Changing Europe: A Historical Perspective*. Helsinki: National Defence College, 1994.

Pesu, Matti and Tuomas Iso-Markku. *Finland as a NATO Ally: First Insights into Finnish Alliance Policy*. Helsinki: Finnish Institute of Internal Affairs (FIIA), December 2022. www.fiia.fi/en/publication/finland-as-a-nato-ally. Accessed 16 January 2024.

Pirhonen, Jouko. *Laivastotoiminta Saaristotaistelussa Suomen Olosuhteissa*. Tampere: Juvenes Print, 2015.

Puolustusvoimat [Finnish Defence Force]. 'Uudistettu Taistelutapa—Perusteet' [Army Doctrine 2015—Basics]. 18 February 2013, 7:29. www.youtube.com/watch?v=2crAx8kibis. Accessed 17 March 2023.

———. 'Ennätysmäärä vapaaehtoisia naisia suoritti varusmiespalveluksen—kotiutuneita yli 1000' [A Record Number of Women Concluded Military Service—More than 1,000 Joining Reserve]. 30 December 2022. https://puolustusvoimat.fi/-/ennatysmaara-vapaaehtoisia-naisia-suoritti-varusmiespalveluksen-kotiutuneita-yli-1000. Accessed 16 January 2024.

———. 'Kansainväliset Harjoitukset ja Koulutukset' [International Training and Exercises]. Undated. https://puolustusvoimat.fi/kansainvaliset-harjoitukset. 24 March 2023.

———. 'Pohjanmaa-Luokka Vahvistaa Suomen Puolustuskykyä' [Pohjanmaa-Class Corvettes Will Strengthen Finland's Defence]. Undated. https://puolus-tusvoimat.fi/web/kansainvalinen-kriisinhallinta/-//1951215/pohjanmaa-luokka-vahvistaa-suomen-puolustuskykya. Accessed 21 March 2023.

Reuters. 'Finland Buys More Arms from United States as It Boosts Defence'. *Reuters*, 29 November 2022. www.reuters.com/world/europe/finland-buys-more-arms-united-states-it-boosts-defences-2022-11-29/. Accessed 16 January 2024.

Rytkönen, Anne-Pauliina. 'Suomi Hankkii Käytettyjä Panssarivaunuja Hollannista' [Finland Acquires Used Main-Battle Tanks from the Netherlands]. *YLE News*, 16 January 2014. https://yle.fi/a/3-7034964. Accessed 16 January 2024.

Salonius-Pasternak, Charly and Henri Vanhanen. *Finnish-Swedish Defence Cooperation: What History Suggests about Future Scenarios.* Helsinki: Finnish Institute of International Affairs (FIIA), June 2020. www.fiia.fi/julkaisu/finnish-swedish-defence-cooperation?read. Accessed 16 January 2024.

Schultz, Teri. 'In Defense, Finland Prepares for Everything'. *DW*, 10 April 2017. www.dw.com/en/finland-wins-admirers-with-all-inclusive-approach-to-defense/a-40806163. Accessed 16 January 2024.

Selden, Zachary. 'Will Finland and Sweden Joining NATO Deepen the Alliance's Problems?'. *War on the Rocks*, 31 May 2022. https://warontherocks.com/2022/05/will-finland-and-sweden-joining-nato-deepen-the-alliances-problems/. Accessed 16 January 2024.

*Shephard Media.* 'Finland Purchases David's Sling Long-Range Air Defence System from Israel'. 13 November 2023. www.shephardmedia.com/news/landwarfareintl/finland-signs-for-davids-sling-long-range-air-defence-system/. Accessed 16 January 2024.

Stefan-Gady, Franz. 'Breaking the Mannerheim Line: Soviet Strategic and Tactical Adaptation in the Finnish-Soviet Winter War'. *War on the Rocks*, 8 February 2023. https://warontherocks.com/2023/02/breaking-the-mannerheim-line-soviet-strategic-and-tactical-adaptation-in-the-finnish-soviet-winter-war/. Accessed 16 January 2024.

The Security Committee (Finland). 'Concept of Comprehensive Security—Building National Resilience in Finland'. Undated. https://turvallisuuskomitea.fi/concept-of-comprehensive-securi-ty-building-national-resilience-in-finland/. Accessed 2 March 2023.

Tuunainen, Pasi. 'Motti Tactics in Finnish Military Historiography since World War II'. *International Bibliography of Military History* 33, no. 2 (2013): 121–147. https://doi.org/10.1163/22115757-03302003.

Uusi-Suomi. 'Jussi Niinistö: "Liikekannallepanon Hitaus On Puolustusjärjestelmämme Akilleen Kantapää"' [Jussi Niinistö: Inertia as the Achille's Heel of Finland's Defence System]. 12 August 2017. www.uusisuomi.fi/uutiset/jussi-niinisto-liikekannallepanon-hitaus-on-puolustusjar-jestelmamme-akilleen-kantapaa/eef0efca-e087–320f-8117-a9fa8b61cf88. Accessed 16 January 2024.

Valtioneuvosto [Finnish Government]. 'The Lockheed Martin F-35A Lightning II Is Finland's Next Multi-Role Fighter'. 10 December 2023. https://valtioneuvosto.fi/-/lockheed-martin-f-35a-lightning-ii-on-suomen-seuraava-monitoimihavittaja?languageId=en_US.

Valtioneuvoston Kanslia [Finnish Government]. 'Suomen Turvallisuus- ja Puolustuspolitiikka 2012' [Finland's Security and Defence Policy]. May 2012. https://vnk.fi/documents/10616/622970/J0512_Suomen+turvallisuus-+ja+puolustuspolitiikka+2012.pdf/b534174a-13bc-4684-beb0a093 be30ce2a? version=1.0. Accessed 17 March 2023.

———. 'Valtioneuvoston Puolustusselonteko 2021' [Government's Defence Report 2021]. 9 September 2021. https://julkaisut.valtioneuvosto.fi/handle/10024/163407. Accessed 16 January 2024.

Vanhanen, Henri. 'Artikla 42.7:llä On Vielä Matkaa EU:n Todelliseksi Turvatakuuksi' [A Long Way Ahead for the Article 42.7 to Be Considered as EU's 'Security Guarantee']. *The Ulkopolitist*, 28 November 2018. https://ulkopolitist.fi/2018/11/28/artikla-42-7lla-on-viela-matkaa-eun-todelliseksi-turvatakuuksi/. Accessed 16 January 2024.

Vehviläinen, Olli. *Finland in the Second World War: Between Germany and Russia.* Hampshire: Palgrave Macmillan, 2002.

*YLE News.* 'Poll: Citizens' Willingness to Defend Finland, Support for NATO Hit All Time High'. 1 December 2022. https://yle.fi/a/74-20006876. Accessed 16 January 2024.

———. 'Finnish Military to Procure Surveillance Drones Valued at €14M'. 11 April 2022. https://yle.fi/a/3-12400277. Accessed 16 January 2024.

# 6    Retail path-dependence

Indonesia's post-authoritarian
defence planning

*Evan A. Laksmana*

Indonesia plans for its defence almost like how its people pay for cigarettes—by small retail pieces, rather than wholesale boxes. Indonesia has never had the ability to fully standardise its weapons systems and equipment and has settled instead on diversifying platforms and suppliers in smaller numbers. Rather than, for example, acquiring five squadrons of frontline fighter aircraft, Indonesia acquires one to three squadrons of different fighter jets from different suppliers over a long period of time. In other words, Indonesia develops a long-term modernisation plan but implements it in retail fashion. Much like smoking, there are various negative repercussions of this behaviour. While no single state can hold hostage Indonesia's arms supply, it has lost what little resources it had on expensive maintenance, repair and training costs, not to mention inter-operability challenges. It also made platform standardisation too unwieldy so that the country is effectively locked into a cycle of never having the fiscal space to buy no more than a few expensive platforms from different sources.

What explains Indonesia's retail-oriented defence planning? This chapter argues that Indonesia's retail defence planning stems not from resource constraints, but more broadly from the deeper structural features of its civil-military relations and operational experience. In this regard, Indonesia's persistent internal security threats as well as the ability of the military (*Tentara Nasional Indonesia* or TNI) to develop and implement its own defence planning free from civilian intervention following the end of authoritarian rule in 1998 help explain the persistence of the retail outlook. Indonesia's fractured and under-institutionalised post-authoritarian defence policymaking process further exacerbated the problem. Consequently, Indonesia has been unable to achieve major military innovations thus far.

The following sections expand and elaborate these arguments further. The first section anchors the conceptual roots of Indonesia's retail defence planning within the broader defence planning literature. It makes the case for why Indonesia's retail-oriented process can be classified as a form of 'resource-based' defence planning. The second section describes the evolution and challenges of Indonesia's civil-military relations and defence planning process following the end of authoritarian rule in 1998. It describes the nature and origins of the fractured and under-institutionalised defence planning process. The third section describes Indonesia's defence posture and dependencies as well as the absence of major military

DOI: 10.4324/9781003398158-6

innovations. It also notes the need to differentiate strategic alignment from defence partnerships. The penultimate section analyses the implications of Indonesia's retail approach for its historical lack of major military innovations. The concluding section summarises the key arguments and offers some policy reflections for other small and medium powers.

## Indonesia's retail approach to defence planning

Defence planning is generally seen as the deliberate process of deciding on a nation's future military forces, force postures and force capabilities.[1] In peacetime, this process is concerned with managing and addressing the conflict between ends (particular missions or operations) and means (capabilities and assets) under the influence of constraints (especially resources).[2] There are many different approaches to defence planning and recent studies have provided different models to simplify the complex and often discreet if not secretive processes.[3] In this context, Indonesia's defence planning model can perhaps be broadly classified under the rubric of 'resource-based planning'. This approach begins with the status quo and budget pressure or a tight fiscal environment; it then considers available resources as a limit and looks at priorities, trade-offs and substitution opportunities to meet the most possible potential demands from a finite resource base.[4] This model represents a more incremental approach to obtain capabilities 'on the margins but still in meaningful ways'.[5] In many ways, the idea here is perhaps to make 'good enough' provisions for national security.[6]

In Indonesia, as we shall see in more detail later, various constraints, rather than future strategic concerns, have been the primary feature of defence planning. The constraints come in several strains. First, *financial resources*. While Indonesia's defence budget has grown by almost ten times from around USD 1.5 billion in 1999 to about USD 10.1 billion in 2022, most of the funding has been allocated to routine expenditure and personnel costs (between 50 and 60 per cent on average).[7] Accounting for other activities like military exercises or operational contingencies, that leaves only roughly one to two billion dollars per year to be divided by the three different armed services. Given the enormity of the modernisation needs across all major platforms (most of which are past their life cycle), such an amount will hardly be sufficient for serious future-oriented defence planning. Consequently, defence policymakers can only rely on small pots of funding on an annual basis for procurement.

Second, *platform diversity*. Indonesia has had the largest number of foreign suppliers (33 different countries) for its major arms and equipment among Southeast Asian countries for decades now.[8] While this structural feature has allowed defence policymakers to claim that no single country can veto or dominate Indonesia's arms supply and force employment, it also means that the cost of maintenance, repair and overhaul as well as training, education and exercises to operate a multitude of different weapons systems has been significantly high. This does not even account for the operational costs of ensuring intra-service inter-operability, let alone across the three different armed services or with international partners.

Consequently, in the absence of a genuine organisation-wide push for platform standardisation, Indonesian defence policymakers are placed on a path-dependent trajectory of procuring small numbers of platforms from different suppliers, some of whom they were already committed to before.

Third, *operational demands*. The Indonesian military has not been seriously geared towards external defence requiring the full use of complex and expensive weapons and systems for decades. In fact, more than 70 per cent of Indonesia's 370 military operations conducted between 1945 and 2020 were geared towards internal security threats, from unrest and insurgencies to secessionism and counterterrorism.[9] What this means is that most of the operational needs of the military are not found in expensive, complex weapons systems they must import or develop themselves. Instead, they needed to rely on internal security functions and planning for their future needs, such as intelligence gathering down to the village level. Such operational history and experience (the social processes of militaries experiencing and giving meaning to a variety of tasks and activities during missions that are conducted over time by a significant share of its personnel) has been shown to have lasting effects on civil-military relations, and presumably, defence planning as well.[10] Indonesia's internal security history in this regard acts as another constraint in the broader defence planning process.

Surely these constraints are not unique to Indonesia; other small and middle powers face resource crunches on a regular basis. However, the nature of Indonesia's operational history and civil-military relations has allowed such constraints to persistently affect defence planning processes. Because of these constraints, nonetheless, Indonesia has little choice but to embark on what I call long-term retail defence planning, where the defence ministry develops long-term modernisation or acquisition plans but implements them in retail fashion where it acquires small numbers of expensive and complex platforms from different suppliers. Furthermore, the retail purchases have also often been done in a transactional way *sans* a wider strategic framework or broader capability development plans, nor have they been integrated efforts across defence units. In short, Indonesia's military modernisation plan resembles a long-term shopping list of assets of what needs to be replaced or modernised, and which supplier will give the best financing terms.

The notion of retail planning under a resource-driven approach is best encapsulated under the Defence Ministry's Minimum Essential Force (MEF) blueprint issued back in 2010 (discussed later) to provide the TNI with the bare minimum set of capabilities to execute its daily operational tasks. In this regard, retail planning is perhaps closer to what others have called 'task-based planning', where the need for armed forces to be able to perform '*elementary* tactical and operational tasks' becomes the guiding principle of defence planning.[11]

I will discuss why and how Indonesia has practised this form of planning in the following sections. For now, let me note that Indonesia's retail planning persists in a path-dependent trajectory because it is rooted within the overall context of the country's broader strategic policymaking ecosystem. Scholars have noted in particular how civil-military history has affected forms and practices of defence planning.[12] Where past civil-military relations have been deeply conflictual, as has been the case in Indonesia historically, one can expect the regime to remain fearful of coups and therefore to prioritise their prevention in the development of military organisational practices.[13] As a consequence, Indonesian political leaders may have been hesitant to provide all of the resources for and support defence planning practices that would allow the military to quickly obtain major complex weapons that could be used to destabilise the regime. Further, when there is no consensus that the country is facing immediate external threats, it can be hard to impose strategic discipline on defence planning.[14] We shall see how all these constraints were manifested over the past two decades in the next section.

### Post-authoritarian transition: national ambitions, alliances and dependencies

As noted earlier, Indonesia's threat landscape for much of its history has been internally focused, rather than externally. As such, domestic political changes are often the primary catalysts in major military changes.[15] It is not surprising therefore that as far as Indonesia's defence planning is concerned, the end of the Cold War was not as crucial as the end of authoritarian rule in 1998. That transition mattered for defence planning in several ways. First, the violent nature of the transition—where nationwide riots killed hundreds if not thousands accompanied by a splintered military with one faction on the cusp of launching a coup—meant that internal security concerns became more amplified. Indeed, shortly after the transition, Indonesia was racked by a multitude of such challenges, from the destructive separation of East Timor, the raging conflicts in Aceh and Papua, to the rise of violent religious extremism and terrorism. Maintaining territorial integrity and strengthening the coercive arm of the state became a first-order priority for much of the first post-authoritarian decade. It is not surprising therefore that the new TNI Law of 2004 prescribed a range of non-conventional warfighting tasks under Military Operations Other Than War (MOOTW), ranging from humanitarian assistance and disaster relief (HADR) to civic action and counterterrorism, still to the military.

Within the context of defence planning, such a threat environment was exacerbated by the damaging consequences of the 1996 Asian financial crisis which saw many of the military's big-ticket procurement items such as warships and fighter jets shelved for almost a decade. Any budgetary room and fiscal space were allocated to economic recovery, not defence spending. It was not until the MEF blueprint was issued in 2010 (codified in 2015) that the military started to engage in long-term modernisation and acquisition build-up again. The notion here is to gradually (over 15 years) modernise the hardware of the TNI so as to not cause destabilising effects in the region, in a way that would not hurt the economic recovery, but also allow the organisation to fulfil its basic operational tasks.[16]

The civil-military contexts, however, continue to be significant in shaping the defence planning process. At the broader domestic level, the political and military leaders struck a 'grand bargain' during and shortly after the authoritarian transition: the military would support the civilian leaders and overall democratisation process while the civilians in turn promised to protect the military's corporate interests, including its promotion policies, budget, business interests and others.[17] While that bargain toned down the initially violent transition, it left a path-dependent legacy where the military was essentially left alone to its own defence planning processes 'free' from civilian intervention. Successive post-authoritarian Presidents have been reluctant to significantly reverse course in this area.[18]

Furthermore, while the democratisation process theoretically increased the role of the national legislature (*Dewan Perwakilan Rakyat* or DPR) in providing oversight over the military, the implementation of that role has largely been performative. Senior retired officers were assigned to DPR's Commission 1, overseeing defence and foreign affairs, which remained understaffed and lacking in professional expertise around defence planning and policies. Their budgetary role was further curtailed by a court decision in 2014 that allowed the defence ministry to only reveal the top two-line budget items, thus reducing the DPR's ability to meaningfully and publicly question detailed budget items. Overall, the Commission chose to work 'cooperatively' with the military, which stabilised civil-military relations but stalled significant military reforms and undermined serious budgetary control.[19]

At the organisational level, the military's command and control structure was separated from the defence ministry, which is now tasked with overall policy management and budgeting responsibilities. However, military officers continue to dominate key positions within the defence ministry bureaucracy, despite having several civilian defence ministers since 1999.[20] The military high command therefore can exercise 'tacit control' over key positions within the ministry, including those tasked with defence planning and acquisition.[21] As part—or perhaps a consequence of—this dual chain system, defence planning has been understood to be taken up by the respective armed services. As discussed later, they would draw up the list of operational requirements to be submitted to the defence ministry who would then submit the necessary budgetary requests. For over a decade, however, the ministry often compiles and submits without fully exercising independent authority over what those requests contain. Current defence minister Prabowo Subianto has in recent years tried to reverse this process and sought to determine all procurement requirements, plans and suppliers, causing significant tensions with the TNI.[22]

But there are broader sets of organisational fault lines that hinder a more effective defence planning process. For one thing, Indonesia's defence planning system remains under-institutionalised as it continues to be changed or revised every few years; there are no less than 51 different regulations recorded between 1945 and 2020 concerning defence planning and budgeting systems.[23] These changes have only become less frequent over the past decade (at least five regulatory revisions). What these changes mean is that Indonesia's integrated and systematic defence planning process is a relatively new arrival as a by-product of the post-authoritarian transition. In other words, Indonesia issued a long-term defence planning document (the MEF in 2010) without having a clearly institutionalised defence planning ecosystem in place—and with more changes coming along the way.

For another, the defence planning process remains fractured and inconsistent. The defence ministry unit tasked with strategic documents (e.g., Defence White

Paper or Strategic Defence Review) is different from the one tasked with acquisition planning. More recently, new actors have been added to the acquisition process mix. Any final defence procurement must now get the full consent and approval from six different actors (in different ways and capacities): the defence ministry, the military, the National Development Planning Agency or BAPPENAS (tasked with overall long-term planning), the finance ministry, the DPR and the National Defence Industrial Committee or KKIP (created in 2010). Both BAPPENAS and the finance ministry could effectively veto the defence procurement plans, even if the entire defence planning process from the bottom up has been followed to the letter.

The bottom-up process starts within the service headquarters, from each unit up to logistics staff and then planning staff.[24] Once the acquisition planning documents and requests are consolidated within the defence ministry, trilateral meetings and concurrence between the ministries of defence, finance and the BAPPENAS are necessary to finalise the budgetary items and financing. The defence ministry then provides a consolidated five-year acquisition list based on the MEF document (and subject to annual revisions as necessary). The defence ministry's Directorate General for Defence Planning formulates the itemised budget line and financing strategy while its Defence Facilities Agency selects the necessary vendors or suppliers of the requested items. Possible defence industrial participation falls under a different unit within the defence ministry—the Directorate General of Defence Potential. The KKIP as a separate agency outside of the defence ministry provides yet another different planning process in terms of coordinating national defence industrial policies.

In short, Indonesia's post-authoritarian transition has opened the door for the military to exercise control over its own defence planning process. However, the policy ecosystem that has only been established over the past decade has fractured the acquisition and planning process in a way that exacerbated bureaucratic haggling and politics. Consequently, Indonesia has only been able to implement a long-term modernisation plan in retail fashion and an inconsistent manner. This section demonstrates in general how the path-dependent trajectory of civil-military relations and operational demands shape defence planning as well.

### Defence posture, dependencies and stagnation

As the previous sections show, Indonesia's military ambitions have been focused on internal security, rather than external ones. For decades, among the most common force planning goals has been the extent to which the military can be deployed in two major trouble hotspots within the country simultaneously. This outlook is reflected even in its post-authoritarian 'flashpoint-based' defence which projected the country's top threats as revolving around foreign powers supporting domestic armed separatists and the persistent threat of terrorism.[25] These concerns are rooted in the military's guerrilla warfare-style doctrinal foundation of 'territorial defence' as part of Total People's Defence and Security.[26] Territorial defence in Indonesian military parlance implies the enduring need for the Army's territorial command structure that mirrors civilian governments across the country down to the village level.[27] In other words, Indonesia's defence posture remains army-heavy, rather than naval-centric. If anything, Indonesia's post-authoritarian transition has further expanded the territorial command structure, although the need to create jobs to reduce promotional logjams was also responsible for this trend alongside the internal threats noted earlier.[28]

This is not to say that Indonesia does not have external security challenges. Issues surrounding border security, maritime incursions and regional contingencies such as Taiwan or the South China Sea are all within the thought processes of Indonesian defence policymakers. After all, Indonesia's defence doctrine calls for the ability to stop the enemy from advancing from beyond the country's outer layer (its exclusive economic zone [EEZ] and airspace) into its territorial waters and further. This anti-access impulse has been one of the constant features of Indonesia's layered defence strategy, although it remains underdeveloped as a concept of operations.[29] These external threats, however, have not redefined Indonesia's defence planning orientation, let alone its ecosystem. They do, however, provide an additional public rationale for and justification to push or accelerate pre-existing long-term procurement plans like the MEF, especially the threat posed by China's recent aggressive maritime behaviour. In other words, external threats provide the additional push for—not the overhaul of—the retail defence planning.

Given this outlook, Indonesia has never felt the need to fundamentally rely on treaty alliances with great powers to address its security needs. If anything, alliance entanglement is seen as a source of domestic political instability, and therefore, internal insecurity.[30] ASEAN therefore has remained the only cornerstone of Indonesian foreign policy as a strategic buffer between competing great powers. But Indonesia develops defence partnerships in different ways. It seeks to fulfil specific military needs (e.g. training and education) from advanced armed forces such as the United States and Australia. Other partners like South Korea, Turkey and France are seen as potentially reliable defence industrial partners. Indonesia is also willing to invest in decades of joint exercises with neighbouring countries like Malaysia and Singapore. As noted earlier, it has also traditionally eschewed having a small number of external foreign suppliers of arms. There is no basis therefore to assume strategic spillovers from a sector-specific alignment like defence to a broader cross-sector (strategic) alignment.

Defence partnership is also not equivalent or interchangeable with strategic alignment in general beyond defence. Indonesia's broader strategic alignment (mutual policy coordination over common interests and concerns across multiple policy domains in an integrated manner) with a single great power on an exclusive basis remains absent.[31] To some extent, the defence partnerships follow a similar logic but for different reasons with different countries. The major difference between strategic and defence partnerships is the role of ASEAN. There is no such alternative for the defence sector as ASEAN cannot provide more than a bare minimum of defence goods like norms and simple joint exercises.

Taken together, Indonesia's ultimately internally anchored defence posture, its resistance to formal alliance entanglements and its under-developed defence capability generation infrastructure (from training to education and research and development as well as defence industrial bases) means that the country's defence policymakers instinctively focused on retail procurement plans while simultaneously resisting broader strategic alignment overtures. In recent years, however, there has been increasingly a wider privately held gap between foreign ministry and defence ministry officials over the strategic utility of ASEAN in Indonesia's broader strategic calculus. Defence officials are more likely to pay lip service to that ASEAN is still a key feature of its 'diplomacy is the first line of defence' position, while complaining privately that the grouping could not stop the deterioration of the regional environment. Diplomats meanwhile seemed unable or unwilling to seriously contemplate non-ASEAN options over a wide range of regional security challenges.

## Consequences for military innovation

Indonesia's retail approach to defence planning has consequences for the country's serious lack of major military innovations—major fundamental changes in how a military thinks, plans and trains for and conducts its operations in a way that significantly if not drastically improves the efficiency of military power generation.[32] Technological innovation is only but one subset or element of major military innovations. That does not mean that Indonesia has never witnessed military changes; if anything, the TNI organisation changed too often creating under-institutionalised major practices and policies all around. What is absent however is a major military innovation implemented as a breakthrough policy based on systematic assessment and planning in response to major external military threats. Given Indonesia's threat landscape and history above, it is arguably the case that the military never felt the need to engage in major military innovations to begin with.

Furthermore, Indonesia's under-developed economic and defence industrial bases, which created a path-dependence over-reliance on external suppliers of arms and various offset policies and 'localisation' efforts, exacerbated this lack of urgency.[33] As noted earlier, the challenge for Indonesia's defence planning here is to manage the balance between political autonomy and the cost of diversity of arms suppliers. As many of these major platforms require a complex set of maintenance, repair and training infrastructure, they effectively lock in Indonesia's future procurement plans. Consequently, Indonesia today can only afford to procure small numbers of advanced platforms from different suppliers. Analysts have noted this increasingly 'opportunistic' transactionalism in Indonesia's defence procurement plans.[34] Without commonality and standardisation plans, Indonesia also cannot afford to engage in major technological innovations or even obtain the highest level of advanced platforms. Indonesia therefore often settles on 'modernisation plus' schemes where they obtain the next generation item, enough to meet their daily operational task but not enough to be operating the most sophisticated set of arms of the day.[35] In other words, retail defence planning has contributed to the lack of major military innovations in Indonesia.

Indeed, Indonesia over the past 20 years has never systematically planned for and successfully implemented major military innovations. If there ever were such innovations in its 77-year history, it would have been during the 1940s and 1950s when Indonesia fused various doctrinal modifications of its guerrilla warfare campaign into 'new' counter-insurgency tactics while modifying Western concepts to successfully engage in its first-ever joint operations against rebels.[36] Nonetheless, as noted earlier, Indonesia's Total People's Defence doctrine remained unchanged. Indonesia's fundamental premise and posture of its Territorial Command have also remained a constant, even though its size and levels vary over time.

Technologically, Indonesia over the past two decades largely relied on off-the-shelf procurements and offset programmes, as noted earlier. If there were limited domestic R&D activities, they were focused on a small number of non-strategic platforms like drones or modifications to current platforms like rockets or guided missiles (many of which remain under-tested). In this realm, Indonesia's joint venture with South Korea in its 4.5 generation fighter jet KF-21 *Boramae* is perhaps the furthest it has ever gone in terms of technological innovation. But Indonesia was not involved in developing advanced features of the design and development or major production of the aircraft.

Overall, for much of Indonesia's history, it has managed to engage in a few conceptual or doctrinal modifications, but the military has never developed major technological or operational innovations let alone major military innovations. Indonesia's retail approach to defence planning is both a cause and a consequence of this trend. That said, many of the underlying challenges behind Indonesia's military stagnation appear to be organisational and political in nature, rather than caused by a lack of threats alone. In this regard, we return to the importance of Indonesia's civil-military relations and fractured post-authoritarian strategic policymaking ecosystem over the past decade.

## Conclusion

The previous sections have described why and how Indonesia has embarked on long-term retail defence planning where the defence ministry develops long-term modernisation or acquisition plans but implements them in retail fashion, where it acquires a small number of expensive and complex platforms from different suppliers. The chapter has highlighted the path-dependent effects of Indonesia's civil-military history and operational history and experience. Indonesia's internal security challenges and fractured defence policymaking following the end of authoritarian rule in 1998 further complicate attempts to develop a systematic and outwardly focused defence planning system. Furthermore, the absence of personal presidential interest in managing defence policy exacerbated the growing domestic political role of the military and the army's dominant outlook and strategic thinking. While the diversity of military platforms has hamstrung military effectiveness, defence resources and technological innovation, it has allowed the defence ministry under Prabowo Subianto to seek new arms deals with a wide range of international partners in recent years.

Indonesia's retail defence planning is a useful reminder for other small and medium powers seeking to enhance their military capabilities. While policymakers are not wrong in seeking out modern arms and equipment, they should not only focus on technological solutions. Nor is a larger defence budget the only answer. Instead, they should focus on improving and institutionalising all the necessary defence planning policy processes, from proper budgetary allocations to personnel management and management expertise. They should also consider the balance between arms autonomy and diversity, which makes arms modernisation more than a defence policy problem as it needs a whole-of-government approach to make it sustainable and efficient in the long run. In other words, broader questions surrounding civil-military relations, domestic and bureaucratic politics, defence diplomacy and industrial bases are part of the equation as well. Finally, Indonesia should also be seen as the case where strategic alignments are not equivalent to sector-specific (i.e. defence) alignments. Conflating the two types of alignments will muddy our broader understanding of defence partnerships as part of defence planning.

Finally, in terms of how Indonesia's retail compares to other small and medium powers (see Figure 6.1), the country suffers from significant financial constraints as well as the burdens of operational history and civil-military relations. These conditions set the parameters and drivers for Indonesia's defence planning processes. Despite the transition from authoritarian rule more than two decades ago, the military has been left to its own devices to determine its procurement requirements and defence planning processes relatively free from civilian control. Domestic dynamics in other words contributed to the military's own search for defence planning processes and mechanisms.

| GENERAL COMMONALITIES | SMPS ADJUST THEIR APPROACHES BASED ON SYSTEMIC CHANGE OUT OF THEIR CONTROL | VARIATIONS WITHIN THESE COMMONALITIES | SMPS REACT TO CHANGES IN THE RELEVANT SECURITY COMPLEX(ES) THE SMP IS PART OF | | | | |
|---|---|---|---|---|---|---|---|
| | *Alliances, Dependencies and National Ambitions* | | | | | | |
| | SMPS FACE CONSTRAINTS IN SAFEGUARDING NATIONAL SECURITY | | FINANCIAL | DEMOGRAPHIC | CULTURAL | GEOGRAPHICAL | CIVIL/MILITARY RELATIONS |
| | SMPS NEED TO POSITION THEMSELVES VIS-À-VIS GLOBAL AND REGIONAL POWER CONSTELLATIONS | | BALANCING | HEDGING | BANDWAGONING | NON-ALIGNING | NOT APPLICABLE |
| | SMPS HAVE TO ENGAGE WITH OTHER STATES TO INCREASE THEIR DEFENCE CAPABILITIES | | SECURITY GUARANTEES | TRANSACTIONAL | FORCE INTEGRATION | SPECIALISATION | OTHER |
| | SMPS HAVE ARMED FORCES FOR DIFFERENT PURPOSES | | INSTRUMENT OF DIPLOMACY | PRESTIGE PROJECT | REDUCE STRATEGIC RISK (DEFEND) | DOMESTIC STABILITY | OTHER |
| | *Approaches, Processes, Methods and Techniques* | | | | | | |
| | SMPS TAKE A MORE CONSTRAINED APPROACH TO DEFENCE PLANNING | | THREAT-NET ASSESMENT BASED | PORTFOLIO-BASED | TASK-BASED | MOBILISATION | RESOURCE-BASED |
| | SMPS HAVE LIMITED RESOURCES TO DESIGN AND IMPLEMENT DEFENCE PLANNING PROCESSES | | TAILORED ANALYTICAL PROCESS | OUTSOURCING STRATEGY | ACCEPT DILUTION / FAIR SHARE | EXTERNAL TEMPLATE | OTHER |
| | SMP EMPLOY MULTIPLE METHODS AND TECHNIQUES | | COPY GREAT POWERS | INFORMAL CONSENSUS | TAILORED METHODS | AD HOC | OTHER |
| | SMPS RELY ON DIFFERENT PROCESSES TO DECIDE ON FORCE POSTURE | | ANALYTICAL RESULTS | (BUREAUCRATIC) POLITICS DECIDES | MILITARY DECIDES | GREAT POWER DECIDES | OTHER |
| | *Military Innovation* | | | | | | |
| | SMPS FACE CHALLENGES IN KEEPING PACE WITH TECHNOLOGICAL ADVANCES | | BUILD DOMESTIC TECH/INDUSTRY | TECH TRANSFER AGREEMENTS | ACQUIRE READY INNOVATIONS | PARTNER WITH GREAT POWER | OTHER |
| | SMPS PURSUE VARIOUS MILITARY INNOVATION PATHS AND PATTERNS | | EMULATION / SPECULATION | ADAPTATION / EXPERIMENTATION | INNOVATION / IMPLEMENTATION | OTHER / OTHER | LIMITED SCOPE INNOVATION |
| | SMPS FIND THEMSELVES AT DIFFERENT STAGES OF ORGANISATIONAL CHANGE | | EXPLORATION | MODERNISATION | TRANSFORMATION | OTHER | OTHER |

*Figure 6.1* Structured focused comparison framework for Indonesia.

*Source:* Author

Further, as Indonesia's threat landscape remains internal-dominant, its policy-makers never felt the need to forge formal alliances and rely instead on sector-specific defence partnerships over the years. Consequently, national ambitions are limited and defence planning narrow, as manifested in the 2010 MEF framework. Indonesia's retail defence planning as a form of task-based and resource-based planning also contributed to the country's overall lack of major military innovations. Nevertheless, recent attempts are revitalising the domestic defence industrial base through offset arrangements but have yet to be properly evaluated following the passing of the 2012 Defence Industrial Law. If anything, Indonesia's fractured strategic policymaking ecosystem has made such assessments more challenging.

Finally, Indonesia's retail approach, as we have seen earlier, also highlights the conceptual limitations of the technologically anchored framework of military innovation and adaptation as used in this special issue. For one thing, the framework replicates the collective conceptual ambiguity in the literature of military innovation where scholars use different terms—from 'emulation', 'adaptation', to 'innovation' in different ways to refer to different analytical constructs.[37] Scholars have tried to break down the different levels and scope of military change and innovation to better specify the different terms employed.[38] Military emulation, for example, is not necessarily an 'early phase' prior to innovation. For another, the technological premise of the framework ignores the reality that other changes and innovations that matter for military effectiveness and operational performance are rarely technological; they could be conceptual or doctrinal or organisational. Furthermore, the framework also starts with the premise that states and their armed forces engage in military innovation—and presumable defence planning—in response to some form of 'structural' threats or challenges in the international system (i.e. foreign enemies). The Indonesian case above clearly shows the limitations of this starting point. Military change—and therefore the planning for it—is driven differently in countries where internal threats and political instability predominate. The causal direction and linearity of the framework are also harder to apply in an analysis that covers decades. Overall, without properly specifying the different levels, types and scope conditions of military change and innovation, scholars are left struggling to see how the framework could better explain their specific empirical case. In other words, an analytical framework of military innovation and defence planning premised on technology and external threats (derived from the Western experience) is unlikely to be the most constructive path to analyse defence planning models in non-Western countries like Indonesia.

## Notes

1 Paul Davis, 'Defense planning when major changes are needed', *Defense Studies* 18, no. 3 (2018): 375.
2 R.G. Coyle, 'A mission-orientated approach to defense planning', *Defense Analysis* 5, no. 4 (1989): 353.
3 See, for example, Stephan Frühling, *Defence planning and uncertainty: Preparing for the next Asia-Pacific war* (Abingdon: Routledge, 2014); Stuart E. Johnson et al., *New challenges, new tools for defense decision making* (Santa Monica, CA: RAND

Corporation, 2003); Henrik Breitenbauch and André Ken Jakobsson, 'Defence planning as strategic fact: Introduction', *Defence Studies* 18, no. 3 (2018): 253–261.

4  Michael J. Mazarr et al., *The US Department of Defense's planning process: Components and challenges* (Santa Monica, CA: RAND Corporation, 2019), 18.

5  Mazarr et al., *The US Department of Defense's planning process*, 19.

6  Colin Gray, *Strategy and defence planning: Meeting the challenge of uncertainty* (Oxford: Oxford University Press, 2014), 26.

7  See details in Curie Maharani, Reine Prihandoko and Wendy Prajuli, 'Assessing the Risk of Technological Disaster in Indonesian Armed Forces', *The Indonesian Quarterly* 50, no. 1 (2022): 19.

8  See details in Evan A. Laksmana, 'Why is Southeast Asia rearming? An empirical assessment', in *US policy in Asia: Perspectives for the future,* eds. Rafiq Dossani and Scott W. Harold (Santa Monica, CA: RAND Corporation, 2018), 128–129.

9  Details are in Evan A. Laksmana, Curie Maharani and Iis Gindarsah, *75 tahun TNI: evolusi ekonomi pertahanan, operasi, dan organisasi militer Indonesia, 1945–2020* [75 years of the TNI: The Evolution of Defence Economics, Military Operations, and Organisational Structures, 1945–2020] (Jakarta: Centre for Strategic and International Studies, 2020), chp. 4.

10  Christoph Harig, Nicole Jenne and Chiara Ruffa, 'Operational experiences, military role conceptions, and their influence on civil-military relations', *European Journal of International Security* 7, no. 1 (2022): 3.

11  Italics mine. See Frühling, *Defence planning and uncertainty*, 112.

12  Breitenbauch and Jakobsson, 'Defence planning as strategic fact', 258.

13  Caitlin Talmadge, 'Different threats, different militaries: Explaining organizational practices in authoritarian armies', *Security Studies* 25, no. 1 (2016): 124; For Indonesia's conflictual civil-military history, see Marcus Mietzner, *Military politics, Islam, and the state in Indonesia: From turbulent transition to democratic consolidation* (Singapore: Institute of Southeast Asian Studies, 2009).

14  Gray, *Strategy and defence planning*, 138.

15  See details in Laksmana, Maharani and Gindarsah, *75 tahun TNI*.

16  See *Peraturan Menteri Pertahanan Republik Indonesia Nomor 39 tahun 2015 tentang Kebijakan Pembangunan Minimum Essential Force Tentara Nasional Indonesia* [Minister of Defence Regulation No. 39 of 2015 on the Development Policy of the Indonesian Armed Forces' Minimum Essential Force]. See a brief discussion on MEF in Yuddy Chrisnandi, 'The political dilemma of defence budgeting in Indonesia', *Revista UNISCI* 15 (2007): 9–18.

17  For this bargain, see Marcus Mietzner, *The politics of military reform in post-Suharto Indonesia: Elite conflict, nationalism, and institutional resistance* (Washington, DC: East-West Center, 2006).

18  President Abdurrahman Wahid was effectively impeached in 2001 partly because of the political crises he ignited by deeply intervening into the inner workings of military promotions. See details in Tatik S. Hafidz, *Fading away?: The political role of the army in Indonesia's transition to democracy, 1998–2001* (Singapore: Institute of Defence and Strategic Studies, 2006).

19  See Jefferson Ng and Yudha Kurniawan, 'The parliament and cooperative oversight of the Indonesian armed forces: Why civil–military relations in Indonesia is stable but still in transition', *Armed Forces & Society* (2022): 1–27.

20  See Aditya Batara Gunawan, 'Civilian control and defense policy in Indonesia's nascent democracy', in *Reforming civil-military relations in new democracies: Democratic control and military effectiveness in comparative perspectives,* eds. Aurel Croissant and David Kuehn (Cham: Springer, 2017), 129–150; Muhamad Haripin, Adhi Priamarizki and Sigit S. Nugroho, 'Quasi-civilian defence minister and civilian authority: The case study of Indonesia's Ministry of Defence during Joko Widodo's presidency', *Asian Journal of Comparative Politics* 8, no. 1 (2023): 164–183.

21 See Laksmana, Maharani and Gindarsah, *75 tahun TNI*, 127.
22 See Institute for the Policy Analysis of Conflict, *Civil-military relations in Indonesia After Jokowi* (Jakarta: Institute for the Policy Analysis of Conflict, 2023), 8.
23 See Laksmana, Maharani and Gindarsah, *75 tahun TNI*, 54.
24 Details in this paragraph are from Laksmana, Maharani and Gindarsah, *75 tahun TNI*, 127–130.
25 See Evan A. Laksmana, 'The enduring strategic trinity: Explaining Indonesia's geopolitical architecture', *Journal of the Indian Ocean Region* 7, no. 1 (2011): 103–104.
26 See Guy J. Pauker, *The Indonesian doctrine of territorial warfare and territorial management* (Santa Monica, CA: RAND, 1963).
27 See details in International Institute for Strategic Studies, 'Indonesia', in *The Military Balance 2021* (Abingdon: Routledge, 2022), 236.
28 See Siddharth Chandra and Douglas Kammen, 'Generating reforms and reforming generations: Military politics in Indonesia's democratic transition and consolidation', *World Politics* 55, no. 1 (2002): 96–136; Evan A. Laksmana, 'Reshuffling the deck? Military corporatism, promotional logjams and post-authoritarian civil-military relations in Indonesia', *Journal of Contemporary Asia* 49, no. 5 (2019): 806–836.
29 See Evan A. Laksmana, 'Indonesia and anti-access warfare: Preliminary policy thoughts', *The Indonesian Quarterly* 48, no. 4 (2020): 303–321.
30 Franklin B. Weinstein, *Indonesian foreign policy and the dilemma of dependence: From Sukarno to Soeharto* (Jakarta and Kuala Lumpur: Equinox Publishing, 2007); Daniel Novotny, *Torn between America and China: Elite perceptions and Indonesian foreign policy* (Singapore: Institute of Southeast Asian Studies, 2010).
31 For the distinction between alliances and alignment, especially in terms of policy coordination, see Thomas S. Wilkins, '"Alignment", not "alliance"—the shifting paradigm of international security cooperation: Toward a conceptual taxonomy of alignment', *Review of International Studies* 38, no. 1 (2012): 53–76.
32 See the discussion of major military innovations in Michael C. Horowitz, *The diffusion of military power: Causes and consequences for international politics* (Princeton, NJ: Princeton University Press, 2010), 22–23.
33 See Curie Maharani and Ron Matthews, 'The role of offset in the enduring gestation of Indonesia's strategic industries', *Defence and Peace Economics* (2022): 1–22.
34 Ristian Supriyanto, 'Indonesia's opportunistic approach to arms procurement', *East Asia Forum*, 1 July 2021, www.eastasiaforum.org/2021/07/01/indonesias-opportunistic-approach-to-arms-procurement/.
35 See Richard A. Bitzinger, 'A new arms race? Explaining recent Southeast Asian military acquisitions', *Contemporary Southeast Asia* 32, no.1 (2010): 50–69.
36 See David J. Kilcullen, *The political consequences of military operations in Indonesia 1945–99: A fieldwork analysis of the political power-diffusion effects of guerilla conflict* (PhD diss., University of New South Wales, 2000).
37 See the discussion of this collective ambiguity in David Morgan-Owen, Aimée Fox, and Alex Gould, 'Sources of military change: Emulation, politics, and concept development in UK defence', *The British Journal of Politics and International Relations* (2023): 2–4.
38 See the conceptual discussion in Evan A. Laksmana, 'Threats and civil—military relations: Explaining Singapore's "trickle down" military innovation', *Defense & Security Analysis* 33, no. 4 (2017): 348–350.

## Bibliography

Bitzinger, Richard A. 'A new arms race? Explaining recent Southeast Asian military acquisitions', *Contemporary Southeast Asia* 32, no. 1 (2010), 50–69.
Breitenbauch, Henrik and Jakobsson, André Ken. 'Defence planning as strategic fact: introduction', *Defence Studies* 18, no. 3 (2018), 253–261.

Chandra, Siddharth and Kammen, Douglas. 'Generating reforms and reforming generations: military politics in Indonesia's democratic transition and consolidation', *World Politics* 55, no. 1 (2002), 96–136.

Chrisnandi, Yuddy. 'The political dilemma of defence budgeting in Indonesia', *Revista UNISCI* 15 (2007), 9–18.

Coyle, R. G. 'A mission-orientated approach to defense planning', *Defense Analysis* 5, no. 4 (1989), 353–367.

Davis, Paul. 'Defense planning when major changes are needed', *Defense Studies* 18, no. 3 (2018), 374–390.

Frühling, Stephan. *Defence planning and uncertainty: preparing for the next Asia-Pacific war*. Abingdon: Routledge, 2014.

Gray, Colin. *Strategy and defence planning: meeting the challenge of uncertainty*. Oxford: Oxford University Press, 2014.

Gunawan, Aditya Batara. 'Civilian control and defense policy in Indonesia's nascent democracy'. In *Reforming civil-military relations in new democracies: democratic control and military effectiveness in comparative perspectives*, edited by Aurel Croissant and David Kuehn, 129–150. Cham: Springer, 2017.

Hafidz, Tatik S. *Fading away? The political role of the army in Indonesia's transition to democracy, 1998–2001*. Singapore: Institute of Defence and Strategic Studies, 2006.

Harig, Christoph, Jenne, Nicole and Ruffa, Chiara. 'Operational experiences, military role conceptions, and their influence on civil-military relations', *European Journal of International Security* 7, no. 1 (2022), 1–17.

Haripin, Muhamad, Priamarizki, Adhi and Nugroho, Sigit S. 'Quasi-civilian defence minister and civilian authority: the case study of Indonesia's Ministry of Defence during Joko Widodo's presidency', *Asian Journal of Comparative Politics* 8, no. 1 (2023), 164–183.

Horowitz, Michael C. *The diffusion of military power: causes and consequences for international politics*. Princeton, NJ: Princeton University Press, 2010.

Institute for the Policy Analysis of Conflict. *Civil-military relations in Indonesia after Jokowi*. Jakarta: Institute for the Policy Analysis of Conflict, 2023.

International Institute for Strategic Studies. *The military balance 2021*. Abingdon: Routledge, 2022.

Johnson, Stuart E., Libicki, Martin C. and Everton, Gregroy. *New challenges, new tools for defense decisionmaking*. Santa Monica, CA: RAND Corporation, 2003.

Kilcullen, David J. *The political consequences of military operations in Indonesia 1945–99: A fieldwork analysis of the political power-diffusion effects of guerrilla conflict*. PhD diss., University of New South Wales, 2000.

Laksmana, Evan A. 'The enduring strategic trinity: explaining Indonesia's geopolitical architecture', *Journal of the Indian Ocean Region* 7, no. 1 (2011), 95–116.

———. 'Threats and civil—military relations: explaining Singapore's "trickle down" military innovation', *Defense & Security Analysis* 33, no. 4 (2017), 347–365.

———. 'Why is Southeast Asia rearming? An empirical assessment'. In *US policy in Asia: perspectives for the future*, edited by Rafiq Dossani and Scott W. Harold, 106–137. Santa Monica, CA: RAND Corporation, 2018.

———. 'Reshuffling the deck? Military corporatism, promotional logjams and post-authoritarian civil-military relations in Indonesia', *Journal of Contemporary Asia* 49, no. 5 (2019), 806–836.

———. 'Indonesia and anti-access warfare: preliminary policy thoughts', *The Indonesian Quarterly* 48, no. 4 (2020a), 303–321.

———., Maharani, Curie and Gindarsah, Iis. *75 tahun TNI: evolusi ekonomi pertahanan, operasi, dan organisasi militer Indonesia, 1945–2020* [75 years of the TNI: the evolution of defence economics, military operations, and organisational structures, 1945–2020]. Jakarta: Centre for Strategic and International Studies, 2020b.

Maharani, Curie and Matthews, Ron. 'The role of offset in the enduring gestation of Indonesia's strategic industries', *Defence and Peace Economics* 34, no. 7 (2022), 981–1002.

Maharani, Curie, Prihandoko, Reine and Prajuli, Wendy. 'Assessing the risk of technological disaster in Indonesian armed forces', *The Indonesian Quarterly* 50, no. 1 (2022), 6–29.

Mazarr, Michael J., Best, Katharina L., Laird, Burgess, Larson, Eric V., Linick, Michael E. and Madden, Dan, *The US Department of Defense's planning process: components and challenges*. Santa Monica, CA: RAND Corporation, 2019.

Mietzner, Marcus. *The politics of military reform in post-Suharto Indonesia: Elite conflict, nationalism, and institutional resistance*. Washington, DC: East-West Center, 2006.

———. *Military politics, Islam, and the state in Indonesia: from turbulent transition to democratic consolidation*. Singapore: Institute of Southeast Asian Studies, 2009.

Morgan-Owen, David, Fox, Aimée and Gould, Alex, 'Sources of military change: emulation, politics, and concept development in UK defence', *The British Journal of Politics and International Relations* (2023), 1–22.

Ng, Jefferson and Kurniawan, Yudha. 'The parliament and cooperative oversight of the Indonesian armed forces: why civil—military relations in Indonesia is stable but still in transition', *Armed Forces & Society* (2022), 1–27.

Novotny, Daniel. *Torn between America and China: elite perceptions and Indonesian foreign policy*. Singapore: Institute of Southeast Asian Studies, 2010.

Pauker, Guy J. *The Indonesian doctrine of territorial warfare and territorial management*. Santa Monica, CA: RAND, 1963.

Supriyanto, Ristian. 'Indonesia's opportunistic approach to arms procurement', *East Asia Forum*, 1 July 2021, available at: www.eastasiaforum.org/2021/07/01/indonesias-opportunistic-approach-to-arms-procurement/.

Talmadge, Caitlin. 'Different threats, different militaries: explaining organizational practices in authoritarian armies', *Security Studies* 25, no. 1 (2016), 111–141.

Weinstein, Franklin B. *Indonesian foreign policy and the dilemma of dependence: from Sukarno to Soeharto*. Jakarta & Kuala Lumpur: Equinox Publishing, 2007.

Wilkins, Thomas S. '"Alignment", not "alliance"—the shifting paradigm of international security cooperation: toward a conceptual taxonomy of alignment', *Review of International Studies* 38, no. 1 (2012), 53–76.

# 7 Emirati defence planning

## The overriding importance of the political-cultural system

*Ash Rossiter and Athol Yates*

### Introduction

The United Arab Emirates (UAE) offers an intriguing case for the comparative study of defence planning of small and medium powers. This is because, we argue, it has some very unusual characteristics that have resulted in it achieving something that other Arab countries, and very few rich but demographically small countries, can do—the ability to translate investment in its armed forces into actual military prowess. For one analyst, it has exemplified unusual levels of military effectiveness and sophistication in hostile campaigns.[1] The country's ability to generate military power and its propensity to use its armed forces abroad are important issues to unpack.[2] Yet these issues can only be properly understood through understanding the circumstances and drivers behind the country's decisions and the processes for translating ambitions for its military into desired outcomes.

The most commonly quoted reason for the UAE's defence abilities is its wealth. There is no doubt that its enormous hydrocarbon reserves mean that Emirati leaders do not face the same harsh resource constraints that many small states face. This has obviously allowed it to buy the latest materiel—and large amounts of it. The UAE keeps much information about its armed forces confidential, but analysts estimate that the country spends $22 billion a year on defence, which is about the same as Turkey.[3] However, wealth alone is far from the key determinant of military ability. If it was, then states such as Qatar and Saudi Arabia, which spend similar amounts on their respective militaries, would reflect the military competence of the UAE.

The narrative that follows makes the argument that the UAE's political and cultural environment plays a much more pervasive role in Emirati defence planning matters than structural realities such as its wealth, population and institutions. This is not to say that these structural factors do not matter; rather, the choices that the UAE makes based on them and the ways it goes about trying to implement these choices are highly conditioned by the political and cultural characteristics of the country.

### The UAE's national ambitions, alliances and dependencies

Up until the 1990 Gulf War, the UAE's military ambitions were limited to deterring aggression through a strategy of making friends with everyone, and seeking

DOI: 10.4324/9781003398158-7

security support in times of crisis, such as following the 1986 intensification of attacks against UAE interests by Iraq and Iran.

Like other small states in the region,[4] the Iraqi invasion of Kuwait in 1990 led the UAE to adjust its defence posture. It not only initiated a modernisation of its materiel and forces, it also immediately removed over ten thousand expatriates serving in its force who were seen as unreliable, imposed a rank ceiling for the remaining expatriates and removed them completely from combat and combat support formations. The UAE also prioritised formalising security guarantees for out-of-region powers—the US, UK and France. Finally, a decision was made to introduce an offset policy designed to foster the development of a national defence-industrial base. These developments were a significant departure from the pre-1990 arrangements and reflected a more proactive approach to defence. This was driven from then to the present by one person: Sheikh Mohammed bin Zayed Al Nahyan (MBZ), who in 1990 was appointed the Deputy Chief of Staff of the UAE Armed Forces, and in 1992 the Chief of Staff, before in 2004 becoming the Deputy Supreme Commander and in 2021 the Supreme Commander and President of the UAE.[5] This centralisation of authority in the UAE is not new; it is a continuation of the Sheikhly political system.[6]

Since the 1990s and up to the present day, the UAE's national ambitions have been pretty constant. They feature one main threat driver (countering Iranian activity throughout the region); one strategic objective (preserving the regional status quo) and one capability aim (becoming an increasingly self-reliant military force and demonstrating to others that the UAE takes its own defence seriously).[7]

Before examining how successfully these aims have been pursued, it is important to identify how shocks have influenced national ambitions and acted as a driver for change. This is because shocks are often very significant to small nations which have less ability to resist such impacts and chart their own future than larger nations. Despite frequent characterisation that major military changes in the UAE are driven by shocks, this is not the case. This misunderstanding arises because, while some significant military changes have occurred at the same time as major shocks, the linkages are not causal. Two examples illustrate this—the creation of the Presidential Guard and the introduction of national service.

The Presidential Guard has been the most noted development in the UAE's force structure in the past 20 years. Its formation in 2011 coincided with the Arab Spring, resulting in these events being linked. Despite its name, the Presidential Guard was not established as a praetorian guard for the protection of the country's political leadership (though a subunit within it is responsible for close personal protection of the political rulers). Rather than arising out of the Arab Spring, it was established as part of a range of national security reforms in the second half of the 2000s and early 2010s designed to modernise and improve national security coordination, the armed forces and internal security. Other institutions established over this period include the National Security Council, the National Electronic Security Authority and the Critical Infrastructure and Coastal Protection Authority.[8] The formation of the Presidential Guard can be better understood as an example of the Emirati leadership's desire to create areas of excellence that strive to be 'best in class'. In this case, it was to establish a high-readiness, multi-purpose, fighting capability that can operate across land, sea and air domains.[9] The Presidential Guard is today a rapid reaction, deployable and self-sufficient force consisting of a combined-arms mechanised brigade group, with amphibious, reconnaissance, assault and special force formations.

A second example of correlation but not causation was the establishment of national service in 2014 just before the Yemen military intervention. The programme introduced universal conscription for men between the ages of 18 and 30.[10] While some have interpreted the initiative as a move to enhance the country's defence preparedness in readiness for the Yemen campaign, a deeper look shows that it was done for reasons more closely linked to nation-building, and in particular the shaping of the Emirati youths' minds and building a national identity.[11]

Another common characterisation of the motivation behind the UAE's military transformation has been that it was to support an increasingly assertive foreign policy agenda. The example usually pointed to is Yemen. There is no doubt that the UAE's greater military capabilities have allowed it to successfully undertake operations there. However, the UAE's military expedition was initially a small activity that expanded into something much bigger.[12] A major and long-lasting military intervention in Yemen was not an objective at the outbreak of the conflict.[13] It should not be seen as some sort of hegemonic play by the UAE to take over a country nor dominate the strategically important Gulf of Aden and the Red Sea. A better way to understand the UAE's military activities abroad is to recognise that what is of paramount importance regionally for the UAE—as it is for other small states in turbulent areas—is stability. Stability allows the UAE to focus on advancing its domestic and foreign priorities without the continual concern of unexpected foreign interference or being collaterally affected by geopolitical tensions and conflicts. Given this understanding and the decline in pre-1990 regional balancing arising from Iraq, Iran and Saudi Arabia, plus regionally active players of Syria and Egypt, then the reasons for the UAE's military interventions abroad can be better characterised as a nation stepping forward (albeit for self-interested reasons frequently) to uphold stability.

Those who argue that the UAE now has the military capability to shape the region overlook the UAE's demographic constraints and what they mean for deployable forces. Despite the country having a population of about ten million, only one million of these are nationals. The UAE Armed Forces since the late 1990s have allowed only nationals to serve in combat and combat support formations, which means that their pool of service personnel is limited. (Admittedly, there are thousands of expatriates in support positions, such as pilot instructors, maintenance technicians and physical trainers, with some dual nationals in command posts such as the head of the Presidential Guard. The use of expatriates in these positions allows more Emiratis to serve in combat and combat support elements.[14])

Currently, the regular component of the UAE Armed Forces numbers around 65,000 and to increase this would cause shortages of Emiratis needed for both the control and staffing of other key UAE government agencies and priority areas of the private sector. To put the UAE's military burden (defined as the ratio of military members per 1,000 people in the population) into a global context, the UAE has the world's highest military burden. In 2016, the UAE's burden was at 63, North Korea's at 47, Singapore's at 12.5 and Jordan's at 12.3.[15] In terms of Emirati boots on the ground, this means that the UAE land forces can only deploy abroad a brigade group of 3,000–4,000 soldiers for an extended period although it can surge to two brigade equivalents. Thus, the demographic constraint means that regardless of the proficiencies of the UAE Armed Forces, the UAE's direct military footprint is hugely limited.

Engagement of out-of-region defence partners has been an important element in building UAE military capability. The US has been particularly important, as reflected in the expansive 1994 US-UAE Defence Cooperation Agreement, which was renegotiated in 2019. The relationship is not only about equipment but also about transferring knowledge. This can be seen in the case of US support for the formation and ongoing development of the Presidential Guard. In 2010, the UAE leadership reached out to the US—and General Jim Mattis in particular who was CENTCOM (Central Command) commander at the time—to assist in the creation of the new force. Whilst standing up the Presidential Guard was an attempt to provide the UAE with 'best in class' specialised mechanised, deployable, assault, amphibious and special force capabilities that would allow it to do more for its own security, it had the initial effect of making the Emiratis more dependent on the US in terms of training and advisory requirements.[16] This type of dependency is not necessarily problematic; quite the opposite. The tighter relationship that emerged from American assistance with the Presidential Guard's establishment deepened the alliance and helped to foster conditions for the signing of a bilateral security guarantee.[17] It should be stressed, however, that the Presidential Guard was not set up by the Emiratis to enhance US commitment to the country—this was just a valued by-product.

Although the United States remains its key security provider, the UAE is striving to achieve greater freedom of manoeuvre by pairing up with other partners. This should be taken to mean, however, that the UAE is displaying strong hedging behaviour. The US remains by some distance its closest security relationship. In the realm of national security-related technologies, the US is still the UAE's main source of weapons and will be so for some time to come. However, it is increasingly looking to other providers. The signing of the Abraham Accords, for example, has opened the door for UAE-Israeli security cooperation and for the former to acquire cutting-edge national security technologies from the latter. In terms of military expertise which can be transferred to the UAE, the UAE has a long history of using UK contracts and seconded personnel. In more recent decades, expertise has been sought from other nations including France, Australia and New Zealand. Some countries are seen to provide expertise in niche areas. An example is South Korea which has had a training element in the UAE since 2011. Known as the *Akh* team, meaning 'brother' in Arabic, its goal has been to improve the combat strength of the UAE Armed Forces by supporting education and training of UAE special warfare units in special warfare, counterterrorism, high-altitude descent and maritime special operations. The use of foreign expertise is not uncommon for small, wealthy states.[18]

While self-reliance has long been a general goal, in the last decade its importance has become elevated. Achievement towards this self-reliance can be seen in the UAE's military intervention in Yemen, which was undertaken with minimal support from its traditional military partners—at least in the war's early phases.[19] There are two reasons for the growing importance attached to self-reliance. First, some of the UAE's traditional partners are increasingly viewed as unreliable, in terms of commitment to the region, the value of their security guarantees and as dependable equipment and ongoing support suppliers. Secondly, the UAE needs to demonstrate that it is taking more care of its own security responsibilities. This desire ratcheted up during the Trump administration, which asked its allies to do more for their own security in their respective regions under a revamped version of the Nixon Guam Doctrine.[20] Increasing self-reliance through building up indigenous military capabilities was viewed by the US as a quid pro quo for retention of security guarantees.

In equipping its military, the UAE is seeking to avoid dependencies. One long-term strategy it has used is to buy from multiple countries. A more recent initiative is to develop and build systems in the UAE, using its own intellectual property through the control of joint ventures and by establishing wholly owned enterprises. While impressive gains have been made, the UAE's nascent defence industrial sector is a long way from being able to provide its military with even a small portion of its defence needs.

Despite its efforts to reduce its military dependencies and achieve greater self-reliance, the UAE's security will, to a significant extent, rest on its external partnerships. This does not mean, however, that it participates in the kind of traditional bandwagoning or hedging behaviour that one might expect from a small state, for two reasons. First, it has the resources to be an attractive partner to the great and the small. It therefore does not have to sit back and accept what it is given. Instead, it has many potential suitors and can pick and choose to some extent. Also, because it is less constrained by resources it can possess overlapping capabilities from various partners—multiple jet fighter variants, for example.

A second reason why it does not exhibit traditional hedging or bandwagoning behaviour is that it has low levels of confidence in the commitments made to it by its current or prospective security partners. The UAE thus displays a propensity to continuously revise its relations based on the situation of the day. It will readily seek new partners and quickly bury old grievances with erstwhile adversaries if there are advantages to be gained. There are no long-term friends, but no long-term enemies at the same time. This, what we term 'fluid pragmatism', can create a confusing picture at times to those who expect to see, and look for, patterns of behaviour in the country's external relations.

## Emirati military innovation and adaptation

Historically, a key driver of UAE innovation has been the real and perceived need to emulate other militaries that exhibit excellence. Previous scholarship has identified the importance of isomorphism in influencing how military organisations take shape.[21] Research shows how as a strategy for survival states often try to, as best they can, mimic the militarily most capable states in the international system. Countries may partly base the design of their militaries not entirely on their own needs but on how the most significant powers in the world structure theirs. The UAE has shown a propensity to emulate the best practices of principally the US and British and, to a lesser extent, French, Australian and South Korean militaries in the past few decades.

A 2023 example of this, which is having a significant force structure impact, was the order to establish a new service—the UAE National Guard. Following observations of the French/Italian Gendarmerie, which is a military organisation charged with civil responsibility, the UAE Supreme Commander decided that such a service would be useful in the UAE. In early 2023, elements of the military were brought together with elements of public security police, civil defence and critical infrastructure protection. Again, this decision to reorganise significant parts of the military and wider national security apparatus was not a reaction to any particular event but part of a continuous effort to achieve greater effectiveness by emulating 'best in class' practices abroad.

Once a model to emulate has been decided upon by the leadership, the military will be directed to skip the speculation and experimentation phases and move directly to implementation. This does not necessarily mean, however, that the existing force structure is transformed; rather, new layers are added to the extant organisation in a process that can best be thought of as accretion.

Such emulation or mimicry is rationally logical, but there is also a constructivist element to this as well. Possessing a military with all the 'standard' elements becomes a marker that the state has become a modern state.[22] In this way, decisions about force structure and weapons acquisition are taken for intersubjective reasons, that is, what such things communicate to other states. There were certainly signs of this in the early years of the UAE Armed Forces, after Britain's withdrawal in 1971.[23]

The dominance of external sources of military innovation is facilitated by the presence of both senior retired Emirati and Western advisers who have a direct relationship with the Supreme Commander. These very senior Western advisors are from the UAE's main English-speaking security partner nations, and a number have been granted Emirati citizenship. An example of the former is Major General (ret.) Khalid Abdullah Al Buainain Al Mazroeui, former head of the UAE Air Force and Air Defence, and examples of the latter are Major General Michael Hindmarsh, a retired Australian officer, who commands the Presidential Guard and retired US Army lieutenant colonel Stephen Toumajan, promoted to the rank of major general in the Emirati military, who formerly led the UAE's Joint Aviation Command and National Search and Rescue Centre. These advisors and their ideas about how the UAE Armed Forces should innovate and in what directions are often fed into the top of the political-military hierarchy and then come down from above for implementation.[24] There are also a number of expatriate advisors at service, brigade and sometimes unit level who also inject ideas.[25]

The adviser-to-leadership route can only be considered an indirect source of innovation. Decisions of the force structure and procurement are made at the apex of the political-military system, whether this is the insistence that jiu-jitsu is taught to all Emirati service personnel, acquisition choices for modern fighter jets, or the introduction of the annual public military spectacle known as Union Fortress display.[26] All significant—and many modest and even small decisions—are taken by essentially one person—MBZ.[27] Subordinates—senior officers or otherwise— may be asked to provide a menu of options, but the decision ultimately rests with MBZ.[28] The single most important reality for understanding mechanisms and processes for innovation in the UAE military is that this single person makes key decisions unconstrained by elite involvement, let alone institutions.[29]

It stands to reason then that much of the force structure or doctrine innovation comes from the leadership's observations of other states and subsequent decrees that certain aspects need to be emulated and adopted. Emulation is the germ of the idea, but aspects are also adapted to suit local circumstances. Existing force structure is no barrier to innovation. Simply, the armed forces are required to restructure to meet this new direction if it is decreed. The magnitude of change can be transformative and rapid, as long as there is sustained leadership attention on implementing the change by MBZ or his delegates.

An important thrust of innovation reflects the manpower realities of the UAE. This has the effect of placing emphasis on upskilling Emiratis and technological solutions to make up for the lack of manpower. On the former, the UAE relies heavily on its American partner; the US training mission within the UAE and on the US mainland ranges from aviation and missile defence training to special operations training and artillery, reconnaissance and manoeuvre exercises.[30]

In the absence of numbers, the UAE Armed Forces is largely a platform-centric and technology-driven military. But how does it decide which technologies to buy-in or develop? First, the military does not lead in technological innovation. Rather for major systems it has traditionally waited to receive what the political leadership believes it should possess, although this may be initiated by a senior officer bringing forward an idea to the leadership. With the restructuring of the Ministry of Defence in the second half of the 2010s, 'best practice' separation of capability and procurement has been adopted. This has seen the Ministry defining what capability it needs based on strategic assessments with Tawazun, a state-run and leadership-controlled entity, tasked with developing the detailed requirements and actually procuring the materiel. Even with this system, the role of the senior political leadership appears not to have been diminished, as all important decisions, including procurements above a few million dirhams, require their approval.

Industry tries to anticipate the UAE's emerging technology innovation requirements, but there is little explicit direction given. Industry may hear that the political leadership is talking firmly about autonomous systems and then respond. Innovation in this way is responsive to hearing the 'mood music' of the day. Because the potential financial rewards are so high, companies are willing to develop concepts and prototype systems in areas in which they perceive there is leadership interest.

We can also think of innovation in the UAE defence scene not in terms of outcomes but as a desirable process to undertake, whatever the result. 'We need to innovate . . . we need to have new ideas', may best summarise the prevailing sentiment.[31] The pressure for innovation has been further fuelled by the UAE government announcement that 2015 was to be the 'year of innovation'.[32] To be seen to be innovating is as important as any actual innovation.

Even though the concept of innovation is prized and efforts are made towards continuous improvement in the military, the UAE faces several challenges in translating this innovation into desired outcomes. The first obstacle is a cultural predisposition towards risk aversion.[33] The risk-reward equation for attempting something new is heavily weighted towards making a mistake,[34] which creates huge disincentives for risk-taking, even though this is crucial in the process of innovation.[35]

A further problem is that of meritocracy. Promotion to senior positions is often based on considerations beyond an individual's assessed competency. Tribal and family status and connections still matter in the military, though there are signs that this is much reduced than before.

Another issue is that the UAE Armed Forces is a highly fragmented organisation. It is therefore difficult to enact a change that is implemented across the board. Few individuals have a clear picture of the military as a whole, which significantly reduces consultation with different parts of the military about what may or may not be a suitable innovation and stymies the implementation of change. Moreover, senior officers view themselves as little more than temporary custodians of their particular fiefdom. There are few incentives for commanders to attempt to innovate within their own units and very little ability to pollinate other units outside of their command with successful practices. Typically, they will maintain the status quo until told otherwise. For these reasons, senior military leaders within the service branches are generally passive. (The exceptions are those who are empowered by the leadership to make changes and/or have a greater appetite for risk-taking in a culture that is generally risk-averse.[36]) Passivity affects the propensity for organic innovation to occur.

A clear distinction exists between presentational innovation and actual innovation in the UAE. As mentioned previously, much of the attempted change in the UAE Armed Forces derives from attempts to emulate the perceived 'best in class' practices of others. But whilst the accoutrements and artefacts can be imported, real, substantive change does not occur, especially if it runs counter to the dominant national cultural norms in the UAE. Take, for example, efforts to create a process for learning lessons. This is an extremely difficult process to put into practice as there is a cultural aversion to observations about operational experiences—good or bad—being pushed up the system. Regardless of whether a lessons-learned process is instituted, the role of the military will remain to passively wait to be told what the lessons were by the political leadership and whether any change therefore needs to occur.

In sum, the cultural and political system of the UAE does not allow for bottom-up innovation or sideways discussion of innovation in any significant way. Rare is the individual who attempts change. Even when radical change is successfully brought about, it may not be enduring if it runs counter to national cultural norms. Such change will only be enduring for as long as ongoing leadership continually focuses on it and empowers the custodian of the change. In addition, there is no institutional mechanism for continual improvement.

### Approaches, processes, methods and techniques for UAE's defence transformation

Defence planning is in good part the attempted management of risk in an unavoidably uncertain future. Risk management methods cannot reduce this uncertainty; the future has yet to happen. But defence establishments do create institutional processes for attempting to understand the emerging threat environment.

Historically, the UAE has not adopted a risk management approach to defence planning. It has not engaged in systematic, institutionalised over-the-horizon defence planning as it is prized and practised by, say, NATO member states.[37] However, this has been identified as a weakness and as such, elements (such as writing a National Defence Strategy) and processes (such as a strategic assessment review) are slowly being introduced as part of the ongoing, nearly decade-long reform of the Ministry of Defence. When introduced, the approach will be markedly different from that of Western states. This is because the Western approach devolves many decisions to the defence ministry which is culturally inappropriate in the UAE as it would take away decision-making from the political leadership.

Will this change in the future? Increasingly, strategic planning is being put in place, but its importance may be limited as plans and ideas will still go up to the top, and national leadership will be the ultimate determinant of their fate.

What we are left with is a defence planning process that is neither consistent, regular, nor institutionalised. It is sporadic and driven from the top down. The system described throughout this chapter is one whereby decisions are made by a small number of figures on issues about which they may have limited information.

Decisions may also be made with little consultation. Rapid and swift movements can be made at the helm. The leadership can decree that state institutions are to move in new directions.[38] However, it also means that there is little consideration given to the implementation of the policy and mitigating unintended consequences. This can lead to confusion, a lack of coordination and expedient decisions in implementing the change. In addition, it also leads to the accretion of organisations and functions as the new direction does not see a rationalisation in the existing arrangements and the elimination of duplications and overlap.

Historically, advisors to the senior political leadership and the use of consultants have had an outsized impact on the leadership's decisions. This is because they have been able to bring forward ideas to the leader without any organisational impact assessment of their costs and benefits. This is changing now with the 2022 establishment of the Office of the Supreme Commander which is functioning as a strategic evaluator and coordinator of potential changes, as well as acting as the Supreme Commander's eyes to allow him to see what is actually occurring on the ground. This will continue the system of power remaining with the Supreme Commander, but may introduce more rigour to policy and other change proposals.

It is not helpful really to think of defence planning in the UAE in terms of threats or capabilities. Rather defence planning reflects a period of time and does not shift or change until another big leadership intervention comes. The force structure of that particular moment is of its time. Threat perception is important, but trying to achieve 'best in class' and attempting to bring about greater effectiveness are more important drivers for how the UAE leadership shapes what the military looks like and what it is intended for. Please refer to Figure 7.1. to see the case study of the UAE plotted on the structured comparative framework and to Figure 7.2 to see where the UAE falls on the innovation pathways framework.

| GENERAL COMMONALITIES | SMPS ADJUST THEIR APPROACHES BASED ON SYSTEMIC CHANGE OUT OF THEIR CONTROL | SMPS REACT TO CHANGES IN THE RELEVANT SECURITY COMPLEX(ES) THE SMP IS PART OF | | | | |
|---|---|---|---|---|---|---|
| | *Alliances, Dependencies and National Ambitions* | | | | | |
| | SMPS FACE CONSTRAINTS IN SAFEGUARDING NATIONAL SECURITY | FINANCIAL | DEMOGRAPHIC | CULTURAL | GEOGRAPHICAL | OTHER |
| | SMPS NEED TO POSITION THEMSELVES VIS-À-VIS GLOBAL AND REGIONAL POWER CONSTELLATIONS | BALANCING | HEDGING | BANDWAGONING | NON-ALIGNING | PRAGMATISM |
| | SMPS HAVE TO ENGAGE WITH OTHER STATES TO INCREASE THEIR DEFENCE CAPABILITIES | SECURITY GUARANTEES | TRANSACTIONAL | FORCE INTEGRATION | SPECIALISATION | OTHER |
| | SMPS HAVE ARMED FORCES FOR DIFFERENT PURPOSES | INSTRUMENT OF DIPLOMACY | PRESTIGE PROJECT | REDUCE STRATEGIC RISK (DEFEND) | DOMESTIC STABILITY | REGIONAL STABILITY |
| | *Approaches, Processes, Methods and Techniques* | | | | | |
| | SMPS TAKE A MORE CONSTRAINED APPROACH TO DEFENCE PLANNING | THREAT-NET ASSESMENT BASED | PORTFOLIO-BASED | TASK-BASED | MOBILISATION | OTHER |
| | SMPS HAVE LIMITED RESOURCES TO DESIGN AND IMPLEMENT DEFENCE PLANNING PROCESSES | TAILORED ANALYTICAL PROCESS | OUTSOURCING STRATEGY | ACCEPT DILUTION / FAIR SHARE | EXTERNAL TEMPLATE | OTHER |
| | SMP EMPLOY MULTIPLE METHODS AND TECHNIQUES | COPY GREAT POWERS | INFORMAL CONSENSUS | TAILORED METHODS | AD HOC | OTHER |
| | SMPS RELY ON DIFFERENT PROCESSES TO DECIDE ON FORCE POSTURE | ANALYTICAL RESULTS | POLITICS DECIDES | MILITARY DECIDES | GREAT POWER DECIDES | RULER DECIDES |
| | *Military Innovation* | | | | | |
| | SMPS FACE CHALLENGES IN KEEPING PACE WITH TECHNOLOGICAL ADVANCES | BUILD DOMESTIC TECH/INDUSTRY | TECH TRANSFER AGREEMENTS | ACQUIRE READY INNOVATIONS | PARTNER WITH GREAT POWER | OTHER |
| | SMPS PURSUE VARIOUS MILITARY INNOVATION PATHS AND PATTERNS | EMULATION / SPECULATION | ADAPTATION / EXPERIMENTATION | INNOVATION / IMPLEMENTATION | OTHER | OTHER |
| | SMPS FIND THEMSELVES AT DIFFERENT STAGES OF ORGANISATIONAL CHANGE | EXPLORATION | MODERNISATION | TRANSFORMATION | OTHER | OTHER |

*Figure 7.1* Structured focused comparison framework for the UAE.

*Source:* Authors

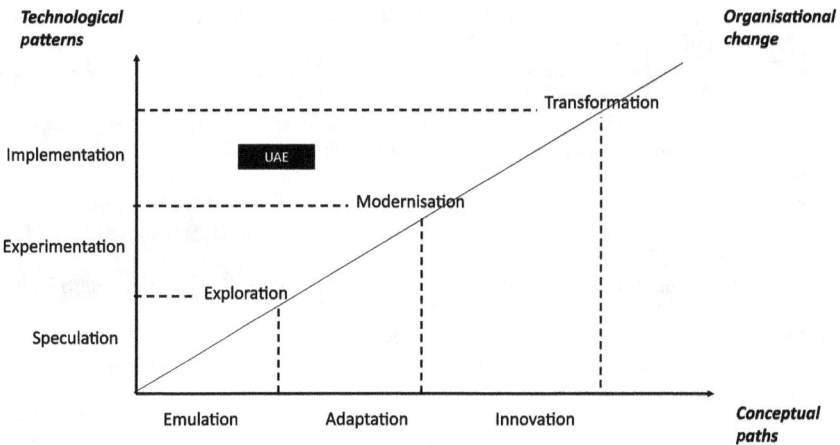

*Figure 7.2* The UAE's military innovation trajectories.

*Source:* Figure created by authors based on Michael Raska, Military Innovation of Small States—Creating a Reverse Asymmetry (Routledge, 2016)[39]

## Conclusion

If the UAE were to invest more in its strategic planning capabilities, it might be better able to map its priorities to its resourcing. Certainly, maturing institutions for strategic, operational and tactical lessons learned—most immediately from the Yemen war, while these lessons are still fresh—would help to inform future force planning. Desirable as this may be, they butt up against the political and cultural characteristics of the country which play an outsized role at the input and output ends of innovation.

Emirati defence planning is driven by the political leadership of the country with the sources of innovation being emulation of external 'best in class' practices. The low institutional capacity for capturing adaptions means that such innovations will invariably not endure unless there is sustained oversight and interest by the leadership. This means that change occurs mainly at the macro level. When change is implemented it is usually substantive rather than bottom-up or incremental.

## Notes

1 David B. Roberts, 'Bucking the Trend: The UAE and the Development of Military Capabilities in the Arab World', *Security Studies* 29, no. 2 (2020): 301–334.
2 K.M. Pollack, *Sizing Up Little Sparta: Understanding UAE Military Effectiveness* (Washington, DC: American Enterprise Institute, 2020).
3 Figures given in Craig Whitelock and Nate Jones, 'UAE Relied on Expertise of Retired U.S. Troops to Beef Up Its Military', *Washington Post*, 18 October 2022, www.washingtonpost.com/investigations/interactive/2022/uae-military-us-veterans/.
4 H. Edström, D. Gyllensporre and J. Westberg, *Military Strategy of Small States: Responding to External Shocks of the 21st Century* (Abingdon: Routledge, 2018).
5 David D. Kirkpatrick, 'The Most Powerful Arab Ruler Isn't M.B.S. It's M.B.Z.', *New York Times*, 2 June 2019, www.nytimes.com/2019/06/02/world/middleeast/crown-prince-mohammed-bin-zayed.html; Athol Yates, 'Challenging the accepted understanding of the executive branch of the UAE's Federal Government', *Middle Eastern Studies* 57, no. 1 (2021): 119–133.
6 K.C. Ulrichsen, *The United Arab Emirates: Power, Politics and Policy-Making* (Basingstoke: Taylor & Francis, 2016).
7 Melissa Dalton and Hijab Shah, *Evolving UAE Military and Foreign Security Cooperation: Path Toward Military Professionalism* (Washington, DC: Carnegie Endowment for International Peace, 2021).
8 Athol Yates, 'The UAE's National Security Machinery of Government', in *Facets of Security in the United Arab Emirates*, ed. W. Guéraiche and K. Alexander (Abingdon, OX: Routledge, 2022), 23–34.
9 Presidential Guard, *Vision Statement [Wall Plaque]* (Majalis, Presidential Guard Building: Al Manhal Palace, Abu Dhabi, 2012).
10 J.B. Alterman and M. Balboni, *Citizens in Training: Conscription and Nation-Building in the United Arab Emirates* (Washington, DC: Center for Strategic & International Studies, 2017).
11 Eleonora Ardemagni, '"Emiratization of Identity": Conscription as a Cultural Tool of Nation-Building', *Gulf Affairs*, no. 'Identity and Culture in the 21st century Gulf' (2016): 6–9.
12 M. Knights, *25 Days to Aden: The Unknown Story of Arabian Elite Forces at War* (Dubai: Profile, 2023).
13 Michael Knights, 'Lessons from the UAE War in Yemen', *Lawfare*, 18 August 2019, www.lawfareblog.com/lessons-uae-war-yemen.

14 Athol Yates, 'Western Expatriates in the UAE Armed Forces, 1964–2015', *Journal of Arabian Studies* 6, no. 2 (2016): 182–200.

15 Athol Yates, *The Evolution of the Armed Forces of the United Arab Emirates* (Solihull: Helion & Co, 2020).

16 For a recent press release on CENTCOM's role with PG, see: Capt. Joshua Hays, 'UAE Armed Forces and U.S. Marines complete defense training in UAE', *US Central Command*, 26 January 2021, www.centcom.mil/MEDIA/NEWS-ARTICLES/News-Article-View/Article/2483392/uae-armed-forces-and-us-marines-complete-defence-training-in-uae/.

17 Joshua L. Schulze, 'The Efficacy of US Security Cooperation in the Gulf' (PhD diss., Monterey, CA: Naval Postgraduate School, 2021).

18 For a historical overview of this practice, refer to Ash Rossiter, *Security in the Gulf: Local Militaries before British Withdrawal* (Cambridge: Cambridge University Press, 2020).

19 Knights, 'Lessons from the UAE War in Yemen'.

20 Richard M. Jennings, 'The Thrust of the Nixon Doctrine', *Military Review* 52, no. 2 (1972): 3.

21 Joelien Pretorius, 'The Security Imaginary: Explaining Military Isomorphism', *Security Dialogue* 39, no. 1 (2008): 99–120.

22 Alessio Patalano, '"A Symbol of Tradition and Modernity": Itō Masanori and the Legacy of the Imperial Navy in the Early Postwar Rearmament Process', *Japanese Studies* 34, no. 1 (2014): 61–82.

23 Athol Yates and Ash Rossiter, 'Military Assistance as Political Gimmickry? The Case of Britain and the Newly Federated UAE', *Diplomacy & Statecraft* 32, no. 1 (2021): 149–167.

24 For how foreign advisers influence state policy, see Paul Drake, 'The Money Doctors: Foreign Advisers and Foreign Debts in Latin America', *NACLA Report on the Americas* 31, no. 3 (1997): 32–36.

25 Whitelock and Jones, 'UAE Relied on Expertise of Retired U.S. Troops'.

26 Yates, *The Evolution of the Armed Forces of the United Arab Emirates*.

27 Kirkpatrick, 'The Most Powerful Arab Ruler'; Yates, 'Challenging the accepted understanding'.

28 On the evolution of the position of the ruler in the Arab Gulf states, see James Onley and Sulayman Khalaf, 'Shaikhly Authority in the Pre-Oil Gulf: An Historical—Anthropological Study', *History and Anthropology* 17, no. 3 (2006): 189–208.

29 This point is made extensively by Z. Barany, *Armies of Arabia: Military Politics and Effectiveness in the Gulf* (Oxford: Oxford University Press, 2021).

30 Jeremy M. Sharp, *The United Arab Emirates (UAE): Issues for U.S. Policy* (Washington, DC: US Congressional Research Services, 2023).

31 Mohammed Dulaimi, 'The Climate of Innovation in the UAE and Its Construction Industry', *Engineering, Construction and Architectural Management* 29, no. 1 (2022): 141–164.

32 The theme for 2015 implored federal government entities to enhance cooperation and formulate policies that provide opportunities fostering innovation. See, for example, the Ministry of Economy's website: 'Our Participation in Government Initiatives', *United Arab Emirates Ministry of Economy,* updated 18 September 2023, www.moec.gov.ae/en/-/year-of-innovation.

33 Yacov Tsur, Menachem Sternberg and Eithan Hochman, 'Dynamic Modelling of Innovation Process Adoption with Risk Aversion and Learning', *Oxford Economic Papers* 42, no. 2 (1990): 336–355.

34 Jamal Mohammad Ibrahim Abdulla Alnassai, 'A Study on the Barriers to Entrepreneurship in the UAE', *Journal of Risk and Financial Management* 16, no. 3 (2023): 146–160.

35 Ana García-Granero et al., 'Unraveling the Link Between Managerial Risk-Taking and Innovation: The Mediating Role of a Risk-Taking Climate', *Journal of Business Research* 68, no. 5 (2015): 1094–1104.

36  Pollack identifies the risk taking dimension of Arab cultures in K.M. Pollack, *Armies of Sand: The Past, Present, and Future of Arab Military Effectiveness* (Oxford: Oxford University Press, 2019). Yates identifies the importance of risk aversion in limiting UAE's military capabilities. Yates, *The Evolution of the Armed Forces of the United Arab Emirates.*

37  Thomas-Durell Young, 'Questioning the "Sanctity" of long-term defense planning as practiced in Central and Eastern Europe', *Defence Studies* 18, no. 3 (2018): 105–121.

38  C.M. Davidson, *From Sheikhs to Sultanism: Statecraft and Authority in Saudi Arabia and the UAE* (London: C. Hurst Limited, 2021).

39  Michael Raska, 'A Structured-Phased Evolution: The Third Generation Force Transformation of the Singapore Armed Forces', in *Military Innovation in Small States: Creating a Reverse Asymmetry*, ed. M. Raska (New York: Routledge, 2016), 130–162.

## Bibliography

Alnassai, Jamal Mohammad Ibrahim Abdulla. 'A Study on the Barriers to Entrepreneurship in the UAE'. *Journal of Risk and Financial Management* 16, no. 3 (2023): 146.

Alterman, J.B., and M. Balboni. *Citizens in Training: Conscription and Nation-Building in the United Arab Emirates*. Washington, DC: Center for Strategic & International Studies, 2017.

Barany, Z. *Armies of Arabia: Military Politics and Effectiveness in the Gulf*. Oxford: Oxford University Press, 2021.

Dalton, Melissa, and Hijab Shah. *Evolving UAE Military and Foreign Security Cooperation: Path Toward Military Professionalism*. Washington, DC: Carnegie Endowment for International Peace, 2021.

Davidson, C.M. *From Sheikhs to Sultanism: Statecraft and Authority in Saudi Arabia and the UAE*. London: C. Hurst Limited, 2021.

Drake, Paul. 'The Money Doctors: Foreign Advisers and Foreign Debts in Latin America'. *NACLA Report on the Americas* 31, no. 3 (1997): 32–36.

Dulaimi, Mohammed. 'The Climate of Innovation in the UAE and Its Construction Industry'. *Engineering, Construction and Architectural Management* 29, no. 1 (2022): 141–164.

Edström, H., D. Gyllensporre, and J. Westberg. *Military Strategy of Small States: Responding to External Shocks of the 21st Century*. Abingdon, OX: Routledge, 2018.

García-Granero, Ana, Óscar Llopis, Anabel Fernández-Mesa, and Joaquín Alegre. 'Unraveling the Link between Managerial Risk-Taking and Innovation: The Mediating Role of a Risk-Taking Climate'. *Journal of Business Research* 68, no. 5 (2015): 1094–1104.

Knights, M. *25 Days to Aden: The Unknown Story of Arabian Elite Forces at War*. Dubai: Profile, 2023.

Patalano, Alessio. '"A Symbol of Tradition and Modernity": Itō Masanori and the Legacy of the Imperial Navy in the Early Postwar Rearmament Process'. *Japanese Studies* 34, no. 1 (2014): 61–82.

Pollack, K.M. *Armies of Sand: The Past, Present, and Future of Arab Military Effectiveness*. Oxford: Oxford University Press, 2019.

———. *Sizing Up Little Sparta: Understanding UAE Military Effectiveness*. Washington, DC: American Enterprise Institute, 2020.

Presidential Guard. *Vision Statement [Wall Plaque]*. Abu Dhabi: Majalis, Presidential Guard Building: Al Manhal Palace, 2012.

Pretorius, Joelien. 'The Security Imaginary: Explaining Military Isomorphism'. *Security Dialogue* 39, no. 1 (2008): 99–120.

Roberts, David B. 'Bucking the Trend: The UAE and the Development of Military Capabilities in the Arab World'. *Security Studies* 29, no. 2 (2020): 301–334.

Rossiter, Ash. *Security in the Gulf: Local Militaries before British Withdrawal*. Cambridge: Cambridge University Press, 2020.

Sharp, Jeremy M. *The United Arab Emirates (UAE): Issues for U.S. Policy*. Washington, DC: US Congressional Research Services, 2023.

Tsur, Yacov, Menachem Sternberg, and Eithan Hochman. 'Dynamic Modelling of Innovation Process Adoption with Risk Aversion and Learning'. *Oxford Economic Papers* 42, no. 2 (1990): 336–355.

Ulrichsen, K.C. *The United Arab Emirates: Power, Politics and Policy-Making*. Basingstoke: Taylor & Francis, 2016.

Yates, Athol. 'Western Expatriates in the UAE Armed Forces, 1964–2015'. *Journal of Arabian Studies* 6, no. 2 (2016): 182–200.

———. *The Evolution of the Armed Forces of the United Arab Emirates*. Solihull: Helion & Co., 2020.

———. 'Challenging the Accepted Understanding of the Executive Branch of the UAE's Federal Government'. *Middle Eastern Studies* 57, no. 1 (2021): 119–133.

———. 'The UAE's National Security Machinery of Government'. In *Facets of Security in the United Arab Emirates*, edited by W. Guéraiche and K. Alexander, 23–34. Abingdon, OX: Routledge, 2022.

———, and Ash Rossiter. 'Military Assistance as Political Gimmickry? The Case of Britain and the Newly Federated UAE'. *Diplomacy & Statecraft* 32, no. 1 (2021): 149–167.

Young, Thomas-Durell. 'Questioning the "Sanctity" of Long-term Defense Planning as Practiced in Central and Eastern Europe'. *Defence Studies* 18, no. 3 (2018): 357–373.

# 8 Leveraging dependencies

## Defence planning in the Sultanate of Oman

*Nikolas Gardner*

## Introduction

Over the past 50 years, oil and gas revenues have vaulted the states of the Gulf Cooperation Council (GCC) from the pre-industrial to the information age. Economic growth in these states has been accompanied by the rapid development of political, economic and military institutions, some of which have taken centuries to evolve in other societies. This is certainly true of the Sultanate of Oman. Since 1970 Oman's gross domestic product has increased by more than tenfold, facilitating the establishment of modern infrastructure and institutions, including capable armed forces.[1] This chapter examines the development of these armed forces, focusing in particular on the twenty-first century. It will consider the extent to which Oman's allies, as well as the threat perceptions of its leaders, have influenced their size, force structure and capabilities. The rationale for Omani defence planning can be difficult to discern. Oman does not publish a national defence strategy or other documents pertaining to national security. Nor are decisions about military spending and deployments subject to public debate. For the large majority of the period under examination in this chapter, these decisions were made personally by Sultan Qaboos bin Said al Said, who ruled the country from 1970 until his death and succession by his cousin Haitham bin Tarik al Said in 2020. Given Sultan Qaboos's profound influence over Oman's development, his passing had the potential to disrupt the country's foreign and domestic policies. The extent of this disruption, however, remains unclear. As Prime Minister and Supreme Commander of the armed forces, Sultan Haitham retains personal control over Oman's defence and national security policies. As was the case under Qaboos, the Sultan's decisions on these issues are classified. Nonetheless, by examining sources such as military cooperation agreements between Oman and its allies, as well as Oman's military spending and procurement choices, it is possible to infer the rationale for Oman's alliances and the extent to which its armed forces have innovated or adapted in response to changing threats. In addition, while the opacity of Omani governance makes it difficult to determine how defence-related decisions are made, we can at least identify informal processes and the development of the institutional expertise that can shape such decisions in the future. In the process, we can assess the extent to which Sultan Haitham's accession marks a fundamental change in the traditional patterns and practices of Omani defence planning.

DOI: 10.4324/9781003398158-8

## Oman's alliances, dependencies and national ambitions

Oman has long depended on allies and partners to contribute to its defence. Their interest in doing so stems from Oman's status as a pivot state in the international system. The country is pivotal primarily, if not entirely, due to its location.[2] Situated along the maritime trade route linking Europe and Africa with Asia, Oman has long-held strategic significance. Its importance to the European powers increased as they established trade relationships and then colonies in India and further to the east. In 1798, Britain signed the first of a series of treaties with the rulers of Oman in an effort to deny its rivals access to the subcontinent. Oman's location grew even more important in the twentieth century. While aircraft facilitated travel throughout Britain's global empire, they required refuelling. This led the Royal Air Force (RAF) to establish an air base on Masirah Island, just off the Omani coast, in the 1930s. Even as the British Empire contracted after the Second World War, the massive expansion of oil production in the Middle East added to the significance of Oman, which overlooks the Strait of Hormuz, the most important oil transit chokepoint in the world.[3] Concern about the spread of communism in the region led Britain to increase its military support to Oman during the Cold War. In 1958, the British government initiated an assistance programme that aimed to develop Omani transportation, educational and medical infrastructure, as well as its armed forces.[4] When Sultan Said bin Taimur proved unable to contain a communist-backed insurgency in the Dhofar region, however, British authorities supported Qaboos in replacing his father in 1970.[5] A graduate of the Royal Military Academy at Sandhurst, Qaboos was more aggressive in his efforts to develop the country and fight the insurgency. He was also more reliant on British assistance. Hundreds of British military personnel served in Oman, leading Omani forces in a campaign that defeated the insurgents by the end of 1975.[6]

Afterwards the British remained, training Omani personnel in the operation and maintenance of advanced weapon systems, including fighter aircraft and naval patrol vessels. This involved dozens of active duty 'loan service personnel' seconded from the British armed forces, as well as hundreds of contractors, most of whom were recently retired members of the British services. British personnel also managed the institutional development of the Omani armed forces. Even at the end of the Dhofar campaign, few Omani officers had the experience to serve in command and staff positions.[7] As a result, British officers occupied the most senior posts in all three of the Omani armed services through the 1980s. For example, seconded British officers commanded Oman's navy and air force until 1990.[8] While Omanis have long since assumed command of their own armed forces, the British presence remains. Along with an unknown number of British contractors, 70–90 loan service personnel have consistently been seconded to Oman over the past two decades.[9] The most senior is an active-duty major-general who provides advice to Omani leaders, including the Sultan.

The relationship between the United States and Oman is not as venerable as the Anglo-Omani alliance. It is hardly new, however, extending back to 1833 when the US and Oman signed 'a treaty in support of friendship and navigation'.[10] In recent decades, the US has increasingly recognised the strategic significance of

Oman's location. By the mid-1970s, the end of Britain's imperial commitments in Southeast Asia and the defeat of the Dhofar insurgency reduced the value of the Masirah Island air base for the RAF. As the British prepared to transfer the base to Oman's air force, however, the US expressed interest in it. In 1980 the US and Oman signed the Oman Facilities Access Agreement, which allowed American military forces to use the Masirah base. The same year, it served as a staging point for *Operation Eagle Claw*, the ill-fated attempt to rescue American hostages held in Iran. Subsequently, US maritime surveillance aircraft used the base for refuelling as they monitored Soviet activities in the Indian Ocean.[11] While the need for such activities diminished with the end of the Cold War, Oman took on renewed significance to the United States at the beginning of the twenty-first century, when US forces used multiple Omani facilities to support their operations in Afghanistan and Iraq. During *Operation Iraqi Freedom*, Oman allowed more than 3000 American troops to operate out of the country, in return for which the US provided more than $80 million in security assistance.[12] The US maintains its access to multiple Omani military bases today, thanks to the renewal of the Facilities Access Agreement in 2010. These include airfields as well as storage facilities. In addition, in 2017 the US Defense Logistics Agency opened a material processing centre at the port of Salalah, which provides 'material aggregation, trans-shipment, short-term storage and delivery functions for visiting U.S. Navy vessels and other U.S. customers'.[13]

American and Omani forces also conduct joint military exercises on a regular basis. For three weeks each year, for example, the US and Omani land forces train together in the Inferno Creek exercise, which aims to increase interoperability between the two armies by exposing each to the tactics and decision-making processes of the other. Similarly, the US and Omani air forces conduct a biannual exercise, known as Accurate Test, that 'improves proficiency, as well as increases integration between US and Omani military forces'.[14] Moreover, the US has built relationships with senior Omani military officers through its International Military Education and Training (IMET) programme. As of 2021, all of the Omani service chiefs as well as the Chief of Staff of the armed forces were IMET graduates, meaning that they had attended US professional military education institutions, building relationships with their American counterparts in the process.[15]

While Oman has relied primarily on the US and UK as security partners, it has also established military ties with its neighbours. A member of the Gulf Cooperation Council (GCC), Oman has contributed to the organisation's joint and multinational military component, the Peninsula Shield Force (PSF), since its establishment in the mid-1980s. Organised to counter aggression on the part of larger regional powers such as Iran and Iraq, the newly established force proved unable to deter Saddam Hussein from invading Kuwait in 1990. Afterwards, Sultan Qaboos advocated for the expansion of the PSF to a standing army of 100,000 troops, but encountered opposition from Saudi Arabia, which would have been expected to contribute the largest number of personnel. Moreover, after the US-led eviction of Iraqi forces from Kuwait in 1991, other Gulf leaders became less concerned with countering conventional military threats.[16]

Military cooperation among the Gulf states has increased in the twenty-first century. In 2003, the PSF deployed to Kuwait to defend the country prior to the US invasion of Iraq, although it did not participate in the invasion itself. In 2011, elements of the force were deployed to Bahrain to support its government in suppressing internal protests. In 2013, the GCC agreed to establish a unified military command based in Riyadh.[17] In addition, Gulf military personnel have served in the armed forces of their GCC allies. For example, UAE personnel have recently been seconded to the Omani armed forces.[18] Nonetheless, despite Sultan Qaboos's apparent enthusiasm for regional military cooperation, Oman has not participated in recent military operations alongside its GCC partners. The Sultanate did not commit military or security personnel to the GCC intervention in Bahrain in 2011.[19] In 2015, Oman was the only GCC member that did not participate in Operation Decisive Storm, the Saudi-led intervention in Yemen.[20]

Moreover, Oman has a long history of military cooperation with Iran alongside its ties with its Arab neighbours. This relationship extends back to the 1970s, when the Shah sent thousands of Iranian troops to aid Qaboos in defeating the Dhofar insurgency. Cooperation between the two armed forces has continued in the twenty-first century. In 2009, they committed to work together to combat smuggling in the Gulf of Oman and in 2010 they signed an agreement to cooperate in patrolling the Strait of Hormuz and to hold joint military exercises, which have taken place on multiple occasions since then.[21] The most recent, involving the Omani Navy, the Islamic Republic of Iran Navy and the Islamic Revolutionary Guard Corps Navy, took place in October 2022.[22] Given that Oman has consistently maintained closer diplomatic relations with Iran than the rest of the GCC, this level of cooperation is not entirely surprising. Nonetheless, Oman's willingness to maintain it even during periods of tension between Iran and other GCC members has complicated the process of establishing collective defence arrangements with its Arab neighbours. Despite advances in GCC military cooperation over the past two decades, Oman has shown a decided preference for defence partnerships with great powers.

China may be the exception. Oman's location along international maritime trade routes and its proximity to Middle East oil production have made the Sultanate a magnet for Chinese investment, with a Chinese consortium planning to invest $10.7 billion to develop the port of Duqm, as well as multiple industrial projects in its vicinity.[23] These investments will help diversify Oman's economy, so it is not surprising that it has welcomed them. Chinese naval vessels have also visited Oman in recent years and in 2022 Chinese leaders indicated their desire 'to raise the level of cooperation in joint drills, military technology, logistics support and personnel training'.[24] In October 2023, Chinese and Omani military representatives apparently engaged in more focused discussions regarding the establishment of a Chinese military 'presence' in Oman.[25] But the Sultanate has been markedly more enthusiastic about establishing economic ties with China than military ones. In fact, Oman appears to be pursuing a deliberate strategy to offset Chinese influence by deepening military cooperation with Beijing's rivals. In addition to the aforementioned Facilities Access Agreement with the US military, Oman has allowed the Indian Navy to use its ports.[26]

Oman has also taken full advantage of Britain's desire to re-establish a measure of its former global influence following Brexit. Designated a 'strategic hub' in Britain's 2021 *Integrated Review of Security, Defence, Development and Foreign Policy*, Oman has recently experienced an increase in British military presence on a scale not seen in decades.[27] In 2017, the Ministry of Defence (MOD) announced the construction of a Joint Logistics Support Hub in close proximity to the Chinese-funded port at Duqm. In 2018, this new facility supported the largest military exercise by British and Omani air, land and naval forces in nearly 20 years.[28] The following year saw the establishment of the Ras Madrakah Joint Training Area to support military exercises involving British and Omani forces. Britain's military footprint in Oman has continued to expand even after the passing of Qaboos and his succession by Sultan Haitham in early 2020. Later that year, the UK announced additional investment to triple the size of the logistics hub at Duqm, which would include dry docks capable of accommodating the Royal Navy's two aircraft carriers. In 2022, the British Army also announced that it was increasing its presence at the Ras Madrakah facility from six weeks to eight months each year.[29]

Overall, Oman's strategic behaviour can be understood as a form of hedging, in the sense that the Sultanate seeks to strengthen economic relations with a rising China while maintaining strong defence relationships with the US and the UK. This is a strategy increasingly common among the Arab Gulf states. In terms of military alliances and partnerships, however, Oman is thus far balancing with the western powers against China. While the American and British military presence in Oman remains relatively small, they have the effect of keeping the Chinese military at arm's length while also enabling Oman to play a neutral role in disputes in its own neighbourhood. Oman maintains ties with both Iran and its GCC allies without becoming dependent on either. Thus, a degree of dependency on the western powers enables Oman to maintain its military and diplomatic freedom of action with respect to China and on a regional level.

### Military innovation and adaptation

Oman's close relationship with the western powers has also influenced the ability of its armed forces to adapt to new threats and technologies. In the absence of transparent governance and indigenous professional literature, the extent of military innovation in Oman can be difficult to measure. By examining the force structure and procurement decisions of the armed forces over the past two decades, however, it is possible to gain insights into how they have adapted to changing threats and new technologies. Financial considerations do not appear to have constrained Oman's procurement choices in the period under examination. In real terms, Oman's annual military spending rose from $2.56 billion in 2003 to a high of $9.88 billion in 2015.[30] It declined afterwards, but in 2022, it remained at $6.43 billion.[31] As a percentage of Gross Domestic Product (GDP), Omani military spending has been higher than that of its GCC counterparts and significantly higher than that of most western states. From 2009–13 it stood at approximately 8% of GDP, while from 2015–19 it rose to approximately 13% of GDP, before declining to approximately 6% in 2022.[32] These dramatic fluctuations reflect changes in the price of oil, which have a significant effect on Oman's overall GDP. Nevertheless, it is clear that Oman has devoted significant resources to defence over the past two decades. Despite this level of spending, however, the size and structure of Oman's armed forces have remained quite stable. The reported strength of the army has remained constant at 25,000 personnel, while that of the navy has remained at 4,200. The armed forces also include the 'Royal Household', comprising an armoured brigade and special forces units, which has maintained a strength of 6,400 personnel. The only service that has grown is the air force, which expanded from 4,100 to 5,000 personnel in 2007.[33]

Rather than funding the expansion of the armed forces and the development of new capabilities, the primary purpose of Omani defence spending has been to replace its core inventory of air and naval weapon systems, such as combat aircraft as well as patrol ships and support vessels.[34] In 2002, for example, the Omani Ministry of Defence (MOD) ordered 12 F-16 combat aircraft from American manufacturer Lockheed-Martin to allow the Royal Air Force of Oman (RAFO) to begin replacing its existing fleet of 24 Anglo-French Jaguar fighters, originally acquired in the 1970s and 80s. In 2011, Oman ordered an additional 12 F-16s, enabling the RAFO to phase out the Jaguars entirely.[35] In 2012, the MOD signed a contract with Singapore Technologies Marine to design and build four Al-Ofouq class patrol vessels to replace older ships in service since the early 1980s.[36] In 2013, the MOD ordered the National Advanced Surface to Air Missile System (NASAMS) made by Norwegian manufacturer Konigsburg and US-based Raytheon to update Oman's existing air defence system, which was originally installed in the 1970s.[37]

The Omani armed forces have not focused solely on replacing outdated equipment. They have also invested in new systems that enhance their capabilities. For example, in 2002 the MOD ordered 16 Agusta-Westland Super Lynx helicopters. Able to deploy from naval vessels, the helicopters gave the Royal Navy of Oman (RNO) an aviation capability it had previously lacked.[38] In addition to the aforementioned patrol vessels, the RNO took delivery of three Khareef-class corvettes in 2013 and 2014. Built by BAE systems in the UK, the vessels are the largest and most advanced in the Omani fleet, equipped with stealth technology to evade detection as well as surface-to-surface and surface-to-air missiles.[39] Although Oman had already ordered F-16s to replace its older combat aircraft, the MOD ordered an additional 12 Eurofighter Typhoons in late 2012.[40] Manufactured by BAE in conjunction with European companies Airbus and Leonardo, the Typhoon has more advanced capabilities than the F-16. In addition, in late 2022, the US government approved the sale of 48 Raytheon Joint Standoff Weapons to Oman to improve the air-to-ground strike capabilities of RAFO combat aircraft.[41]

Some of these purchases might be construed as payments on what one scholar has termed Oman's 'insurance policy' with the western powers. Admittedly, a desire to maintain a strong security relationship with the UK has influenced Omani procurement decisions in the past. In 1974, when Sultan Qaboos purchased Oman's first integrated air defence system, comprising aircraft, missiles and radar, he chose a package of British-made equipment over French and American alternatives as a reward for British assistance in combating the Dhofar insurgency.[42] It is conceivable that the decision to purchase Typhoons, despite earlier efforts by French President Nicolas Sarkozy to convince the Sultan to buy the French-made Rafale combat aircraft, stemmed from a continued desire to sustain Oman's security relationship with Britain.[43] Even if this was the case, however, it is unlikely that Oman would choose British equipment if it was not of comparable quality to the alternatives. This is certainly true of the Typhoon. While states such as the UAE and India have recently chosen the Rafale over the Typhoon, others such as Austria, Saudi Arabia and Kuwait have chosen the Typhoon. Qatar has purchased both.[44] Although they have different capabilities, both are among the most advanced combat aircraft available on the market today, and there are legitimate reasons for choosing either. Oman has generally favoured British and American manufacturers, but the companies it has patronised, like BAE and Lockheed-Martin, have world-leading reputations for producing state-of-the-art technology. The close security relationship that the UK and US have with the Sultanate may facilitate the sales of their products, but it is unlikely that Oman would buy products that were irrelevant to its defence requirements.

Moreover, Oman has been willing to shop elsewhere to secure systems that meet its needs, even when such purchases have no value as 'insurance'. For example, the decision to purchase patrol vessels from a company based in Singapore, which can offer Oman no security guarantee, suggests that their quality and capabilities were more important than their country of origin. In 2016, the MOD also signed a contract to purchase 172 Pars III infantry fighting vehicles from Turkish manufacturer FNSS.[45] In addition, in 2021, it apparently reached an agreement with the Turkish Ministry of Defence to procure Bayraktar TB2 drones.[46] While Turkey has developed a close security relationship with Qatar, there is little evidence that it is doing the same with Oman. Turkey and Oman signed a Memorandum of Understanding to promote military cooperation in 2001, but it has not resulted in any interaction between the armed forces of the two states. Oman's primary interest appears to be to acquire specific weapon systems rather than an additional ally.

One result of the Pars III purchase, however, was a 2022 agreement between the MOD and FNSS to establish a maintenance and repair facility for the vehicles in Oman.[47] This agreement is part of a broader effort initiated under the leadership of Sultan Haitham to develop indigenous technical expertise that Oman currently lacks. While the maintenance facility will be established by FNSS, it will apparently involve Omani firms, in an effort to 'develop capabilities of local companies'. At the same time as revealing this agreement, the Omani government also announced a plan to transfer military aircraft engine maintenance responsibilities from the British manufacturer Rolls Royce to the RAFO itself. In addition, it announced the establishment of a Spanish-run 'earth observation center', specialising in the analysis of satellite images of the earth.[48] The ultimate aim of all of these projects is to develop the expertise necessary to facilitate military adaptation and innovation. In this respect, Oman lags behind its Saudi and Emirati neighbours, both of which are developing indigenous defence industries. Projects like these, however, show that Oman's new leader intends to follow suit, even if it is not in the immediate future.

How then, should we characterise military innovation and adaptation in the Omani armed forces? As demonstrated in the preceding section, Oman has established enduring relationships with the US and UK that enable it to maintain an independent foreign policy in its own region. But its military procurement decisions are not driven solely by a desire to sustain these relationships. Over the past two decades, Oman has systematically updated and enhanced its military capabilities, particularly those of its navy and air force. It has often chosen American or British technology, but it has also purchased weapons systems produced elsewhere when the capabilities of those systems match Omani requirements. Admittedly, Oman's approach to military innovation is not particularly sophisticated in comparison to other small states such as Israel and Singapore.[49] To use Michael Raska's terminology, while it has begun to develop indigenous military expertise, Oman lacks a mature 'defense-innovation ecosystem', comprising research and development capabilities as well as political, institutional and cultural conditions conducive to the development of new ideas and approaches. As a result, it has engaged in what Raska calls emulation, importing tools that have proven effective in other military organisations. Recent initiatives implemented under Sultan Haitham show a recognition of the value of indigenous technical expertise. At present, however, the armed forces lack the capabilities to engage in independent speculation about, experimentation with, or implementation of emerging concepts and technologies.[50] This has not stopped Oman from modernising its military capabilities, but it has done so by following the lead of its western allies (see Figure 8.1).

## Approaches, processes, methods and techniques

Oman's approach to defence planning bears the mark of Sultan Qaboos, its ruler, defence minister and supreme commander from 1970 until 2020. Oman has a parliament that includes an elected assembly comprising 86 members, but rather than possessing the power to pass laws, it can only vote on issues raised by the

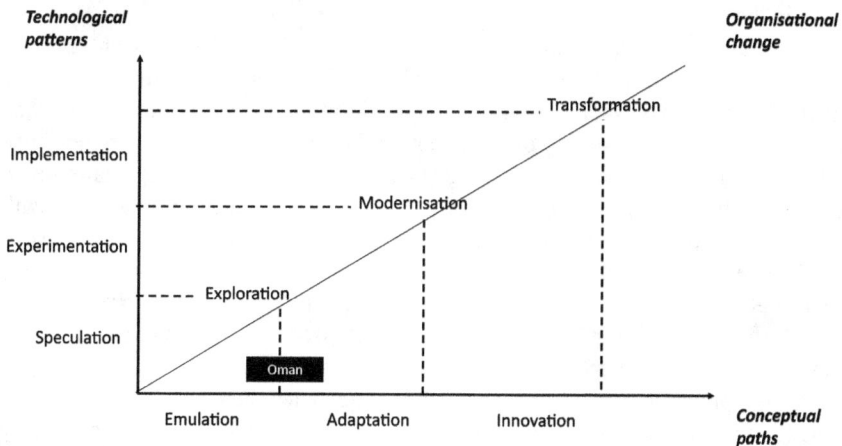

*Figure 8.1* Oman's military innovation trajectories.

*Source:* Figure created by the author based on Michael Raska, Military Innovation of Small States—Creating a Reverse Asymmetry (Routledge, 2016)[51]

Sultan.[52] Thus, throughout his reign, Qaboos maintained full executive authority over defence policy and plans. The most significant threat the Sultanate faced since Qaboos took power was the Dhofar insurgency. While it was defeated by the end of 1975, the Sultan later faced coup attempts in 1994 and 2004.[53] It is therefore not surprising that one of the risks that Qaboos prepared to address was internal opposition. As a result, even today the armed forces include the Royal Guard and the Sultan's Special Forces, which together comprise more than 6,000 personnel directly under the command of the Sultan.[54]

Yet Qaboos also prepared to meet conventional military threats. He made the decision to acquire an integrated air defence system in the 1970s even though his British advisors pointed out that Oman's only hostile neighbour at the time, the People's Democratic Republic of Yemen, did not possess an air force capable of threatening it.[55] When a more serious threat emerged in the 1990s, in the form of Saddam Hussein's Iraq, Qaboos advocated for the expansion of the GCC's Peninsula Shield Force and acquired capabilities, such as US and British tanks and armoured personnel carriers, to contribute to it. In the twenty-first century, Oman has maintained and updated its conventional military capabilities. It is significant, however, that it has invested primarily in upgrading the capabilities of the air force and navy. Aside from the infantry fighting vehicles recently purchased from Turkey, Oman's army still uses equipment acquired in the 1990s. This prioritisation of resources likely reflects the emergence of new threats in the region since the US-led invasion of Iraq overthrew Saddam Hussein in 2003. While Oman has enjoyed stable relations with its neighbours since then, non-state actors such as ISIS and the Houthis have emerged as threats to regional stability. In addition, Oman sits astride sea lanes that have seen an increase in piracy and smuggling. Such threats are better addressed using air and naval power rather than conventional land forces. Nevertheless, according to the most recent edition of *The Military Balance*, after investing in new air and naval weapon systems, Oman 'is now looking to do the same in the land domain'.[56]

Thus, Oman's approach to defence planning resembles what Stephan Frühling has called 'portfolio planning'.[57] Its armed forces are constituted to respond to a diverse range of risks including internal opposition, malign activities by non-state actors and even the possibility of a conventional military conflict in the region. The likelihood of such a conflict involving Oman may seem remote, especially given its success in maintaining and even strengthening military and diplomatic ties with its neighbours on both sides of the Gulf. The past two decades, however, have seen GCC forces deploy to Kuwait to defend against the Iraqi Army. They have also seen internal instability in Bahrain and civil war in Yemen. Therefore, preparations for such an eventuality should not be dismissed simply as Omani efforts to purchase an Anglo-American security guarantee. In light of the recent instability of the region, preparation to meet a range of contingencies seems justified.

It is important, however, not to minimise the importance of Oman's security relationships with the western powers in influencing its defence planning processes. Sultan Qaboos had almost complete control over defence-related decisions, and while Sultan Haitham has appointed his brother Shihab as Deputy Prime Minister for Defence Affairs, there are still no institutional checks on the ruler's authority.[58] But there is evidence to suggest that the United States and particularly the United Kingdom have been able to influence defence-related decisions in Oman over the past two decades, even if it is only indirectly. For example, Oman's participation in the American IMET programme has meant that the most senior Omani military officers, who are in a position to advise the Sultan on key military decisions, are familiar with American operational concepts and strategic priorities in the Middle East and the Indian Ocean. With dozens of officers seconded to the Omani armed forces over the past two decades, British input is more direct. Most significantly, the senior British loan service officer in Oman is in a position to advise the Sultan personally. Until the late 1980s, the highest-ranking military position in the Omani armed forces, the Chief of the Defence Staff (CDS), was held by a British officer.[59] As noted previously, however, a British major-general has remained in Oman in a mentoring role, advising his Omani counterparts and the Sultan himself on defence-related matters.

The UK also exerts influence over Omani defence planning through the Saif Sareea exercises, an ongoing series of large-scale joint training initiatives involving the British armed forces and Omani military and civilian personnel, the first of which took place in 1986. While Oman participates in exercises with other states such as the US, UAE and Iran, the Saif Sareea events are on a much larger scale. The most recent took place in 2018, involving 5,500 members of the British armed forces alongside more than 80,000 Omani military and civilian personnel. The exercise 'was based on a sophisticated scenario involving a multi-agency response . . . to unrest and a territorial threat'.[60] These exercises allow the British armed forces to assess their own expeditionary capabilities and the impact of desert conditions on equipment and human performance. But they also allow the UK to influence Oman's preparations to address security threats by exposing the Omani government and armed forces to hypothetical but realistic conflict situations.

While Sultan Haitham does not have the same close ties to the British armed forces as his cousin Qaboos, Oman's new leader has remained receptive to British advice. In 2021, the UK armed forces supported Oman in conducting a 'security and defence review'.[61] While the results of this initiative are unknown, it is evident that Britain continues to play a key advisory role in Oman's decisions regarding its armed forces. As *The Military Balance* has observed: 'Working substantially with a single partner through a government-to-government arrangement, as Muscat has done with the UK, may not suit other countries, but the durability of the Oman-UK partnership should not be underestimated'.[62] Overall, Oman's political system provides little opportunity for public debate over defence policy and planning. Although Qaboos gradually introduced political reforms that limited his own authority and Haitham seems to be doing the same, the Sultan maintains control over all matters related to national security. Nevertheless, it is clear that the US and particularly the UK have been able to influence Oman's approach to defence planning in a variety of ways. While the impact of this advice is difficult to measure, it is also clear Oman's rulers have willingly accepted it and continue to do so.

## Conclusion

Since well before the turn of the twenty-first century, Oman has relied heavily on its western allies to ensure its security. Even as its economic ties with China have grown in recent years, Oman has bolstered its military ties with the US and particularly the UK. Omani defence spending has been robust over the past two decades, with the US and UK being the principal beneficiaries. Given the length and intensity of the Dhofar insurgency, Sultan Qaboos understandably viewed internal opposition as one of the principal security risks that he faced. But Oman's strong ties with the western powers are not simply a means of ensuring internal security. They also enable Oman to maintain its neutrality in regional disputes, allowing it to preserve its long-standing relationship with Iran and to resist entanglement in disagreements within the GCC. Oman lacks the capability to engage in independent military innovation, but it has modernised its armed forces by emulating the procurement decisions of its allies. Its close ties with the US and UK may have encouraged Oman to purchase American and British weapon systems, but Oman has also sought out manufacturers in other countries in order to acquire capabilities that suit its needs. Moreover, its acquisitions have been relatively modest in comparison to certain of its GCC counterparts. Qatar, for example, increased the number of combat aircraft in its air force from 12 to 96, produced by three different manufacturers, in just a decade.[63] In comparison, Oman's acquisitions appear to be more consistent with the requirements and capabilities of its armed forces.

While it is still too early to make a definitive judgement, the accession of Sultan Haitham appears to represent continuity rather than disruption in Oman's foreign and defence policies. Haitham has largely maintained the same diplomatic and defence partnerships as his predecessor. That said, rather than relying solely on emulation, Oman's new ruler is seeking to develop the expertise necessary to enable indigenous military innovation. Oman's allies have supported this process. Britain and the US have undoubtedly influenced many of Oman's defence-related decisions. But they have also helped the Omani armed forces develop the capability to make these decisions independently. For example, Oman's professional military education institutions are based on the British model and are among the most competitive in the GCC.[64] Like the American colleges attended by senior Omani officers, these institutions aim to encourage critical thinking about defence policy, planning and procurement. At present, Oman remains dependent on the US and particularly the UK for assistance in these areas. It lags behind its neighbours in developing the expertise necessary to carry out defence planning and innovation independently. Moreover, neither its resource-based economy nor its highly centralised systems of military and political decision-making seem particularly conducive to the growth of these capabilities. Nonetheless, its leaders have demonstrated the ability to make rational decisions about procurement, and the desire to develop the ability to innovate independently, even if this process takes time. Figure 8.2 depicts Oman's approach to defence planning in relation to the three themes of this volume. Its relatively recent transformation into a modern state and the extensive authority of its rulers have shaped this approach significantly. Oman has hedged between great powers to develop its economy but has balanced with

| GENERAL COMMONALITIES | VARIATIONS WITHIN THESE COMMONALITIES | SMPS ADJUST THEIR APPROACHES BASED ON SYSTEMIC CHANGE OUT OF THEIR CONTROL | | SMPS REACT TO CHANGES IN THE RELEVANT SECURITY COMPLEX[ES] THE SMP IS PART OF | | |
|---|---|---|---|---|---|---|
| | | *Alliances, Dependencies and National Ambitions* | | | | |
| | | SMPS FACE CONSTRAINTS IN SAFEGUARDING NATIONAL SECURITY → FINANCIAL / DEMOGRAPHIC / CULTURAL / GEOGRAPHICAL / OTHER | | | | |
| | | SMPS NEED TO POSITION THEMSELVES VIS-À-VIS GLOBAL AND REGIONAL POWER CONSTELLATIONS → BALANCING / HEDGING / BANDWAGONING / NON-ALIGNING / OTHER | | | | |
| | | SMPS HAVE TO ENGAGE WITH OTHER STATES TO INCREASE THEIR DEFENCE CAPABILITIES → SECURITY GUARANTEES / TRANSACTIONAL / FORCE INTEGRATION / SPECIALISATION / OTHER | | | | |
| | | SMPS HAVE ARMED FORCES FOR DIFFERENT PURPOSES → INSTRUMENT OF DIPLOMACY / PRESTIGE PROJECT / REDUCE STRATEGIC RISK (DEFEND) / DOMESTIC STABILITY / OTHER | | | | |
| | | *Approaches, Processes, Methods and Techniques* | | | | |
| | | SMPS TAKE A MORE CONSTRAINED APPROACH TO DEFENCE PLANNING → THREAT-NET ASSESSMENT BASED / PORTFOLIO-BASED / TASK-BASED / MOBILISATION / OTHER | | | | |
| | | SMPS HAVE LIMITED RESOURCES TO DESIGN AND IMPLEMENT DEFENCE PLANNING PROCESSES → TAILORED ANALYTICAL PROCESS / OUTSOURCING STRATEGY / ACCEPT DILUTION / FAIR SHARE / EXTERNAL TEMPLATE / OTHER | | | | |
| | | SMP EMPLOY MULTIPLE METHODS AND TECHNIQUES → COPY GREAT POWERS / INFORMAL CONSENSUS / TAILORED METHODS / AD HOC / OTHER | | | | |
| | | SMPS RELY ON DIFFERENT PROCESSES TO DECIDE ON FORCE POSTURE → ANALYTICAL RESULTS / POLITICS DECIDES / MILITARY DECIDES / GREAT POWER DECIDES / RULER DECIDES | | | | |
| | | *Military Innovation* | | | | |
| | | SMPS FACE CHALLENGES IN KEEPING PACE WITH TECHNOLOGICAL ADVANCES → BUILD DOMESTIC TECH/INDUSTRY / TECH TRANSFER AGREEMENTS / ACQUIRE READY INNOVATIONS / PARTNER WITH GREAT POWER / OTHER | | | | |
| | | SMPS PURSUE VARIOUS MILITARY INNOVATION PATHS AND PATTERNS → EMULATION / ADAPTATION / INNOVATION / OTHER; SPECULATION / EXPERIMENTATION / IMPLEMENTATION / OTHER / OTHER | | | | |
| | | SMPS FIND THEMSELVES AT DIFFERENT STAGES OF ORGANISATIONAL CHANGE → EXPLORATION / MODERNISATION / TRANSFORMATION / OTHER / OTHER | | | | |

*Figure 8.2* Structured focused comparison framework for Oman.

*Source:* Author

the US and UK to guarantee its security and to help build armed forces that employ a portfolio-based approach to counter internal and external threats, while serving as a diplomatic tool that supports its independence in regional affairs. Oman's rulers have taken a top-down approach to force development but have received considerable advice and assistance from their western allies. As a result, Oman has acquired weapons systems and developed a force structure similar to these allies. Given that Oman is only beginning to develop its own capacity to innovate, it has relied largely on emulation to modernize its armed forces. These armed forces have ensured Oman's security through the first two decades of this century, fulfilling the ambitions of its rulers. But the extent and character of their development have depended far more on its allies than on internal processes.

## Notes

1 'GDP (Constant 2015 US$)—Oman', *World Bank,* https://data.worldbank.org/indicator/NY.GDP.MKTP.KD?locations=OM (accessed 20 September 2023).
2 Tim Sweijs et al., *Why Are Pivot States So Pivotal? The Role of Pivot States in Regional and Global Security* (The Hague: Hague Centre for Strategic Studies, 2014), 30.
3 US Energy Information Administration, 'The Strait of Hormuz Is the World's Most Important Oil Transit Chokepoint', *U.S. Energy Information Administration,* 20 June 2019, www.eia.gov/todayinenergy/detail.php?id=39932 (accessed 12 February 2023).
4 John Beasant, *Oman: The True-Life Drama and Intrigue of an Arab State* (London and Edinburgh: Mainstream, 2013), 109–10.
5 James Worrall, *Statebuilding and Counterinsurgency in Oman: Political and Diplomatic Relations at the End of Empire* (London: I.B. Tauris, 2014), 72–74.
6 Geraint Hughes, 'A "Model Campaign" Reappraised: The Counter-Insurgency War in Dhofar, Oman, 1965–1975', *Journal of Strategic Studies* 32, no. 2 (2009): 271–305.

7  Nikolas Gardner, 'Defense Sales and British Security Assistance to Oman, 1975–81', *Marine Corps University Journal* 10, no. 1 (2019): 55–56.

8  Nikolas Gardner, 'British Assistance to the Sultan of Oman's Air Force, 1970–1981', in *Air Force Advising and Assistance: Developing Airpower in Client States,* eds. Donald Stoker and Ed Westermann (Warwick: Helion, 2018), 187–204.

9  See International Institute for Strategic Studies, *The Military Balance* (Abingdon: Routledge, 2002–23) for a listing of UK military personnel serving in Oman.

10 'US Security Cooperation with Oman', *US Department of State, Bureau of Political-Military Affairs Fact Sheet*, 15 June 2021, www.state.gov/u-s-security-cooperation-with-oman/.

11 'Use of Royal Air Force (RAF) Base at Masirah, Oman, by USA', UK Foreign and Commonwealth Office File FCO 8/2468, *Arabian Gulf Digital Archive*, www.agda.ae (accessed 12 February); 'Relations between Oman and the USA, 1984 Jan 01–1984 Dec 31', UK Foreign and Commonwealth Office File FCO 8/5502, *Arabian Gulf Digital Archive*, www.agda.ae (accessed 12 February 2023); Matthew Wallin, 'US Military Bases and Facilities in the Middle East', *American Security Project Fact Sheet*, June 2018, 6; Capt. Gregory Ball, '1980—Operation Eagle Claw', *Air Force Historical Support Division*, www.afhistory.af.mil/FAQs/Fact-Sheets/Article/458949/1980-operation-eagle-claw/ (accessed 13 February 2023).

12 Robert Mason, 'The Omani Pursuit of a Large Peninsula Shield Force: A Case Study of a Small State's Search for Security', *British Journal of Middle Eastern Studies* 41, no. 4 (2014): 364.

13 Wallin, 'US Military Bases and Facilities in the Middle East', 6.

14 Capt. Melissa Heintz, 'Accurate Test 22: Tests US, Oman Air Forces Agile Combat Employment Capabilities', *Air Combat Command*, 1 June 2022, www.acc.af.mil/News/Article/3049097/accurate-test-22-tests-us-oman-air-forces-agile-combat-employment-capabilities/ (accessed 21 February 2023); Staff Sgt. Jennifer Milnes, 'Inferno Creek Strengthens US-Omani Partnership', *US Army Central Newsroom*, 31 January 2018, www.usarcent.army.mil/News/Article/1428571/inferno-creek-strengthens-us-omani-partnership/ (accessed 21 February 2023).

15 'US Security Cooperation with Oman'.

16 Mason, 'The Omani Pursuit of a Large Peninsula Shield Force', 362.

17 International Institute for Strategic Studies, *The Military Balance* (Abingdon: Routledge, 2017), 355.

18 International Institute for Strategic Studies, *The Military Balance* (Abingdon: Routledge, 2021), 326.

19 Barany, *Armies of Arabia*, 51.

20 Giorgio Cafiero and Daniel Wagner, 'Oman's Diplomatic Bridge in Yemen', *Atlantic Council MENASource*, 17 June 2015, www.atlanticcouncil.org/blogs/menasource/oman-s-diplomatic-bridge-in-yemen/ (accessed 14 February 2023).

21 Abdullah Baabood, 'The Middle East's New Battle Lines: Qatar, Kuwait and Oman', *European Council on Foreign Relations*, https://ecfr.eu/special/battle_lines/oman (accessed 14 February 2023).

22 'Iran, Oman Hold Combined Naval Exercise in N Indian Ocean', *Mehr News Agency*, 12 October 2022, https://en.mehrnews.com/news/192387/Iran-Oman-hold-combined-naval-exercise-in-N-Indian-Ocean (accessed 14 February 2023).

23 Mordechai Chaziza, 'The Significant role of Oman in China's Maritime Silk Road Initiative', *Contemporary Review of the Middle East* 6, no. 1 (2019): 51.

24 'China, Oman Vow to Promote Strategic Partnership, Military Cooperation', *Xinhuanet*, 29 April 2022, http://eng.chinamil.com.cn/view/2022-04/29/content_10151238.htm (accessed 15 February 2023).

25 Jennifer Jacobs, 'Biden Speaks to Oman's Ruler with Concerns Rising over China's Presence in Region', 10 November 2023, www.bloomberg.com/news/articles/2023-11-10/biden-speaks-to-oman-s-ruler-amid-concern-over-china-s-presence?embedded-checkout=true (accessed 10 January 2024).

26 Camilla Lons, 'Onshore Balancing: The Threat to Oman's Neutrality', *European Council on Foreign Relations*, https://ecfr.eu/article/commentary_onshore_balancing_the_threat_to_omans_neutrality/ (accessed 15 February 2023).

27 HM Government, *Global Britain in a Competitive Age: The Integrated Review of Security, Defence, Development and Foreign Policy* (March 2021), 76.

28 Ministry of Defence, 'UK and Oman Sign Historic Joint Defence Agreement', 21 February 2019, www.gov.uk/government/news/uk-and-oman-sign-historic-joint-defence-agreement (accessed 15 February 2023); Jeremy Binnie, 'UK to Expand Base in Oman', *Janes,* 14 September 2020, www.janes.com/defence-news/news-detail/uk-to-expand-base-in-oman (accessed 15 February 2023).

29 Richard Thomas, 'Oman Becoming Lynchpin for UK Global Military Hub Network', *Army Technology*, 22 November 2022, www.army-technology.com/features/oman-becoming-lynchpin-for-uk-global-military-hub-network/ (accessed 15 February 2023).

30 International Institute for Strategic Studies, *The Military Balance* (Abingdon: Routledge, 2005), 204; International Institute, *The Military Balance* (2017), 396. All figures are US dollars.

31 International Institute for Strategic Studies, *The Military Balance* (Abingdon: Routledge, 2023), 346.

32 International Institute, *The Military Balance* (2023), 346; International Institute for Strategic Studies, *The Military Balance* (Abingdon: Routledge, 2020), 334.

33 International Institute for Strategic Studies, *The Military Balance* (Abingdon: Routledge, 2008), 257.

34 International Institute, *The Military Balance* (2023), 346.

35 'Oman Orders Second Squadron of F-16s', *Times Aerospace,* 21 December 2011, www.timesaerospace.aero/news/defence/oman-orders-second-squadron-of-f-16s (accessed 19 February 2023); International Institute for Strategic Studies, *The Military Balance* (Abingdon: Routledge, 2003), 104.

36 'Oman Receives Final Al-Ofouq Patrol Vessel from ST Marine', *naval-today.com*, 27 June 2016, www.navaltoday.com/2016/06/27/oman-receives-final-al-ofouq-patrol-vessel-from-st-marine/ (accessed 19 February 2023).

37 'Oman Upgrading its Air Defenses', *Defense Industry Daily*, 11 January 2016, www.defenseindustrydaily.com/oman-upgrading-its-air-defenses-07161/ (accessed 19 February 2023).

38 International Institute for Strategic Studies, *The Military Balance* (Abingdon: Routledge, 2002), 99.

39 'Khareef-Class Corvettes', *Naval Technology*, 3 May 2016, www.naval-technology.com/projects/khareef-class/ (accessed 19 February 2023).

40 International Institute for Strategic Studies, *The Military Balance* (Abingdon: Routledge, 2014), 304.

41 'Oman—Joint Stand Off Weapons (JSOW)', *US Defense Security Cooperation Agency*, www.dsca.mil/press-media/major-arms-sales/oman-joint-stand-weapons-jsow (accessed 19 February 2023).

42 Nikolas Gardner, 'The Limits of the Sandhurst Connection: The Evolution of Oman's Foreign and Defense Policy, 1970–77', *Journal of the Middle East and Africa* 6, no. 1 (2015): 56.

43 'France Offers Rafale fighter to Oman—Sources', *Reuters*, 11 February 2009, www.reuters.com/article/france-oman-fighters-idINLB71245420090211 (accessed 19 February 2023).

44 Paul Iddon, 'These Middle East States Are Ordering a Lot of Advanced Fighter Jets', *forbes.com*, 27 September 2022, www.forbes.com/sites/pauliddon/2022/09/27/these-middle-east-states-are-ordering-a-lot-of-advanced-fighter-jets/?sh=1e754681f927 (accessed 19 February 2023); 'The Programme', *Eurofighter Typhoon*, www.eurofighter.com/customers (accessed 23 February 2023).

45 International Institute, *The Military Balance* (2020), 386.

46   Sinem Cengiz, 'Deciphering the Dynamics of Turkiye-Oman Relations', *Arab News,* 4 February 2023, www.arabnews.com/node/2244536 (accessed 20 February 2023).
47   'Turkey's FNSS to Establish Armoured Vehicle Repair Center in Oman', *Daily Sabah,* 7 July 2022, www.dailysabah.com/business/defense/turkeys-fnss-to-establish-armored-vehicle-repair-center-in-oman (accessed 19 February 2023).
48   'Oman Signs Four Pacts for Modernisation of Defence Infrastructure', *Muscat Daily,* 3 July 2022, www.muscatdaily.com/2022/07/03/oman-signs-four-pacts-for-modernisation-of-defence-infrastructure/ (accessed 20 February 2023).
49   Michael Raska, 'The Sixth RMA Wave: Disruption in Military Affairs?', *Journal of Strategic Studies* 44, no. 4 (2021): 473.
50   Michael Raska, 'Strategic Competition for Emerging Military Technologies: Comparative Paths and Patterns', *Prism* 8, no. 3 (2020): 68–69.
51   Michael Raska, 'A Structured-Phased Evolution: The Third Generation Force Transformation of the Singapore Armed Forces', in *Military Innovation in Small States: Creating a Reverse Asymmetry*, ed. M. Raska (New York: Routledge, 2016), 130–162.
52   Yasmina Abouzzohour, 'Oman, Ten Years after the Arab Spring: The Evolution of State-Society Relations', *Arab Reform Initiative*, 9 February 2021, www.arab-reform.net/publication/oman-ten-years-after-the-arab-spring-the-evolution-of-state-society-relations/ (accessed 20 February 2023).
53   'Tribute: Sultan Qaboos Ably Steered Oman in Turbulent Middle East', *onmanorama,* www.onmanorama.com/news/world/2020/01/11/tribute-sultan-qaboos-oman-middle-east.html (accessed 20 February 2023).
54   Barany, *Armies of Arabia*, 91; International Institute, *The Military Balance* (2023), 347.
55   Gardner, 'The Limits of the Sandhurst Connection', 55.
56   International Institute, *The Military Balance* (2023), 346.
57   Stephan Frühling, *Defence Planning and Uncertainty: Preparing for the Next Asia-Pacific War* (Abingdon: Routledge, 2014).
58   'Sultan Appoints Deputy Prime Minister for Defence Affairs', *Economist Intelligence Unit*, 19 March 2020, https://country.eiu.com/article.aspx?articleid=459231829&Country=Oman&topic=Politics&subtopic=Forecast&subsubtopic=Political+stability (accessed 21 February 2023).
59   J.E. Peterson, *Oman's Insurgencies: The Sultanate's Struggle for Supremacy* (London: Saqi, 2013).
60   International Institute, *The Military Balance* (2020), 330; 'Exercise SAIF SAREEA 3', *Army,* 4 October 2018, www.army.mod.uk/news-and-events/news/2018/10/exercise-saif-sareea-3/ (accessed 21 February 2023).
61   International Institute for Strategic Studies, *The Military Balance* (Abingdon: Routledge, 2022), 361.
62   International Institute for Strategic Studies, *The Military Balance* (Abingdon: Routledge, 2019), 331.
63   Jon Lake, 'Orders Accelerate the Qatar Fighter Evolution', *Times Aerospace*, www.timesaerospace.aero/features/defence/orders-accelerate-the-qatar-fighter-evolution (accessed 21 February 2023).
64   International Institute, *The Military Balance* (2021), 327.

## Bibliography

Abouzzohour, Yasmina. "Oman, Ten Years after the Arab Spring: The Evolution of State-Society Relations," *Arab Reform Initiative*, 9 February 2021, www.arab-reform.net/publication/oman-ten-years-after-the-arab-spring-the-evolution-of-state-society-relations/.
Baabood, Abdulla. "The Middle East's New Battle Lines: Qatar, Kuwait and Oman," *European Council on Foreign Relations*, https://ecfr.eu/special/battle_lines/oman (accessed 14 February 2023).

Ball, Capt. Gregory. "1980—Operation Eagle Claw," *Air Force Historical Support Division*, www.afhistory.af.mil/FAQs/Fact-Sheets/Article/458949/1980-operation-eagle-claw/ (accessed 13 February 2023).

Barany, Zoltan. *Armies of Arabia: Military Politics and Effectiveness in the Gulf.* New York: Oxford University Press, 2022.

Beasant, John. *Oman: The True-Life Drama and Intrigue of an Arab State.* London and Edinburgh: Mainstream, 2013.

Binnie, Jeremy. "UK to Expand Base in Oman," www.janes.com/defence-news/news-detail/uk-to-expand-base-in-oman (accessed 15 February 2023).

Cafiero, Giorgio and Daniel Wagner. "Oman's Diplomatic Bridge in Yemen," *Atlantic Council MENASource*, 17 June 2015, www.atlanticcouncil.org/blogs/menasource/oman-s-diplomatic-bridge-in-yemen/.

Cengiz, Sinem. "Deciphering the Dynamics of Turkiye-Oman Relations," *Arab News*, 4 February 2023, www.arabnews.com/node/2244536.

Chaziza, Mordechai. "The Significant Role of Oman in China's Maritime Silk Road Initiative," *Contemporary Review of the Middle East* 6:1 (2019), 51.

"China, Oman Vow to Promote Strategic Partnership, Military Cooperation," *Xinhuanet*, 29 April 2022, http://eng.chinamil.com.cn/view/2022-04/29/content_10151238.htm.

Economist Intelligence Unit. "Sultan Appoints Deputy Prime Minister for Defence Affairs," 19 March 2020, https://country.eiu.com/article.aspx?articleid=459231829&Country=Oman&topic=Politics&subtopic=Forecast&subsubtopic=Political+stability (accessed 21 February 2023).

Eurofighter. www.eurofighter.com/customers (accessed 23 February 2023).

"France Offers Rafale Fighter to Oman—Sources," *Reuters*, 11 February 2009, www.reuters.com/article/france-oman-fighters-idINLB71245420090211.

Frühling, Stephan. *Defence Planning and Uncertainty: Preparing for the Next Asia-Pacific War.* Abingdon: Routledge, 2014.

Gardner, Nikolas. "The Limits of the Sandhurst Connection: The Evolution of Oman's Foreign and Defense Policy, 1970–77." *Journal of the Middle East and Africa* 6:1 (2015), 45–58.

———. "British Assistance to the Sultan of Oman's Air Force, 1970–1981." Donald Stoker and Ed Westermann, eds., *Air Force Advising and Assistance: Developing Airpower in Client States*. Warwick: Helion, 2018, 187–204.

———. "Defense Sales and British Security Assistance to Oman, 1975–81." *Marine Corps University Journal* 10:1 (Spring 2019), 50–62.

"GDP (Constant 2015 US$)—Oman," https://data.worldbank.org/indicator/NY.GDP.MKTP.KD?locations=OM (accessed 20 September 2023).

Heintz, Capt. Melissa. "Accurate Test 22: Tests US, Oman Air Forces Agile Combat Employment Capabilities," *Air Combat Command*, 1 June 2022, www.acc.af.mil/News/Article/3049097/accurate-test-22-tests-us-oman-air-forces-agile-combat-employment-capabilities/

HM Government. *Global Britain in a Competitive Age: The Integrated Review of Security, Defence.* Development and Foreign Policy. March 2021.

Hughes, Geraint. "A 'Model Campaign' Reappraised: The Counter-Insurgency War in Dhofar, Oman, 1965–1975." *Journal of Strategic Studies* 32:2 (April 2009), 271–305.

Iddon, Paul. "These Middle East States Are Ordering a Lot of Advanced Fighter Jets," *forbes.com*, 27 September 2022, www.forbes.com/sites/pauliddon/2022/09/27/these-middle-east-states-are-ordering-a-lot-of-advanced-fighter-jets/?sh=1e754681f927.

International Institute of Strategic Studies. *The Military Balance.* Abingdon, OX: Routledge, 2002.

———. *The Military Balance.* Abingdon, OX: Routledge, 2003.

———. *The Military Balance.* Abingdon, OX: Routledge, 2005.

———. *The Military Balance.* Abingdon, OX: Routledge, 2008.

———. *The Military Balance.* Abingdon, OX: Routledge, 2014.

————. *The Military Balance*. Abingdon, OX: Routledge, 2017.

————. *The Military Balance*. Abingdon, OX: Routledge, 2019.

————. *The Military Balance*. Abingdon, OX: Routledge, 2020.

————. *The Military Balance*. Abingdon, OX: Routledge, 2021.

————. *The Military Balance*. Abingdon, OX: Routledge, 2022.

————. *The Military Balance*. Abingdon, OX: Routledge, 2023.

"Iran, Oman Hold Combined Naval Exercise in N Indian Ocean," *Mehr News Agency*, 12 October 2022, https://en.mehrnews.com/news/192387/Iran-Oman-hold-combined-naval-exercise-in-N-Indian-Ocean.

Jacobs, Jennifer. "Biden Speaks to Oman's Ruler with Concerns Rising over China's Presence in Region", 10 November 2023, www.bloomberg.com/news/articles/2023-11-10/biden-speaks-to-oman-s-ruler-amid-concern-over-china-s-presence?embedded-checkout=true (accessed 10 January 2024).

"Khareef-Class Corvettes," *Naval Technology*, 3 May 2016, www.naval-technology.com/projects/khareef-class/.

Lake, John. "Orders Accelerate the Qatar Fighter Evolution," *Times Aerospace*, www.timesaerospace.aero/features/defence/orders-accelerate-the-qatar-fighter-evolution (accessed 21 February 2023).

Lons, Camilla. "Onshore Balancing: The Threat to Oman's Neutrality," *European Council on Foreign Relations*, https://ecfr.eu/article/commentary_onshore_balancing_the_threat_to_omans_neutrality/ (accessed 15 February 2023).

Mason, Robert. "The Omani Pursuit of a Large Peninsula Shield Force: A Case Study of a Small State's Search for Security." *British Journal of Middle Eastern Studies* 41:4 (2014), 355–367.

Milnes, Staff Sgt. Jennifer. "Inferno Creek Strengthens US-Omani Partnership," *US Army Central Newsroom*, 31 January 2018, www.usarcent.army.mil/News/Article/1428571/inferno-creek-strengthens-us-omani-partnership/.

"Oman—Joint Stand Off Weapons (JSOW)," *US Defense Security Cooperation Agency*, www.dsca.mil/press-media/major-arms-sales/oman-joint-stand-weapons-jsow (accessed 19 February 2023).

"Oman Orders Second Squadron of F-16s," *Times Aerospace*, 21 December 2011, www.timesaerospace.aero/news/defence/oman-orders-second-squadron-of-f-16s.

"Oman Receives Final Al-Ofouq Patrol Vessel from ST Marine," *naval-today.com*, 27 June 2016, www.navaltoday.com/2016/06/27/oman-receives-final-al-ofouq-patrol-vessel-from-st-marine/.

"Oman Signs Four Pacts for Modernisation of Defence Infrastructure," *Muscat Daily*, 3 July 2022, www.muscatdaily.com/2022/07/03/oman-signs-four-pacts-for-modernisation-of-defence-infrastructure/.

"Oman Upgrading Its Air Defenses," *Defense Industry Daily*, 11 January 2016, www.defenseindustrydaily.com/oman-upgrading-its-air-defenses-07161/.

Peterson, J. E. *Oman's Insurgencies: The Sultanate's Struggle for Supremacy*. London: Saqi, 2013.

Raska, Michael. "Strategic Competition for Emerging Military Technologies: Comparative Paths and Patterns." *Prism* 8:3 (2020), 65–81.

————. "The Sixth RMA Wave: Disruption in Military Affairs?." *Journal of Strategic Studies* 44:4 (2021), 456–479.

"Relations between Oman and the USA, 1984 Jan. 1–1984 Dec. 31," UK Foreign and Commonwealth Office File FCO 8/5502, Arabian Gulf Digital Archive, www.agda.ae.

Sweijs, Tim, Oosterveld, Willem T., Knowles, E. and Schellekens, M. *Why Are Pivot States So Pivotal? The Role of Pivot States in Regional and Global Security*. The Hague: Hague Centre for Strategic Studies, 2014.

Thomas, Richard. "Oman Becoming Lynchpin for UK Global Military Hub Network," *Army Technology*, 22 November 2022, www.army-technology.com/features/oman-becoming-lynchpin-for-uk-global-military-hub-network/.

"Tribute: Sultan Qaboos Ably Steered Oman in Turbulent Middle East," www.onmanorama. com/news/world/2020/01/11/tribute-sultan-qaboos-oman-middle-east.html (accessed 20 February 2023).

"Turkey's FNSS to Establish Armoured Vehicle Repair Center in Oman," *Daily Sabah*, 7 July 2022, www.dailysabah.com/business/defense/turkeys-fnss-to-establish-armored-vehicle-repair-center-in-oman.

UK Ministry of Defence, "UK and Oman Sign Historic Joint Defence Agreement," 21 February 2019, www.gov.uk/government/news/uk-and-oman-sign-historic-joint-defence-agreement.

US Department of State, "US Security Cooperation with Oman," *Bureau of Political-Military Affairs Fact Sheet*, 15 June 2021, www.state.gov/u-s-security-cooperation-with-oman/.

US Energy Information Administration, "The Strait of Hormuz Is the World's Most Important Oil Transit Chokepoint," 20 June 2019, www.eia.gov/todayinenergy/detail. php?id=39932.

"Use of Royal Air Force (RAF) Base at Masirah, Oman, by USA," UK Foreign and Commonwealth Office File FCO 8/2468, Arabian Gulf Digital Archive, www.agda.ae.

Worrall, James. *Statebuilding and Counterinsurgency in Oman: Political and Diplomatic Relations at the End of Empire*. London: I.B. Tauris, 2014.

# 9 Turkey

## A rising star with structural problems

*Murat Caliskan*

## Introduction

The Bayraktar TB2, a Turkish combat unmanned aerial vehicle (UAV), has grabbed the world's attention for its role in Ukraine. It has become a legend for Ukrainians to the degree that the chief of Ukraine's air force, Lt. Gen. Mykola Oleshchuk, described the TB2 drones as 'life-giving'[1] and Ukrainian soldiers were inspired to compose a song dedicated to the Turkish drones.[2] This was only the latest of Bayraktar's remarkable performances in a series of wars in Syria, Ethiopia, Libya and Azerbaijan, where the drone played a decisive role and shaped the outcome, improving its performance in each war.[3] Selcuk Bayraktar, the designer of TB2 and also the son-in-law of Turkish President Recep Tayyip Erdogan, stated in a recent interview that 'the demand for the TB2 drone has soared so high that there is now a three-year waitlist to purchase the weapon'.[4] This recent demand has made Turkey one of the world's most prolific users and exporters of armed drones.

Turkey's rise as a weapon producer is not limited to drones, however. It has demonstrably increased efforts to produce indigenous weapon systems ranging from the main battle tank to warships, from fighter planes to infantry rifles. This 'indigenous production' policy has been pursued since the early 1970s, but Turkey has become notably ambitious in the past two decades under *Adalet ve Kalkınma Partisi* (Justice and Development Party-AKP) rule. The Turkish defence budget has grown from US$11.1 billion in 2002 to US$16.7 billion in 2021.[5] Since 2015, the sector doubled its turnover from 4,908 million US$ to 10,159 million US$ in 2021.[6] According to the *Savunma Sanayii Başkanlığı* (Defence Industry Agency-SSB), the indigenous production rate increased from 20% in 2002 to 65% in 2021. Meanwhile, exports increased 13-fold, from US$248 million in 2002 to US$3,224 billion in 2021,[7] making Turkey the 11th largest arms exporter in the world.[8] While Turkey had one company in the Top 100 Global Defence Companies list in 2010, it had three companies (Aselsan, TAI and Roketsan) in 2022. The number even reached seven in 2020—more companies than Israel, Russia, Sweden and Japan combined.[9]

This chapter examines Turkey's defence policy and planning with an emphasis on the past two decades. To achieve this purpose, the first section investigates how alliance strategies and national ambitions influence defence policy and force

DOI: 10.4324/9781003398158-9

development. The second section describes the defence planning approaches, processes and methods that Turkey uses to decide and implement different aspects of defence transformation. The final section discusses how Turkey engages in military innovation and adaptation to carry out its defence transformation. Before proceeding to the first section, it is useful to note that for Turkey, the disruption that is mentioned in the subtitle of this book lies mainly in the regime change and concomitant transformation in cultural/identity aspects rather than in technology.

## Part 1: national ambitions, alliances, dependencies

### *The dramatic change in foreign policy*

For most of the post-1945 period, Turkey was firmly embedded in the Western alliance. In the first decade of AKP rule (2002–2011), Turkish foreign policy—largely driven by a strong commitment to the EU membership process—was based on 'the logic of interdependence', characterised by a 'zero-problem with neighbours' notion and drawing on soft power instruments like cultural proximity and regional economic engagement. With a notable increase in prosperity and improvements in human rights, it began to be seen as a model for an Islamic democratic country.[10] Perhaps, for the first time, EU membership became a potential reality for Turkey. Following the Arab upheavals in 2011, however, the foreign policy orientation has gradually shifted from a Western-anchored middle power mostly using cautious activism to a more interventionist actor using assertive military engagement and coercive diplomacy,[11] with a notable rapprochement with Russia and China. The second decade of AKP rule witnessed regime change attempts in Syria and Egypt, four military operations against Syria, the employment of the aggressive 'Blue Homeland' doctrine against Greece and active involvement in the Libyan and Nagorno-Karabakh wars. The relations with the United States and NATO members, particularly since 2016, have deteriorated to the extent that Turkey's expulsion from NATO is openly discussed in credible outlets.[12]

In parallel with a more autonomous foreign policy and so-called 'strategic autonomy', Turkey improved its relations with non-Western powers, particularly with Russia, and to some extent with China. While Russia has been a traditional threat to Turkey for the last several centuries, the relationship between the two countries took a qualitatively different turn after 2011, not only in economic terms but also in terms of political identity and collective security. Russia ceased to be a rival and even became a security partner, particularly after the 2016 failed coup attempt in Turkey. Today, Turkey is the only NATO member which does not employ sanctions against Russia for the war in Ukraine and the only member which objects to the NATO memberships of Sweden and Finland, a move that is certainly very well received by Russia. China, on the other hand, has significantly increased its influence on the Turkish economy over the past decade. It became a convenient alternative for Turkey after 2016 to solve its political and economic problems while Turkey became an important partner for China's 'Belt and Road Initiative' due to its geostrategic location.[13] In short, Turkey seems to be torn between its traditional alignments with the West and new possibilities to align with non-Western global powers like Russia and China.[14]

From the perspective of alignment strategies, Turkey's recent behaviour can be best defined as a *hedging strategy*. Despite the notable improvement in relations with Russia and China, Turkey has not been completely divorced from its relations with the West, in fact, it even acts to please the US and NATO partners from time to time. According to Ilhan Uzgel, a Turkish professor of IR (International Relations), 'the Erdogan administration is desperately looking for a way out by aligning more with the West rather than Eurasian powers like Russia' to save itself from the desperate situation it created.[15] Turkey's solid commitment to NATO tasks (willingness to protect Kabul Airport as part of the NATO forward security role in 2021; alignment with the Black Sea and the Eastern Europe policies in compliance with NATO; the recent efforts to repair the relations with US Allies like Saudi Arabia, Egypt, Israel and the United Arab Emirates and the normalisation talks with Armenia behind closed doors upon the request of Western countries), show that Turkey pursues a hedging strategy rather than bandwagoning with Russia.[16]

However, the author argues that recent Turkish foreign policy is beyond the power of balance strategies and cannot be explained without considering domestic factors. A 'hedging strategy' instead of a 'balancing' against rising powers, namely Russia and China, as a NATO member doesn't seem to be plausible for several reasons. First, a country that pursues a hedging strategy should normally feel uneasy with both parties, but Turkey has no clear reason to feel concerned about losing the security protection that has been provided by NATO for decades. Additionally, Turkey enjoys the prestige provided by NATO membership as the only Muslim country in the Alliance. Secondly, the partnership with Russia does not present an alternative to Turkey's alliance with the West. It does not have a strong institutional basis and is instead driven by the personal affinity of like-minded leaders. Furthermore, the relationship between the two is rather asymmetric. While the relationship with Turkey provides Russia with an instrument to fragment the solidarity of NATO in addition to economic benefits, Turkey does not seem to benefit much from this relationship, in neither political nor economic terms. Europe remains Turkey's primary trading partner, representing 51% of Turkish foreign trade while its combined trade with Russia and China represents only 14%, and the latter comes with massive trade deficits.[17] Even though US hegemony has relatively declined in the region, the international context does not force Turkey to get closer to the East. Turkey is too embedded in the Western system to leave immediately. The foreign policy pursued by Turkey in the past decade is 'simply beyond its material capabilities'.[18] For this reason, the next section examines domestic factors to understand the change in Turkish foreign policy.

### *The prominent factor: increasing authoritarianism in domestic politics*

The AKP achieved unprecedented economic, political and democratic success in its first decade in power (2002–2011) despite military ultimatums, party closure threats and alleged coup plans and even tamed the bureaucratic leviathan which had held the main power in the Turkish state system.[19] The inclusive domestic policies and favourable international context provided by the EU accession process can be considered as the main drivers behind this success. However, things started to change when the AKP won the third consecutive general election in 2011 with a large majority. Perhaps, for the first time in the history of Turkey, Erdogan, coming from the periphery, had the opportunity to break the vicious cycle caused by the centre-periphery dichotomy. However, he preferred taking an incremental path to authoritarianism, co-opting the unruly elements in the security and judicial apparatus, redefining the bureaucratic centre on his own terms rather than solidifying a consensual democratic centre and establishing an inclusive social order.[20] In other words, Erdogan used the good credit earned in the first decade but only to become the new bureaucratic leviathan with even more authoritarian characteristics.

In many ways, the initial liberal-inclusive-democratic stance of the first decade was gradually replaced by a more nationalist, conservative-authoritarian posture in the second decade. The polarisation of society following the Gezi protests in 2013, the blockage of the 17–25 December 2013 bribery and corruption investigations, in which Erdogan and several ministers were involved, the passing of unconstitutional laws on critical matters such as the immunity of intelligence service members, internet privacy and the structure of the high judiciary in 2014–2015 and the end of the peace process with Kurdish separatists can be considered as the major milestones on the road to the authoritarianism. However, the unsuccessful coup attempt on July 15, 2016, was a decisive moment. The four-year nationwide state of emergency following the coup attempt was used as a tool to crack down on any opposition via the emergency decrees. More than 150 media outlets were shut down and thousands of officials including judges, prosecutors, diplomats, military personnel and police were dismissed from their public service, to be replaced by pro-government people, hired because of loyalty instead of competence.[21] This massive purge enabled Erdogan to control the judiciary, military, media and even opposition parties, thus creating an absolute authoritarian regime. The authoritarianism was institutionalised by the 'Turkish Style Presidential System' after the referendum held on April 16, 2017. The so-called presidential system provided Erdogan authority above all limits of legal-institutional checks and separation of powers, making him a defining element in the Turkish constitutional regime.[22]

Unsurprisingly, foreign policy has also been dominated by Erdogan and his close associates, becoming an extended arm of domestic politics. Erdogan instrumentalises foreign policy mainly to consolidate public support rather than to prioritise national interests.[23] As Kutlay and Onis noted:

> strategic autonomy has been reduced to a discursive tool to legitimise authoritarian practices at home, fragment domestic opposition and accrue popular support through the rhetoric of 'national security'. This explains why the Turkish government has pursued overambitious policies, punching above its weight at the cost of external isolation and going beyond the moderate degree of autonomy compatible with its middle-power credentials.[24]

While this policy generated a populist payoff in the short term, inevitably, there will be adverse consequences in the long term. Turkey has already entered a vicious cycle 'whereby assertive foreign policy and weak economic governance under a highly centralised presidential system reinforced one another to generate a sub-optimal equilibrium'.[25]

### *Consequences for defence planning and force development*

In this context, a strong indigenous defence industry—regardless of whether it is feasible or not—has become a central pillar of the dual transformation at the external-domestic nexus—assertive foreign policy and increasing authoritarianism—providing multiple benefits to Erdogan and the AKP government. The recent assertive foreign policy requires having a strong, indigenous defence industry and reducing the country's dependence on the West. The problem with this logic is that it seems to be putting the cart before the horse. Normally, building a strong defence industry is required before employing an assertive foreign policy. In the case of Turkey, however, it appears to be trying to acquire the capabilities that are required by its assertive foreign policy.

Secondly, and more importantly, investment in the indigenous defence industry has always been a convenient tool to gain political and electoral power in Turkey where the absence of an indigenous defence industry is seen as one of the main causes of the Ottoman defeat during World War I. An indigenous defence industry is not just seen as a way of improving conditions but is in itself seen as a condition of national survival, as proof of national emancipation particularly from the Western powers.[26] The AKP has managed to construct an image that shows how Turkey built an indigenous defence industry during its incumbency and how the party catapulted Turkey to a respectable state at the global level.[27] Indigenous weapon systems provide tangible outcomes and serve AKP image-building efforts to portray Turkey as an increasingly powerful state. Since 2011, the AKP has made pre-election pledges to produce a 'national jet fighter', 'national tank', 'national warship', only to fail to deliver afterwards. However, as long as the AKP remains unaccountable to the judiciary, this failure can always be explained by the fact that it requires more time. For these reasons, Erdogan has been playing to the independent defence industry while rendering nationalism, Islamism and neo-Ottomanism the pillars of his recent domestic policy.

### Part 2: approaches, processes, methods and techniques

The origins of Turkey's indigenous defence policy go back to the 1970s. According to Denis Allen, the British Ambassador to Turkey in 1966, 'the Turkish Armed Forces (TAF) followed United States doctrine and procedure almost on entirely tactical, training and organisational, administrative and logistical matters'.[28] After joining NATO in 1952, the close relations with the US provided Turkey with a security umbrella that was much needed at the time as well as a great deal of the funding, equipment and training of the TAF. However, this also led to the loss of the ability to plan and implement defence policies independently and made Turkey more vulnerable to the demands of the US.[29] US President Lyndon Johnson's blunt and undiplomatic letter regarding the Cyprus question in 1964 and the three-year embargo imposed following Turkey's intervention in Cyprus in 1974 led Turkey to discover the importance of indigenous defence planning and industry. Unable to import basic military needs, Turkey initiated a defence industrialisation plan in the 1980s and established a range of institutions which would be the backbone of the current defence industry.[30] Since then, an independent defence industry has become a universal state policy of Turkey.

Indeed, the underlying goal of the first policy document, 'The Principles of Defence Industry Policy and Strategy', that set out the principles of Turkish defence industrialisation in 1998, and of four strategic plans that were published by the *Savunma Sanayii Müsteşarlığı* (Undersecretariat for Defence Industries— SSM)—now SSB—between 2007 and 2021 was to increase local production through developing national weapon systems and technologies.[31] The yearly meeting held by the Defence Industry Executive Committee (SSİK) in 2004 was an important milestone for the implementation of this policy, in which a series of projects, including a main battle tank, attack helicopter and UAV projects, was cancelled in order to redesign procurement models in a way that enabled the maximum use of national resources and domestic production.[32]

During the Cold War, Turkey's defence planning showed the features of a *net assessment-based planning framework*, which focused on addressing a single, dominant strategic risk, the Soviet Union, as part of the Western Alliance. Turkey remained largely dependent on direct, off-the-shelf procurement from the US in this era. At the end of the Cold War, however, Turkey was one of the few countries that didn't have a sense of enhanced security. On the one hand, it found itself in the middle of a new zone of conflicts, namely in the Balkans, Caucasus-Central Asia and the Middle East, on the other hand, it faced an irregular war started by the *Partiya Karkeren Kurdistan* (PKK—the Kurdistan Workers Party) in its own territory.[33] The security strategy of this era was described as a 'two and a half war' strategy by Sukru Elekdag, a retired diplomat, which required Turkey to have the capacity to fight two major wars with Greece and Syria and a half war with the PKK at the same time.[34] Thus, *a portfolio defence-planning framework* was initiated in the 1990s, which lasts to this day. Although the irregular war with the PKK forced Turkey to a significant amount of direct off-the-shelf procurement, this was gradually replaced by joint ventures, licensed and co-production models in the 1990s and 2000s. Arms import levels did not change significantly in this period, but the quality of imports shifted from the completed major platforms and spare parts to critical technologies at sub-system and component levels.[35] In the following decades, while Turkey has kept its conventional war capacity in accordance with subsequent strategic concepts released by NATO, irregular war against the PKK has been an important factor that influences its defence planning.

A closer look at main defence projects today suggests that Turkey indeed has an ambitious, even aggressive force development plan which aims to produce a wide array of weapon systems. The *Altay* main battle tank, *Kirpi* Mine-Resistant

Tactical Wheeled Vehicle, *Cobra* and *Ejder Yalcın* Armoured Tactical Vehicles, *Samur* Mobile Amphibious Assault Bridge projects for the Land Forces; MILGEM corvette type ship, the LHD Multipurpose Amphibious Assault Ship, LST Amphibious Ship, MOSHIP Submarine Rescue Mother Ship, NTSP New Type Submarine, YTKB New Type Patrol Boats projects for the Naval Forces; TF-X National Combat Aircraft, ATAK Reconnaissance and Tactical Attack Helicopter, HURKUS Basic Training Aircraft, Bayraktar and ANKA Armed UAVs for Air Forces; HISAR and SIPER air defence systems, electronic warfare system-SOJ for Air and Land Forces, weapon systems such as the UMTAS Long-Range Anti-Tank Missile System, OMTAS Middle-Range Anti-Tank Missile System and the MPT-76 National Infantry Rifle are some prominent examples of 750 ongoing projects carried out by Turkish industry. These weapon systems, according to the authorities, have either a full or a high indigenous production rate.[36] However, despite the four-decade policy goal of defence autarky and the notable advances achieved in the past decade, there are complex challenges and constraints emanating from the confluence of political, economic, military and sociocultural factors that impede Turkey's defence transformation, as discussed in the following paragraphs.

### *Organisational/bureaucratic culture*

Turkey has a strategic culture that prioritises security over the economy or the state over individuals. Since its foundation, the 'state' in Turkey has been strong, and perceived largely as sacred. The patrimonial belief that 'the well-being of the society depended upon the well-being of the state',[37] a sense of greatness in belonging to a nation that had ruled an empire and never-ending aspiration to become great again, the Sevres Syndrome, which can be summarised as 'the external world and their internal collaborators are trying to weaken and divide Turkey',[38] and concomitant 'fear of territory loss', have been the main characteristics of Turkish security culture, which not only still has influence on the decisions of the Turkish elite but is also entrenched in the minds of Turkish citizens through formal education, military service and the media. All of this creates an uncertain self-identity with a general feeling of insecurity. But more importantly, this makes the state too strong relative to all other domestic actors and gives the state elites major roles and authority in the name of the state's supreme and even transcendental interests.[39]

Until recently, it has been the military that has controlled the state behind the scenes by conditioning politicians through informal methods.[40] Yet, for the first time in the history of the Turkish Republic, the military's role seems to have been taken over by civilians, namely by the Erdogan regime. Initially, at the beginning of the 2000s, a series of reforms in the context of the EU accession process such as the civilianisation of the *Milli Güvenlik Kurulu* (National Security Council-NSC) diminished the military's influence over civilians. However, just after the failed coup attempt in 2016, the military completely lost its influence due to the radical changes made to the legislation that determines the relationship between the military and government. On top of this, the SSM—renamed SSB—was affiliated with President Erdogan and became the ultimate authority and epicentre of procurement activities.

The rivalry between the military and civilians has been an important constraint for defence transformation. The Turkish military, which perceives itself as the owner of the state, has been reluctant to share its authority over defence planning and opposed any procurement institutions controlled by civilians, which resulted in a lack of communication among defence stakeholders, dual procurement institutions and R&D centres that led to the waste of resources, lack of civilian expertise on defence matters and an overall reductionistic understanding of defence. Since civilians are not involved in the defence planning process from the beginning, it creates numerous problems in the later stages such as long delays in the procurement process. As Mevlutoglu suggested, the average time span between the identification of needs and the acquisition is ten years,[41] which is way longer than the US, which spends less than six years on average for the same phase.[42] As Kurc illustrated in detail, due to the conflicts between military and civilian decision-makers, it took more than 15 years just to make a make/buy decision for the F-16s project in the 1980s, 12 years (1995–2007) for the attack and reconnaissance helicopter project (ATAK) and ten years for the Altay indigenous tank project.[43] An effective civil-military relationship requires the development of a civilian capacity, and the adjustment of the education system and according institutionalisation.[44] Although the recent developments ended the military's supremacy, the problems have yet to be solved as the rivalry between military and civilians appears to be replaced by the rivalry between pro and anti-regime parties.

The supremacy of the state, especially the one that was led by the military, has created an opaque defence sector which has traditionally been unaccountable and uncriticisable. Legislation and legal practices do not allow public institutions to audit defence expenditures properly.[45] 'Secrecy' is generally used as a pretext and *Sayıştay* (Turkish Court of Accounts-TCA) reports on defence expenditures are not examined or audited in the Parliament.[46] Furthermore, although one may find some trivial comments on the procurement process in Turkish media or in the defence studies literature, it is almost impossible to see a critique of the performance, organisational structure, conceptual/doctrinal approach, defence planning methods or practices of the TAF. Even after the military's role was diminished, the opacity of the defence sector remained untouched due to the authoritarian characteristics of the current regime.

### Limited economic resources and technological capabilities

Turkey has an ambitious defence autarky policy, yet a relatively weak economy to support this policy. Because of its historic economic weakness, defence expenditure has been significantly lower than many capable defence industries (see Figure 9.1). South Korea, for instance, which started a similar defence autarky project in the 1980s, has doubled Turkey's defence spending. Similarly, defence-related R&D expenditure—an indicator of innovation—is also significantly lower (see Figure 9.2). These statistics show that the defence expenditures are not sufficient to achieve Turkey's ambitious policy and to close the technology gap between Turkey and its competitors. Moreover, the sustainability of recent growth is also

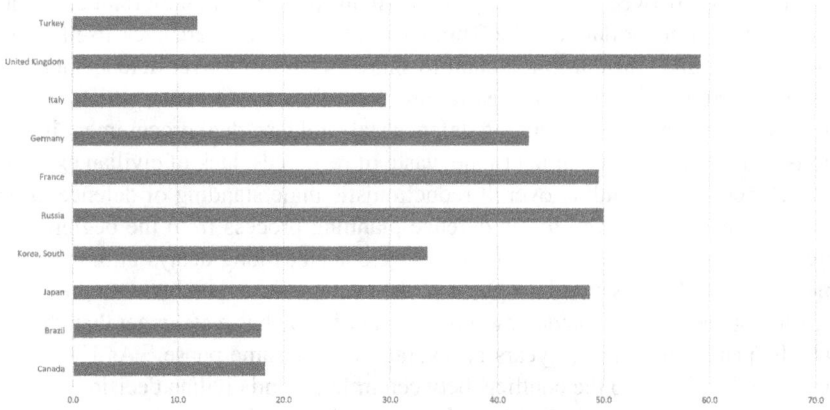

*Figure 9.1* Average military expenditure by country, in constant (2020) US$ billion, 2002–2021.

*Source:* SIPRI[48]

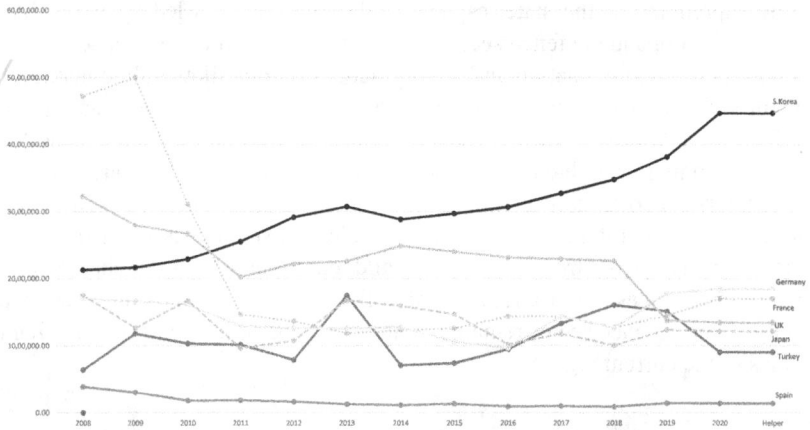

*Figure 9.2* R&D spending by country, in constant (2015) US$ million, 2008–2020.

*Source:* OECD[49]

doubtful, because of the simple fact that Turkey is on the verge of another economic crisis, with an annual inflation rate of 83.45%—by far the highest in OECD
countries—$185.3 billion short-term external national debt, a US$10 billion
current-account deficit—the biggest one in four decades—and the lira having lost
80% of its value against the dollar in five years.[47]

The lack of know-how and over-dependence on foreign supply is another limitation that impedes defence transformation. Despite unprecedented development and a radical increase in the indigenous production rate in the past decade, Turkey is still dependent on foreign suppliers for key technologies such as engines, transmissions, satellite communications, sensors and electronics hardware and software.[50] In general, defence related imports have always been higher than exports (Figure 9.3). Note that the recent decline in imports stems from the embargos imposed by Western countries. The two-month embargo imposed by European countries after Turkey's 'Operation Peace Spring' in northern Syria cost Turkey's industry about $1 billion in production, a high price for an industry with just US$11 billion in revenue.[51] On top of this, the turnover rate per capita—an indicator of labour productivity—for the last five years is between 130–170 thousand US$, which is way lower than the world average of 350 thousand dollars.[52]

Even though the major projects such as the 'attack helicopter', 'main battle tank', 'self-propelled howitzers' are presented as 'national' by the AKP government, they still rely on foreign licences which prevent their export to third countries without the consent of the Western partners. This has been recently revealed on a few occasions. For instance, Turkey could not complete the sale of ATAK helicopters to Pakistan as the US holds the license for the engines and didn't approve the sale.[53] Similarly, Fırtına Howitzers cannot be delivered to Azerbaijan because Germany refused to allow the export license for the engines.[54] With a technology strategy focused on import substitution where export-related activities are of secondary importance; a top-down approach aiming to produce the platform first and indigenisation of critical subsystems later, which carries the risk of becoming even more dependent on foreign supplies, lack of competition that erodes the stimulation

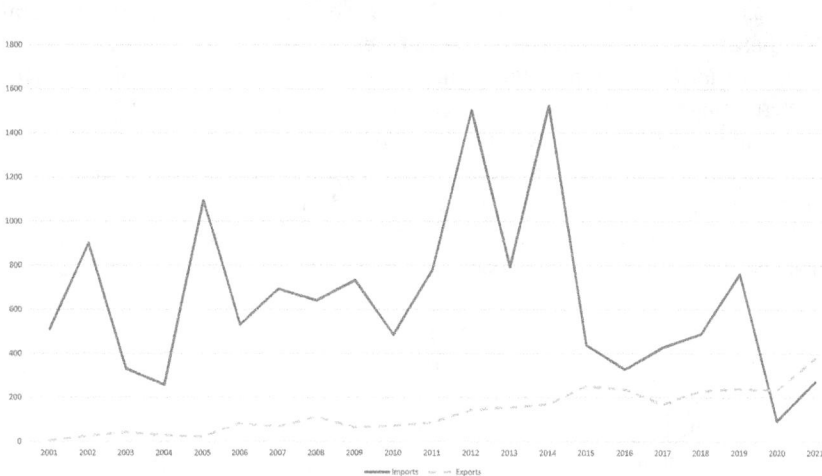

*Figure 9.3* Turkey's defence import–export comparison, in constant (2020) US$ million, 2001–2021.

*Source:* SIPRI[56]

for innovation and insufficient number of qualified personnel,[55] it seems that there is still a long way to go for Turkey to achieve complete self-sufficiency.

### Lack of a coherent defence ecosystem

Turkey uses a domestic version of the US Planning, Programming, Budgeting and Execution System (PPBES) that translates strategic guidance into resource allocation decisions according to the needs of the armed forces. It is a 'cyclic process containing four distinct, but interrelated phases. In addition to establishing the framework and process for decision making on future programmes, the process permits prior decisions to be examined and analysed from the viewpoint of the current environment (threat, political, economic, technological and resources) and for the time period being addressed'.[57] The strategic guidance for the PPBES process is provided by the 'National Military Strategy' that is issued by the Chief of the General Staff (CGS) based on the 'National Security Strategy' that is written by the NSC every five years. In the planning phase, the CGS produces a Strategic Targets Plan (SHP) document for the next 20 years and determines the needs of the TAF under the guidance of the National Military Strategy. In the programming phase, the needs identified in the planning phase are reviewed, prioritised and updated on the basis of the available funds in the budget and programmed in the Ten-Year Procurement Plan (OYTEP). The OYTEP ends with Project Definition Documents (PDD) whose release starts the procurement process through two main channels, namely the Ministry of Defence and SSB—Defence Industry Agency.[58]

It is important to note that the PPBES system in Turkey is a highly 'nationalized' version even though it seems similar in terms of technical and administrative processes. For instance, while the Deputy Secretary of Defense—who could be considered as the equivalent of the deputy Ministry of Defence—manages the overall PPBES process in the US, it is the Chief of General Staff who manages the overall process in Turkey, or while Congress plays a major role in each phase of the PPBES by determining the Department of Defense's (DOD's) authority, by authorising and appropriating the funding or by determining the limits of transfer and reprogramme funds, the role of Parliament in Turkey is rather symbolic. Furthermore, the PPBES system is widely criticised in the US for its 'industrial era' approach and failure to adapt to rapidly changing threats and to leverage commercial technologies.[59] However, there is no such discussion in Turkey mainly because the PPBES itself is regarded as a tool for only regulating rather than shaping, managing or leading procurement activities. It is always easier for decision-makers in Turkey to circumvent the bureaucratic challenges of the PPBES in informal ways given the fact that defence planning is an unaccountable and opaque area. The absence of public criticism despite the long delays in the procurement process is evidence of this.

Taken as a whole, the major issue with Turkey's defence planning mechanism can be described as the absence of a coherent defence ecosystem which includes all relevant stakeholders functioning in synergy. This seems to be confirmed by the SSB in its latest Strategic Plan 2019–2023 by pointing to the need for 'a higher-level institution' with an authority and ability to solve the structural problems of the defence sector and to make a long-term strategic plan with a holistic view.[60] Stemming mainly from the fact that defence planning is largely restricted to the military and that there is a lack of coordination between military and civilian bureaucracy, the current system fails to include the economic, commercial, cultural and sociological aspects of national defence and security, to provide interdisciplinary feedback mechanisms from academia, the defence industry or think tanks and to prioritise among a wide array of requirements.[61]

Another major issue is the lack of proper strategic guidance. Although all documents produced during the defence planning process are classified as 'top secret' and it is difficult to question the whole defence planning process, a review of secondary sources and rarely published official documents might still be helpful. For instance, the defence reform report written by a committee that was tasked by then-President Abdullah Gul in 2014 states:

> In high level documents, the strategic guidance in relation to future security and defence policy is not stated clearly. National Security Strategy and National Military Strategy cannot portray a concrete framework for the force development which must be based on strategic threats and targets. The operational needs are defined independently out of the security context, rather by the weapon systems and platforms. Missions in National Military Strategy must be stated more clearly as in NATO's Level of Ambitions and Mission Statement, the operational needs must be based on capabilities rather than weapon systems/platforms.[62]

This results in procurement decisions that reflect short-term tactical requirements based on almost entirely quantitative analyses without an interdisciplinary perspective and also spreads the already limited financial resources over a large number of ambitious projects.[63] In addition to the chronic limitations discussed in the paragraphs above, there are equally serious limitations like the network of nepotism, the widespread corruption, declining transparency and brain drain that come with increasing authoritarianism, which fall outside of the scope of this chapter.

### Part 3: military innovation and adaptation

Military innovation in Turkey is largely confined to technological innovation. This can be seen not only in the increasing number of defence-related academic publications, websites, TV shows, workshops, think tank publications but also in policy documents published in relation to the defence transformation. Much of the discussion is limited to the technological qualifications of weapon systems and their potential indigenous production. One may argue that despite its long-term policy goal of defence autarky, Turkey appears to have focused solely on indigenous production of existing military systems around the globe, with limited conceptual and organisational aspects. According to Michael Raska's framework,[64] this puts Turkey somewhere between the Tier 3- 'copiers-reproducers' and the Tier-2 'adaptors-modifiers' categories (see Figure 9.4).

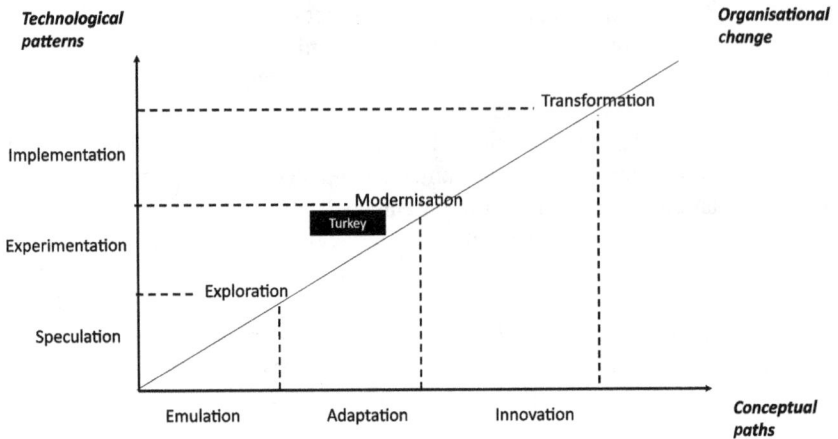

*Figure 9.4* Turkey's military innovation trajectories.

*Source:* Figure created by author based on Michael Raska, Military Innovation of Small States—Creating a Reverse Asymmetry (Routledge, 2016)[65]

In terms of technological innovation, Turkey shows patterns of speculation and experimentation. It is in the process of producing a wide array of weapon systems. However, most of the major projects advertised as the leading examples of indigenous production are in fact based on the design and technologies of other countries, although Turkey produces some selected parts of these weapon systems indigenously. For instance, Turkey's amphibious assault ship—TCG Anadolu—is based on the Spanish Juan Carlo I; a large chunk of Turkey's modern navy vessels, including the Barbaros-class frigates, Yavuz class frigates and Kılıç class fast attack craft, were designed in Germany; the Altay battle tank was co-produced as a variant of K2 Black Panther tank of South Korea while the Fırtına Howitzer was based on the K9 self-propelled guns of South Korea, ATAK helicopters were co-produced by Italy's AgustaWestland and Turkey's first national fighter jet will rely on the British BAE company. Furthermore, as mentioned before, Turkey depends on foreign suppliers for the key technologies. Despite the notable progress in the capabilities of the Turkish defence industry, it has yet to produce high-tech outputs and innovative weapon systems.[66] The existing capabilities do not exceed the technological experimentation phase, which puts the technological innovation of Turkey somewhere between speculation and experimentation.

One exception to this categorisation is the Turkish UAVs. Turkey appears to have passed from the experimentation to the implementation phase of this distinct technology. Up to the early 2000s, despite its investment in UAV technology, Turkey was dependent on the US and Israeli UAVs in its fight against the PKK, which was the main driver for Turkey's UAV interest. However, due to the US refusal to provide Predators and the deteriorating political relationship with Israel, Turkey, in line with its defence autarky project, focused on deriving benefits from this niche capability. Both public—TAI's ANKA and Aksungur—and private sector—Baykar's TB2—products proved effective for a short while. As a result of concerted and top-down efforts of the political leadership, particularly TB2 drones were gradually improved based on the lessons learned in a series of conflicts in Northern Iraq, Syria, Libya, the Caucasus and finally Ukraine. This resulted in the production of a battle-tested and commercially attractive weapon system in high international demand that is capable of destroying Russian Pantsir surface-to-air missile systems in Ukraine, Russian main battle tanks in Armenia and Chinese-made Wing Loong drones in Libya.[67] According to Fukuyama, 'Turkey's use of drones is going to change the nature of land power in ways that will undermine existing force structures' due to its ability to hit tanks in a relatively cheap way without being hit and risking the lives of human pilots.[68] No doubt, Bayraktar's close family ties to President Erdogan made this quick feedback mechanism possible, yet, TB2s emerged as an innovative weapon system in the international market with its favourable price/performance ratio. Furthermore, the use of UAVs in the internal fight against the PKK changed the character of this conflict. According to one recently retired officer: it has become routine for Turkish forces to hit the targets by combat UAVs or by jet fighters after they are marked by UAV lasers while in the past, the artillery guided by a forward observer had been the main method.[69]

However, the progress in technological aspects could not be mirrored in conceptual innovation. The TAF uses the war concepts, doctrines and principles that are emulated and adopted from the US and NATO. Like perhaps many NATO members, the Air-Land Battle Doctrine is in use for the conduct of operational-level conventional war. Although Revolution in Military Affairs (RMA)-oriented innovative concepts have been examined and discussed in various institutions like the Turkish War Academy, there has never been a distinct Turkish RMA-oriented conceptual innovation toward a new theory of war. Despite its 30-year-plus experience in irregular warfare, Turkey has yet to produce a counterterrorism concept/doctrine based on its experiences and lessons learned on the ground. One manual published on irregular warfare in the 1990s, KKT 31–2 Homeland Security Operations (İç Güvenlik Harekatı) was far from reflecting the experiences or lessons learned in the field. Although The Land Forces Command has an Education and Doctrine Command, which is commanded by a four-star general, it rather functions as an administrative unit that regulates educational institutions than a hub of conceptual innovation. In short, Turkey has difficulty translating its battlefield experiences into concepts/doctrines to be used in subsequent operations. All these qualifications reflect characteristics of emulation and limited adaptation in terms of conceptual paths.

In terms of organisational innovation, the most radical change was made in the aftermath of the Soviet Union's collapse. In 1992, in order to increase manoeuvre and firepower capacity, the TAF transformed from a division-based force to a brigade-based force and most brigades were restructured as mechanised/armoured brigades. But this was a change made to remain compatible with other NATO members rather than a change that stemmed from an indigenous need.[70] Besides the continuous and gradual change forced by irregular warfare, the second and more comprehensive transformation attempt came in 2012. Transformation and restructuring units were established in the headquarters of the Chief of General Staff (CGS) and Service Commands. An overall transformation project titled 'TAF 2033 Vision' was started. This project envisaged some important changes such as the creation of a joint command under the CGS, restructuring land forces from four armies to two armies, the creation of a new promotion system based on eligibility,[71] switch from threat-based planning to capability-based planning, building a more professional army—one that is hybrid but not a full all-volunteer system.[72] Some selected projects of TAF 2033, which in fact were coup-proofing measures that extended the authority of the MoD over the military and made the CGS only a war-time commander,[73] were put in action immediately after the 15 July coup attempt. However, since then, 'TAF 2033 Vision' appears to have been laid aside and the transformation units created in the CGS and Service Command HQs lost their functional value. As SSB suggested in Strategic Plan 2019–2023,[74] Turkey has no units like US' DARPA (Defense Advanced Research Projects Agency) or Israel's OTRI (Operational Theory Research Institute) which focus on transformational change and work within an innovation ecosystem that includes academic, corporate and governmental partners.[75] All in all, apart from the notable increase in the number of professional commando brigades fighting against the PKK,[76] the organisational structure of the TAF has remained untouched since the end of the Cold War despite the intensive reform discussions that took place at the beginning of the 2010s. Consequently, Turkey can be considered in the speculation/experimentation and emulation phases in the spectrum of patterns and paths of military innovation.

## Conclusion

The main driver behind the recent growth in Turkish force development is related more to domestic rather than external factors. President Erdogan managed to establish an authoritarian regime in the past decade in Turkey and like all authoritarian regimes, the primary focus of this regime is to consolidate its domestic power even if it is at the expense of national interests in the international system. A strong defence industry has become the central pillar of the new regime—whether it is feasible or not—as it considerably helps consolidate its domestic power in Turkey which already had convenient cultural and structural incentives for an indigenous defence industry. Therefore, recent growth can be considered as the result of top-down stimulation driven mainly by President Erdogan rather than a natural/systemic development process.

However, there are serious questions around the sustainability of this growth as the chronic limitations such as the security notion that prioritises the state over individuals, an opaque defence sector, a weak general economy, a lack of sufficient know-how and the lack of a coherent defence ecosystem have yet to be dealt with. Furthermore, there are also problems such as corruption, declining transparency

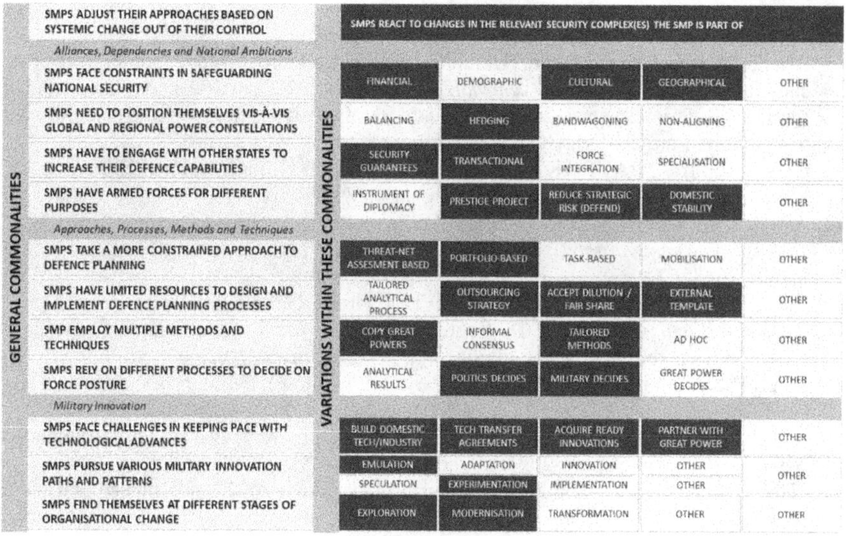
GENERAL COMMONALITIES
VARIATIONS WITHIN THESE COMMONALITIES
SMPS ADJUST THEIR APPROACHES BASED ON SYSTEMIC CHANGE OUT OF THEIR CONTROL
SMPS REACT TO CHANGES IN THE RELEVANT SECURITY COMPLEX(ES) THE SMP IS PART OF
Alliances, Dependencies and National Ambitions
SMPS FACE CONSTRAINTS IN SAFEGUARDING NATIONAL SECURITY
FINANCIAL
DEMOGRAPHIC
CULTURAL
GEOGRAPHICAL
OTHER
SMPS NEED TO POSITION THEMSELVES VIS-À-VIS GLOBAL AND REGIONAL POWER CONSTELLATIONS
BALANCING
HEDGING
BANDWAGONING
NON-ALIGNING
OTHER
SMPS HAVE TO ENGAGE WITH OTHER STATES TO INCREASE THEIR DEFENCE CAPABILITIES
SECURITY GUARANTEES
TRANSACTIONAL
FORCE INTEGRATION
SPECIALISATION
OTHER
SMPS HAVE ARMED FORCES FOR DIFFERENT PURPOSES
INSTRUMENT OF DIPLOMACY
PRESTIGE PROJECT
REDUCE STRATEGIC RISK (DEFEND)
DOMESTIC STABILITY
OTHER
Approaches, Processes, Methods and Techniques
SMPS TAKE A MORE CONSTRAINED APPROACH TO DEFENCE PLANNING
THREAT-NET ASSESMENT BASED
PORTFOLIO-BASED
TASK-BASED
MOBILISATION
OTHER
SMPS HAVE LIMITED RESOURCES TO DESIGN AND IMPLEMENT DEFENCE PLANNING PROCESSES
TAILORED ANALYTICAL PROCESS
OUTSOURCING STRATEGY
ACCEPT DILUTION / FAIR SHARE
EXTERNAL TEMPLATE
OTHER
SMP EMPLOY MULTIPLE METHODS AND TECHNIQUES
COPY GREAT POWERS
INFORMAL CONSENSUS
TAILORED METHODS
AD HOC
OTHER
SMPS RELY ON DIFFERENT PROCESSES TO DECIDE ON FORCE POSTURE
ANALYTICAL RESULTS
POLITICS DECIDES
MILITARY DECIDES
GREAT POWER DECIDES
OTHER
Military Innovation
SMPS FACE CHALLENGES IN KEEPING PACE WITH TECHNOLOGICAL ADVANCES
BUILD DOMESTIC TECH/INDUSTRY
TECH TRANSFER AGREEMENTS
ACQUIRE READY INNOVATIONS
PARTNER WITH GREAT POWER
OTHER
SMPS PURSUE VARIOUS MILITARY INNOVATION PATHS AND PATTERNS
EMULATION
ADAPTATION
INNOVATION
OTHER
SPECULATION
EXPERIMENTATION
IMPLEMENTATION
OTHER
OTHER
SMPS FIND THEMSELVES AT DIFFERENT STAGES OF ORGANISATIONAL CHANGE
EXPLORATION
MODERNISATION
TRANSFORMATION
OTHER
OTHER

*Figure 9.5* Structured focused comparison framework for Turkey.

*Source:* Author

and brain drain that come with increasing authoritarianism. On top of this, military innovation in Turkey, which is an essential factor for a complete indigenous industry, focused largely on technological innovation, with limited conceptual and organisational aspects. As for technological innovation, it focuses on the indigenous production of existing military systems around the globe, whose main function is to promote the current regime with only a secondary aim of having a robust force structure. Please refer to Figure 9.5 to see Turkey's position on the structured comparative framework.

Having said that, as Colin S. Gray explained, defence planning is a complex business as there are so many variables involved and the future security context cannot be known in advance in reliably useful detail.[77] From this point of view, although it is unlikely that Turkey will attain defence autarky in the near future, there is a potential—albeit a tiny one—that the achievement in UAV technology can be disseminated to other fields and that this triggers an overall conceptual and organisational change along with a cultural wind that draws on factors—which normally set a barrier on defence transformation—like the supremacy of the state and an opaque defence sector or cultural factors like 'the aspiration for being great again' in an increasingly heated international context. Considering that an indigenous defence industry and defence expenditures can be used to demonstrate success in the election campaigns despite the heavy economic crisis, this transformation is in the realm of possibility in Turkey even if it is at the expense of impoverishment in other sectors, which is something that may not be imaginable for instance, for a European small/medium power. But only time will tell.

## Notes

1  Brett Forrest and Jared Malsin, 'Ukraine Says It Used Turkish-Made Drones to Hit Russian Targets', *The Wall Street Journal*, 27 February 2022, www.wsj.com/livecoverage/russia-ukraine-latest-news-2022-02-26/card/ukraine-says-it-uses-turkish-made-drones-to-hit-russian-targets-DrigGO7vkGfDzbBuncnA.
2  Clash Report, 'BAYRAKTAR: Official Song [1 Hour Version]', *YouTube*, 13 March 2022, www.youtube.com/watch?v=aEyhV96JW5A.
3  Ash Rossiter and Brendon J. Cannon, 'Turkey's Rise as a Drone Power: Trial by Fire', *Defense & Security Analysis* 38, no. 2 (2022): 210–29, https://doi.org/10.1080/1475179 8.2022.2068562.
4  Maya Carlin, 'How the Turkish-Made TB2 Drone Gave Ukraine an Edge Against Russia', *Business Insider*, 18 September 2022, www.businessinsider.com/how-turkish-baykar-tb2-drone-gave-ukraine-edge-against-russia-2022–9?r=US&IR=T.
5  SIPRI (Stockholm International Peace Research Institute), 'Military Expenditure Database', *SIPRI,* https://milex.sipri.org/sipri, accessed on 28 October 2022.
6  'SASAD 2021 Performance Report 2021', *SASAD*, p. 7, www.sasad.org.tr/uploaded/Sasad-Performans-Raporu-2021.pdf, accessed on 10 March 2023.
7  'Türk Savunma Sanayiinin 2022 Yılı Performansı [Year 2022 Performance of Turkish Defence Industry]', *Savunma Sanayii Başkanlığı*, www.ssb.gov.tr/WebSite/contentlist.aspx?PageID=48&LangID=1, accessed on 23 February 2023.
8  Pieter D. Wezeman, Alexandra Kuimova and Siemon T. Wezeman, *SIPRI Fact Sheet, March 2022, Trends in International Arms Transfers 2021* (Stockholm: SIPRI, March 2022), www.sipri.org/sites/default/files/2022-03/fs_2203_at_2021.pdf.
9  'Top 100 Defense Companies', *DefenseNews*, https://people.defensenews.com/top-100/, accessed on 10 March 2023.
10  Robert Siegel, 'Turkish Democracy: A Model for Other Countries?', *NPR*, 14 April 2011, www.npr.org/2011/04/14/135407687/turkish-democracy-a-model-for-other-countries; Ömer Taşpınar, 'Turkey: The New Model?', *Brookings Institute*, 25 April 2012, www.brookings.edu/research/turkey-the-new-model/.
11  Mustafa Kutlay and Ziya Öniş, 'Understanding Oscillations in Turkish Foreign Policy: Pathways to Unusual Middle Power Activism', *Third World Quarterly* 42, no. 12 (2021): 3052.
12  Mark Wallace and Madeleine Joelson, 'Letter: NATO Should Be Ready to Suspend Turkey's Membership', *Financial Times*, 8 July 2022, www.ft.com/content/f6292e9a-7608-4df8-8a88-b175a873dbec; Steven A. Cook, 'What Erdogan Gets by Being a Spoiler in NATO', *Foreign Policy*, 28 June 2022, https://foreignpolicy.com/2022/06/28/nato-summit-turkey-erdogan-finland-sweden/; Michael Crowley and Steven Erlanger, 'For NATO, Turkey Is a Disruptive Ally', *New York Times*, 30 May 2022, www.nytimes.com/2022/05/30/us/politics/turkey-nato-russia.html.
13  Cüneyt Gürer, 'Turkey-China Relations', *Per Concordiam*, 17 November 2022, https://perconcordiam.com/turkey-china-relations/.
14  Kutlay and Öniş, 'Understanding Oscillations', 3052.
15  İlhan Uzgel, 'Turkey's Foreign Policy in free fall', *Duvar English*, 27 May 2021, www.duvarenglish.com/turkeys-foreign-policy-in-free-fall-article-57636.
16  Davit Safaryan, 'On the Acceleration of Armenian-Turkish and Armenian-Azerbaijani Negotiations', *The Mirror Spectator*, 24 February 2023, https://mirrorspectator.com/2023/02/24/on-the-acceleration-of-armenian-turkish-and-armenian-azerbaijani-negotiations/.
17  Data from Turkish Statistical Institute, 'İSTATİSTİK VERİ PORTALI', *TUIK*, https://data.tuik.gov.tr/Search/Search?text=ihracat, accessed on 1 November 2022
18  Mustafa Kutlay and Ziya Öniş. 'Turkish Foreign Policy in a Post-Western Order: Strategic Autonomy or New Forms of Dependence?', *International Affairs* 97, no. 4 (2021): 1099, doi: 10.1093/ia/iiab094.

19  Bülent Aras, *Turkey's State Crisis, Institutions, Reform and Conflict* (New York: Syracuse University Press, 2022), 48.
20  Aras, *Turkey's State Crisis*, 49.
21  Henri Barkey, 'One Year Later, the Turkish Coup Attempt Remains Shrouded in Mystery', *Washington Post*, 14 July 2017, www.washingtonpost.com/news/democracy-post/wp/2017/07/14/one-year-later-the-turkish-coup-attempt-remains-shrouded-in-mystery/.
22  Aras, *Turkey's State Crisis*, 73.
23  Aras, *Turkey's State Crisis*, 67.
24  Kutlay and Öniş. 'Turkish Foreign Policy', 1103.
25  Kutlay and Öniş, 'Understanding Oscillations', 3064.
26  Anouck Gabriela Côrte Réal-Pinto, 'A Neo-Liberal Exception? The Defence Industry "Turkification" Project', *International Development Policy | Revue internationale de politique de développement* 8 (2017): 300–301, https://doi.org/10.4000/poldev.2316.
27  Hüseyin Bağcı and Çağlar Kurç, 'Turkey's Strategic Choice: Buy or Make Weapons?', *Defence Studies* 17, no. 1 (2016): 19, https://doi.org/10.1080/14702436.2016.1262742.
28  Çağlar Kurç, 'Between Defence Autarky and Dependency: The Dynamics of Turkish Defence Industrialization', *Defence Studies* 17, no. 3 (2017): 262, https://doi.org/10.10 80/14702436.2017.1350107.
29  Réal-Pinto, 'A Neo-Liberal Exception?', 304: Kurç, 'Between Defence Autarky and Dependency', 262.
30  Réal-Pinto, 'A Neo-Liberal Exception?', 305.
31  *Türk Savunma Sanayii Politikası ve Stratejisi Esasları* [The principles of Turkish defence industry policy and strategy], Decision number 98/11173, Resmî Gazete, Issue 23378, 20 June 1998; Savunma Sanayii Başkanlığı [Defence Industry Agency], *Stratejik Plan 2019–2023* (Ankara: Savunma Sanayii Başkanlığı, 2022), www.ssb.gov.tr/WebSite/contentlist.aspx?PageID=43&LangID=1.
32  'Tank ve elicopter ihaleleri iptal edildi' [Tank and Helicopter Tenders were cancelled], *Hurriyet Newspaper*, 14 May 2004, www.hurriyet.com.tr/ekonomi/tank-ve-helikopter-ihaleleri-iptal-edildi-225670.
33  Mustafa Aydın, 'Security Conceptualisation in Turkey', in *Security and Environment in the Mediterranean*, eds. H.G. Brauch et al. (Berlin and Heidelberg: Springer, 2003), 352.
34  S. Elekdağ, '2 1/2 War Strategy', *PERCEPTIONS: Journal of International Affairs* 1, no. 4 (1996): 33–57.
35  Bağcı and Kurç, 'Turkey's Strategic Choice', 8.
36  'Projects', *Turkey Defence Industry Agency*, www.ssb.gov.tr/Default.aspx?LangID=2, accessed on 10 March 2023.
37  Aras, *Turkey's State Crisis*, 47.
38  Michael A. Reynolds, 'Turkey and Russia: A Remarkable Rapprochement', *War on the Rocks*, 24 October 2019, https://warontherocks.com/2019/10/turkey-and-russia-a-remarkable-rapprochement/.
39  Aras, *Turkey's State Crisis*, 46.
40  Steven A. Cook, *Ruling But Not Governing* (Baltimore: The Johns Hopkins University Press, 2007), 104.
41  Arda Mevlutoglu, 'Commentary on Assessing the Turkish Defense Industry: Structural Issues and Major Challenges', *Defence Studies* 17, no. 3 (2017): 285, https://doi.org/10.1080/14702436.2017.1349534.
42  Brendan W. McGarry, *DOD Planning, Programming, Budgeting, and Execution (PPBE): Overview and Selected Issues for Congress* (Washington, DC: Congressional Research Service Report R47178, 11 July 2022), 18.
43  Kurç, 'Between Defence Autarky and Dependency', 266.
44  Turkish Republic Presidency, *Defence Reform Report* (Ankara: Turkish Republic Presidency, 2014), 18.

45 Mehmet Güneş, 'Etkin Bir Kamu Yönetimi İçin Artan Savunma Harcamalarının Denetimi Ve Hesap Verebilirliğin Rolü' [The Audit of Defence Expenditures and the role of Accountability for an Efficient Public Governance], *Gazi Üniversitesi İktisadi ve İdari Bilimler Fakültesi Dergisi* 3, no. 2 (2011): 160.

46 Turkish Republic Presidency, *Defence Reform Report*, 36.

47 'Ahead of a Critical Election Turkey's Economy Is Running on Borrowed Time', *The Economist*, 26 March 2023, www.economist.com/europe/2023/03/26/ahead-of-a-critical-election-turkeys-economy-is-running-on-borrowed-time.

48 SIPRI (Stockholm International Peace Research Institute), 'Military Expenditure Database', *SIPRI*, https://milex.sipri.org/sipri, accessed on 28 October 2022.

49 OECD, 'Research and Development Statistics: Government Budget Appropriations or Outlays for RD (Edition 2021)', *OECD*, https://doi.org/10.1787/043b9ea1-en, accessed on 11 October 2022.

50 Bağcı and Kurç, 'Turkey's Strategic Choice', 2; Mevlutoglu, 'Commentary on the Turkish Defense Industry', 290; Göksel Korkmaz and Mustafa Kemal Topcu, 'Porter's Diamond Model and the Competitiveness of the Turkish Defense Industry', *Journal of Defense Resources Management* 12, no. 1 (2021): 41–74.

51 Ferhat Guruni, 'Turkey's Unpromising Defense Industry', *Carnegie Endowment for International Peace*, 9 October 2020, https://carnegieendowment.org/sada/82936.

52 Korkmaz and Topcu, 'Porter's Diamond Model', 54.

53 Burak Ege Bekdil, Usman Ansari and Joe Gould, 'Pakistan Extends Delayed T129 Helo Deal with Turkey—Again', 15 March 2021, www.defensenews.com/global/mideast-africa/2021/03/15/pakistan-extends-delayed-t129-helo-deal-with-turkey-again/.

54 Mevlutoglu, 'Commentary on the Turkish Defense Industry', 290.

55 Mevlutoglu, 'Commentary on the Turkish Defense Industry', 289–291.

56 SIPRI, 'Arms Transfers Database Importer/Exporter TIV Tables', *SIPRI*, https://armstrade.sipri.org/armstrade/page/values.php, accessed on 7 September 2022.

57 McGarry, 'DOD Planning, Programming, Budgeting, and Execution (PPBE)', 3.

58 Arda Mevlutoglu, 'Türkiye'nin Savunma Reformu: Tespit ve Öneriler [Turkey's Defence Reform: Identifications and Recommendations]', *SETA* 164 (2016): 17–19.

59 McGarry, 'DOD Planning, Programming, Budgeting, and Execution (PPBE)', 18; Jonathan P. Wong, 'Bad Idea: Looking for Easy Solutions for PPBE Reform', *RAND Corporation*, 2 March 2023, www.rand.org/blog/2023/03/bad-idea-looking-for-easy-solutions-for-ppbe-reform.html.

60 Savunma Sanayii Başkanlığı, *Stratejik Plan 2019–2023*, 82–84.

61 Mevlutoglu, 'Commentary on the Turkish Defense Industry', 285–289; Korkmaz and Topcu, 'Porter's Diamond Model', 65–70.

62 Turkish Republic Presidency, *Defence Reform Report*, 18.

63 Mevlutoglu, 'Commentary on the Turkish Defense Industry', 288–89.

64 Michael Raska, 'Strategic Competition for Emerging Military Technologies', *PRISM* 8, no. 3 (2019): 68–69.

65 Michael Raska, 'A Structured-Phased Evolution: The Third Generation Force Transformation of the Singapore Armed Forces', in *Military Innovation in Small States: Creating a Reverse Asymmetry*, ed. M. Raska (New York: Routledge, 2016), 130–162.

66 Özgür Körpe, 'Turks' Pragmatic Solutions the Philosophy Behind Defense Technology', in *Military Innovation in Türkiye An Overview of the Post-Cold War Era*, ed. Barış Ateş (New York: Routledge, 2023), 36, e-book, https://doi.org/10.4324/9781003327127–2.

67 Rossiter and Cannon, 'Turkey's Rise as a Drone Power', 214.

68 Francis Fukuyama, 'Droning on in the Middle East', *Pakistan Defence*, 5 April 2021, https://defence.pk/pdf/threads/droning-on-in-the-middle-east-francis-fukuyama-article.709161/.

69 Interview with a recently retired Turkish Colonel, 23 January 2023, Brussels.

70 'Kara Kuvvetleri Komutanlığı Tarihçesi' [History of Land Forces Command], *Turkish Land Forces Command*, www.kkk.tsk.tr/tarihce.aspx, accessed on 10 March 2023.

71 Barkın Sık, 'TSK silbaştan değişiyor' [TAF transforms from scratch], *Cumhuriyet Newspaper*, 16 December 2013, www.cumhuriyet.com.tr/haber/tsk-silbastan-degisiyor-19283.
72 Murat Gürgen, 'Profesyonel Ordu, hudut ötesinde "Savunma"' [Professional Army, defence beyond the borders], *Habertürk Newspaper*, 30 April 2015, www.haberturk.com/yazi-dizisi/haber/1072029-profesyonel-ordu-hudut-otesinde-savunma.
73 Ayşegül Kars Kaynar, 'Post-2016 Military Restructuring in Turkey from the Perspective of Coup-Proofing', *Turkish Studies* 23, no. 3 (2022): 389–90, https://doi.org/10.1080/14683849.2021.1977631.
74 Turkish Republic Presidency, *Stratejik Plan 2019–2023 Updated Version* (Ankara: Turkish Republic Presidency Defence Industry Agency, 2020), 85.
75 'About DARPA', *Defense Advanced Research Projects Agency*, www.darpa.mil/about-us/about-darpa, accessed on 10 March 2023.
76 'Hulusi Akar TSK'nın başarı bilançosunu anlattı: 2 komando tugayı vardı, şimdi 17', [Hulusi Akar explained TAF's success toll: there was 2 commando brigade, now 17], *Hürriyet*, 20 June 2021, www.hurriyet.com.tr/gundem/hulusi-akar-tsknin-basari-bilancosunu-anlatti-2-komando-tugayi-vardi-simdi-17-41835653.
77 Colin S. Gray, *Strategy and Defence Planning: Meeting the Challenge of Uncertainty* (Oxford: Oxford University Press, 2014), 68.

## Bibliography

Aras, Bülent. *Turkey's State Crisis, Institutions, Reform and Conflict*. Syracuse University Press, New York, 2022.
Aydın, Mustafa. 'Security Conceptualisation in Turkey', In: H.G. Brauch et al. (eds.), *Security and Environment in the Mediterranean*. Springer, Berlin and Heidelberg, 2003.
Bağcı, Hüseyin and Kurç, Çağlar. 'Turkey's strategic choice: buy or make weapons?'. *Defence Studies*, 17:1 (2016). https://doi.org/10.1080/14702436.2016.1262742
Barkey, Henri. 'One year later, the Turkish coup attempt remains shrouded in mystery', *Washington Post*, July 14, 2017, www.washingtonpost.com/news/democracy-post/wp/2017/07/14/one-year-later-the-turkish-coup-attempt-remains-shrouded-in-mystery/
Bekdil, Burak Ege, Ansari, Usman and Gould, Joe. 'Pakistan extends delayed T129 helo deal with Turkey—again', March 15, 2021, www.defensenews.com/global/mideast-africa/2021/03/15/pakistan-extends-delayed-t129-helo-deal-with-turkey-again/
Carlin, Maya. 'How the Turkish-made TB2 drone gave Ukraine an edge against Russia', *Business Insider*, September 18, 2022, www.businessinsider.com/how-turkish-baykar-tb2-drone-gave-ukraine-edge-against-russia-2022-9?r=US&IR=T
Cook, Steven A. *Ruling But Not Governing*. The Johns Hopkins University Press, Baltimore, MD, 2007.
———. 'What Erdogan gets by being a spoiler in NATO', *Foreign Policy*, June 28, 2022, https://foreignpolicy.com/2022/06/28/nato-summit-turkey-erdogan-finland-sweden/
Crowley, Michael and Erlanger, Steven. 'For NATO, Turkey is a disruptive ally', *New York Times*, May 30, 2022, www.nytimes.com/2022/05/30/us/politics/turkey-nato-russia.html
DARPA. 'About DARPA', *Defense Advanced Research Projects Agency*, accessed on March 10, 2023, www.darpa.mil/about-us/about-darpa
*DefenseNews*. 'Top 100 defense companies', accessed on March 10, 2023, https://people.defensenews.com/top-100/
Elekdağ, Sukru. '2 1/2 war strategy'. *Perceptions: Journal of International Affairs*, 1 (1996).
Forrest, Brett and Malsin, Jared. 'Ukraine says it used Turkish-made drones to hit Russian targets,' *The Wall Street Journal*, February 27, 2022, www.wsj.com/livecoverage/russia-ukraine-latest-news-2022-02-26/card/ukraine-says-it-uses-turkish-made-drones- to-hit-russian-targets-DrigGO7vkGfDzbBuncnA
Fukuyama, Francis. 'Droning on in the middle East', *Pakistan Defence*, April 5, 2021, https://defence.pk/pdf/threads/droning-on-in-the-middle-east-francis-fukuyama-article.709161/

Gray, Colin S. *Strategy and Defence Planning: Meeting the Challenge of Uncertainty*, Oxford University Press, Oxford, 2014.

Guruni, Ferhat. 'Turkey's unpromising defense industry', *Carnegie Endowment for International Peace*, October 9, 2020, https://carnegieendowment.org/sada/82936

Güneş, Mehmet. 'Etkin Bir Kamu Yönetimi İçin Artan Savunma Harcamalarının Denetimi Ve Hesap Verebilirliğin Rolü' [The audit of defence expenditures and the role of accountability for an efficient public governance], *Gazi Üniversitesi İktisadi ve İdari Bilimler Fakültesi Dergisi*, 3/2 (2011).

Gürer, Cüneyt. 'Turkey-China relations', *Per Concordiam*, November 17, 2022, https://perconcordiam.com/turkey-china-relations/

Gürgen, Murat. 'Profesyonel Ordu, hudut ötesinde 'Savunma' [Professional Army, defence beyond the borders], *Habertürk Newspaper*, April 30, 2015, www.haberturk.com/yazi-dizisi/haber/1072029-profesyonel-ordu-hudut-otesinde-savunma

*Hurriyet Newspaper*, 'Tank ve helikopter ihaleleri iptal edildi' [Tank and helicopter tenders were cancelled], May 14, 2004, www.hurriyet.com.tr/ekonomi/tank-ve-helikopter-ihaleleri-iptal-edildi-225670

——. 'Hulusi Akar TSK'nın başarı bilançosunu anlattı: 2 komando tugayı vardı, şimdi 17' [Hulusi Akar explained TAF's success toll: There was 2 commando brigade, now 17], June 20, 2021, www.hurriyet.com.tr/gundem/hulusi-akar-tsknin-basari-bilancosunu-anlatti-2-komando-tugayi-vardi-simdi-17-41835653

Kaynar, Ayşegül Kars. 'Post-2016 military restructuring in Turkey from the perspective of coup-proofing'. *Turkish Studies*, 23:3 (2022), pp. 389–390. https://doi.org/10.1080/1468 3849.2021.1977631

Korkmaz, Göksel and Topcu, Mustafa Kemal. 'Porter's diamond model and the competitiveness of the Turkish defense industry'. *Journal of Defense Resources Management*, 12:1 (2021), p. 22.

Körpe, Özgür. 'Turks' pragmatic solutions the philosophy behind defense technology', In: Barış Ateş (ed.), *Military Innovation in Türkiye An Overview of the Post-Cold War Era*. Routledge, 2023. https://doi.org/10.4324/9781003327127-2

Kurç, Çağlar. 'Between defence autarky and dependency: the dynamics of Turkish defence industrialization'. *Defence Studies*, 17:3 (2017). https://doi.org/10.1080/14702436.2017. 1350107262.

Kutlay, Mustafa and Öniş, Ziya. 'Understanding oscillations in Turkish foreign policy: pathways to unusual middle power activism'. *Third World Quarterly*, 42:12 (2021a).

——. 'Turkish foreign policy in a post-western order: strategic autonomy or new forms of dependence?'. *International Affairs*, 97: 4 (2021b). https://doi.org/10.1093/ia/iiab094

McGarry, Brendan W. *DOD Planning, Programming, Budgeting, and Execution (PPBE): Overview and Selected Issues for Congress*. Congressional Research Service Report R47178, Washington, DC, July 11, 2022.

Mevlutoglu, Arda. 'Türkiye'nin Savunma Reformu: Tespit ve Öneriler [Turkey's defence reform: Identifications and recommendations]', *SETA*, August 2016, 164.

——. 'Commentary on assessing the Turkish defense industry: structural issues and major challenges'. *Defence Studies*, 17:3 (2017). https://doi.org/10.1080/14702436.2017.1349534

OECD. 'Research and development statistics: Government budget appropriations or outlays for RD (Edition 2021)', accessed on October 11, 2022, https://doi.org/10. 1787/043b9ea1-en.

Raska, Michael. 'Strategic competition for emerging military technologies'. *PRISM*, 8:3 (2019).

Réal-Pinto, Anouck Gabriela Côrte. 'A neo-liberal exception? The defence industry 'turkification' project'. *International Development Policy | Revue internationale de politique de développement*, 8 (2017). https://doi.org/10.4000/poldev.2316

Reynolds, Michael A. 'Turkey and Russia: A remarkable rapprochement', *War on the Rocks*, October 24, 2019, https://warontherocks.com/2019/10/turkey-and-russia-a-remarkable-rapprochement/

Rossiter, Ash and Cannon, Brendon J. 'Turkey's rise as a drone power: trial by fire'. *Defense & Security Analysis*, 38:2 (2022). https://doi.org/10.1080/14751798.2022.2068 562.

Safaryan, Davit. 'On the acceleration of Armenian-Turkish and Armenian-Azerbaijani negotiations', *The Mirror Spectator*, February 24, 2023, https://mirrorspectator.com/2023/02/24/on-the-acceleration-of-armenian-turkish-and-armenian-azerbaijani-negotiations/

SASAD. 'SASAD 2021 performance report' 2021, p. 7, accessed on March 10, 2023, www.sasad.org.tr/uploaded/Sasad-Performans-Raporu-2021.pdf

Savunma Sanayii Başkanlığı [Defence Industry Agency]. Türk Savunma Sanayiinin 2022 Yılı Performansı [Year 2022 performance of Turkish defence industry], www.ssb.gov.tr/WebSite/contentlist.aspx?PageID=48&LangID=1

———. 'Projects', accessed on March 10, 2023, www.ssb.gov.tr/Default.aspx?LangID=2

———. Strategic Plan 2019–2023, https://www.ssb.gov.tr/WebSite/contentlist.aspx?PageID=43&LangID=1

Siegel, Robert. 'Turkish democracy: A model for other countries?', *NPR*, April 14, 2011, www.npr.org/2011/04/14/135407687/turkish-democracy-a-model-for-other-countries

Sık, Barkın, 'TSK silbaştan değişiyor' [TAF transforms from scratch], *Cumhuriyet Newspaper*, December 16, 2013, www.cumhuriyet.com.tr/haber/tsk-silbastan-degisiyor-19283

SIPRI. Stockholm International Peace Research Institute Arms Transfers database, https://www.sipri.org/databases/armstransfers

SIPRI Fact Sheet, 'Trends in international arms transfers 2021', March 2022, www.sipri.org/sites/default/files/2022-03/fs_2203_at_2021.pdf

Taşpınar, Ömer. 'Turkey: The New Model?'. *Brookings Institute*, April 25, 2012, www.brookings.edu/research/turkey-the-new-model/

*The Economist*. 'Ahead of a critical election Turkey's economy is running on borrowed time', March 26, 2023, www.economist.com/europe/2023/03/26/ahead-of-a-critical-election-turkeys-economy-is-running-on-borrowed-time

TÜİK, *Turkish Statistical Institute*, https://data.tuik.gov.tr

Turkish Republic Presidency, 'Defence reform report', 2014, accessed on March 10, 2023, https://www.memurlar.net/common/news/documents/602402/2014-08-22-savunmareformu.pdf

Türk Kara Kuvvatleri Komutanlığı [Turkish Land Forces Command]. 'Kara Kuvvetleri Komutanlığı Tarihçesi' [History of land forces command], accessed on March 10, 2023, www.kkk.tsk.tr/tarihce.aspx

Uzgel, İlhan. 'Turkey's foreign policy in free fall', *Duvar English*, May 27, 2021, www.duvarenglish.com/turkeys-foreign-policy-in-free-fall-article-57636

Wallace, Mark and Joelson, Madeleine. 'Letter: NATO should be ready to suspend Turkey's membership', *Financial Times*, July 8, 2022, www.ft.com/content/f6292e9a-7608-4df8-8a88-b175a873dbec.

Wong, Jonathan P. 'Bad idea: Looking for easy solutions for PPBE reform', *RAND Corporation*, March 2, 2023, www.rand.org/blog/2023/03/bad-idea-looking-for-easy-solutions-for-ppbe-reform.html

Youtube. 'BAYRAKTAR: Official song' [1 Hour Version], March 13, 2022, accessed on May 15, 2023, www.youtube.com/watch?v=aEyhV96JW5A

# 10 Defence planning in the Netherlands

## Trying to keep all options open

*Lenny Hazelbag, Hans Klinkenberg
and Saskia van Genugten*

### Introduction

The Dutch defence organisation has experienced a number of watershed moments with regard to 'strategic thinking' regarding its purpose, its ambitions as well as its societal value and even its necessity. More than a century ago, the ambition might have been to remain a highly principled, neutral power hammering on parties to abide by the international rule of law. However, that specific strategic choice has been unavailable for the Netherlands since the beginning of World War II. Ever since, this small country on the North Sea has seen its military direction and ambitions firmly linked to, and shaped by, the strategic direction of its most important partners, first and foremost the United States, followed by its larger European counterparts, Germany, the UK and France. While generally seen as a loyal partner, for decades, the Dutch have also proven to be a rather parsimonious defence actor. In particular, after the end of the Cold War, decisions were driven predominantly by cost-efficiency considerations, which went hand-in-hand with a willingness to take the risk of hollowing out the defence organisation in order to be able to spend money elsewhere.

Currently, strategic thinking is pivoting. The Russia-Ukraine War, which started in 2022, is now proving to be the latest watershed moment for the Dutch defence organisation: military budgets and ambitions are rising, leadership is re-emphasising collective defence, the primary consideration of 'cost-efficiency' in decision-making seems to have been weakened and defence cooperation and integration with partners is taken to new heights.

This case study on the Netherlands provides insights into both the changes in 'strategic' thinking about defence requirements within the Ministry of Defence, the processes and methods used to translate this thinking into actual capability procurement, as well as the way the defence organisation engages with military innovation. The chapter traces developments by scrutinising several key official strategy and policy documents, as well as ingrained institutional practices, including an analysis of the more recent adjustments taking place due to the war in Ukraine. The focus will be partly on analysing what has been done in the past, by whom and with what considerations and partly on assessing current thinking and emerging practices within the Ministry of Defence.

DOI: 10.4324/9781003398158-10

## National ambitions, alliances, dependencies

### *From neutrality to great power dependency*

Historically, the ambition of the Netherlands was to be a principled, neutral power in international conflicts. Trying to stay on the sidelines of geopolitical struggles was deemed the most effective strategy to safeguard the international commercial interests of a nation whose economy was largely based on global trade relations.[1] Declaring neutrality came with the additional advantage of claiming a moral high ground, to stand above international conflicts by advocating for the tenets of international law and urging warring parties to abide by them. Based on this double approach, Dutch foreign policy—and by extension its defence policy—has often been characterised as trying to be at the same time 'the merchant and the preacher'.[2] Or, as Joris Voorhoeve, who later became minister of Defence, summarised it: Dutch foreign policy is built around three interlinked parameters: peace, profits and principles.[3]

Despite this historical ambition, the last time the Netherlands semi-successfully declared neutrality was at the outbreak of World War I in 1914. Thereafter, the benefits the Netherlands had accrued in the past from staying aloof no longer materialised: not in terms of economic security, nor in terms of promoting the international rule of law, as all belligerents played with the legal instruments as suited them best.[4] At the start of World War II, the Netherlands tried one last time, but its declaration of neutrality was quickly and decisively overrun by the German *Blitzkrieg* on the Netherlands.

### *Great power guarantees combined with small-state collaboration*

By 1945, the Netherlands had realised it was much better off allying and aligning with a Great Power to receive a credible security guarantee, while at the same time teaming up with its fellow small-state neighbours to generate cross-border collaborations. Nowadays, the Dutch defence organisation lists seven strategic bilateral partners (Belgium, Luxembourg, Germany, France, the UK, the US and Norway). The choice for the US as the Great Power to tag along with was driven by cultural and ideological alignment, as well as by the American role in the liberation of Western Europe and the post-war reconstruction of the continent. With reference to the literature on alliances, the Dutch behaviour qualifies as one of 'balancing'.[5]

In addition to seeking a Great Power security guarantee, the Netherlands started to work more closely with neighbouring countries. Cost efficiency was an increasingly important driver for such cooperation, in particular in the years after the Cold War, when the defence organisation had to absorb large budgetary cuts. These forms of cooperation have helped a small country like the Netherlands keep a wide portfolio of platforms. For decades, the cooperation between the Netherlands, Belgium and Luxembourg was the most prominent, with the flagship of military cooperation the 'BeNeSam', the navy cooperation between the Netherlands and Belgium. This bilateral cooperation includes the integration of operational staff (Admiralty

Benelux), sharing of training and testing facilities and the joint procurement of mine hunters and frigates. The Benelux cooperation is one of the oldest and for a long time was also one of the most advanced examples of defence cooperation between sovereign states.[6]

However, in the past couple of years, the cooperation with Germany has arguably become even more important, and also the most advanced with regard to the integration of (land) forces. This bilateral cooperation has accelerated since the outbreak of the war in Ukraine. Already back in the 1990s, Dutch and German defence ministers decided on the establishment of a binational Headquarters and the integrated 1 Germany-Netherlands Corps. In the 2022 Defence White Paper, further bilateral integration efforts were mentioned and in 2023 a decision was made to start the integration process for the remaining land forces into German military structures. These efforts will reportedly result in a joint force of 50,000 troops, of which around 8,000 come from the Netherlands. The current integration plans do not include the Dutch special forces units.[7] Also in 2022, agreement was reached as part of a 'Common Army Vision' to jointly develop doctrines and jointly procure defence materiel. Plus, the German-Dutch cooperation is being institutionalised in the Regional Plans of the New Force Model of NATO.

Cooperation with France has instead long been channelled through broader European cooperation, in initiatives such as the European Intervention Initiative (EI2). But here as well, the latest geopolitical disruptions have seen the Netherlands warm up to the France-led concept of strategic autonomy and also bilaterally, there is an ambition to do more. In April 2023, the French and Dutch governments signed a declaration of intent, to increase cooperation regarding defence material and the defence industry.[8] And while Brexit has caused confusion regarding the way forward, the UK remains an important partner, with the Netherlands a dedicated participant in the UK-led Joint Expeditionary Force (JEF).

### NATO and European defence cooperation

NATO remains the most prominent post–World War II security framework for the Netherlands. Developments in the past couple of years, from Brexit and Trump to the Russian aggression in Ukraine and the hasty US withdrawal from Afghanistan, have however seen the Netherlands increasingly emphasise the importance of European cooperation—within NATO, but also to be able to act on its own without the US if needed or desired. As a result, in more recent policy documents, the emphasis has shifted from describing NATO 'as the corner stone of Dutch security policies' to positioning NATO and EU/European defence cooperation as 'equally important' or at least complementary.

As international cooperation is key for a small power such as the Netherlands, interoperability with its most important allies is an important consideration in defence planning. This partly explains the choice for air force platforms such as the F-16, F-35, Chinook, Apache and the Patriot, while the land forces value interoperability with their main partner Germany and the navy with the Belgians and the UK.

Only in the last couple of years have Europe-focused resources increased in terms of both manpower and budget. The change is taking off at the policy level but is taking time to trickle down to the more operational level. When it comes to procurement and other planning processes, NATO-related agreements, requests, processes and activities are deeply institutionalised. This includes a heavy reliance on NATO-partners' threat assessments, procurement choices and capability targets (NDPP targets). This does not necessarily mean that the ultimate procurement decisions are always well-aligned with NATO priorities, as often deviating political choices and national priorities interfere in that planning process.

### Applying the Dutch 'cheese grater'

After the demise of the Soviet Union, the Netherlands was one of the first to follow the shift in focus from collective defence to that of crisis management and peace operations. This significant change, which was in line with the thinking in the US and the UK, was written down in the 1991 Defence White Paper [Defensienota 1991] and the related 'Priority Note' [Prioriteitennota].[9] In the years that followed, ministers heading the defence department had to implement budget cuts worth billions of dollars. While in 1990 the Dutch army consisted of almost 261,000 personnel (military professionals, reserve forces and conscripts), at the moment, there are less than 42,000 military professionals and around 6,500 reserves.[10] The budget cuts were applied—to use a good Dutch term—'as a cheese grater': instead of cutting specific abilities and platforms entirely, or making a top-heavy organisation more agile, the organisation remained pretty much the same, but just with less of everything, as can be deducted from Table 10.1.

In this period, conscription was suspended, and a large share of the materiel that was considered purely relevant to collective defence was sold off, mostly against bargain prices. The 913 tanks the Netherlands owned in 1990 were stored or sold over the years, and the number of other armoured vehicles dropped from 2,889 to 752.[18] Frigates were sold to Greece, as well as to the United Arab Emirates, where an old Dutch frigate now sails around as one of the largest Superyachts in the world, owned by a member of Abu Dhabi's Royal Family.

The change in direction and the immense budget cuts affected the perceived purpose of the defence organisation. The choice to participate in an expeditionary mission to promote, protect or re-establish the international rule of law, was a lot more open-ended than agreements around collective defence had been. As a result, the political considerations in decision-making around military deployments grew in importance.[19] Since the Fall of the Berlin Wall, the Dutch defence organisation underwent a transformation from a 'must have' force (collective defence against an existential threat) to a 'nice to have' toolbox (an instrument of foreign policy that could be used for diplomatic-political purposes).

The Russian aggression in Ukraine is now partly reversing that thinking again. The years 1990–2014 could be considered a 'strategic pause' in which Western analysts and decision-makers alike tended to stress uncertainty and had to manage

*Table 10.1* Assets of the Dutch armed forces

| System | 1990[11] | 2009 | 2023[12] |
|---|---|---|---|
| Military personnel (total) | 261.000 | 54.300 | 47.830 |
| Career military personnel | 56.700 | 48.700 | 41.290 |
| Conscripts | 45.900 | 0 | 0 |
| Reservists | 158.400 | 5.600 | 6.540 |
| **Army** | | | |
| Mechanized infantry battalions | 10 \| 9 | 4 | 4 |
| Armed (fighting) vehicles | 2889 | 780 | 745[13] |
| Airmobile battalions | 0 | 3 | 3 |
| Tank battalions | 5 \| 7 | 2 | 0 |
| Tanks (pcs) | 913 | 91 | 0 |
| Artillery battalions | 13 \| 7 | 2 | 1 \| 2 |
| Artillery (pcs) | 481 | 39 | 46 |
| Rocket artillery (pcs) | 23 | 0 | 20 |
| Air defence battery | 9 \| 9 | 2 | 2 \| 3 |
| **Airforce** | | | |
| Fighters (F-16) | 211 | 87 | 0[14] |
| Fighters (F-35) | 0 | 0 | 52 |
| Air defence battery (Patriot) | 12 | 4 | 3 \| 4 |
| Strategic air transport (incl. tankers) | 0 | 3 | 10[15] |
| Tactical air transport | 0 | 4 | 4 \| 5 |
| Assault helicopter (Apache) | 0 | 29 | 28 |
| Heavy utility helicopter (Chinook) | 0 | 11 | 20 |
| Medium utility helicopter (Cougar \| Caracal) | 0 | 17 | 12 \| 14[16] |
| Maritime helicopters | 22 | 21 | 19 |
| Light utility helicopters | 92 | 7 | 0 |
| MQ-9 reaper | 0 | 0 | 4 \| 8 |
| **Navy** | | | |
| Frigates | 14 | 6 | 6 |
| Support ships | 3 | 2 | 1 \| 2 |
| Marine battalions | 2 \| 1 | 2 | 2 |
| Landing platform dock | 0 | 2 | 2 |
| Maritime patrol aircrafts | 13 | 0 | 0 |
| Maritime patrol vessels | 0 | 0 | 4 |
| Submarines | 5 | 4 | 4 |
| Mine countermeasures capability (pcs) | 22 | 10 | 5 \| 6 |

*Source:* Dutch Ministry of Defence[17]

'risks', rather than dealing with any fundamental threats.[20] With regard to the Netherlands, some experts argued that the end of the Cold War and the change in the security environment led 'to a wholesale revision of its defense and security posture', but one in which the Dutch Atlanticist outlook increasingly clashed with its desire to promote the international rule of law. This generated, as Rem Korteweg framed it, 'a pursuit of relevance to US military operations, while lambasting the US for its unilateral approach'.[21]

Overall, between 1990 and 2010 defence spending declined with around 15% in real terms (see Figure 10.1).

2.7 2.6 2.6 2.4 2.3 2 1.9 1.8 1.7 1.8 1.7 1.7 1.7 1.6 1.55 1.5 1.55 1.5 1.5 1.5 1.4 1.3 1.2 1.2 1.15 1.13 1.16 1.15 1.22 1.32 1.41 1.38 1.63 1.7

1990 1991 1992 1993 1994 1995 1996 1997 1998 1999 2000 2001 2002 2003 2004 2005 2006 2007 2008 2009 2010 2011 2012 2013 2014 2015 2016 2017 2018 2019 2020 2021 2022 2023

Years

Defence expenditure as % GDP      % Investment of defence budget      Defence expenditure x 1000

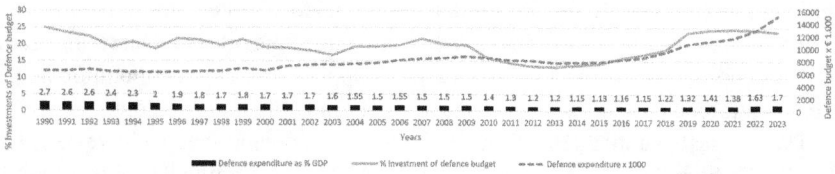

*Figure 10.1* Defence expenditure in the Netherlands.
*Source:* Dutch Ministry of Defence[22]

Despite these cuts, the Netherlands decided to deploy its troops across the globe, often in the framework of NATO or the UN or through ad hoc missions and more sparsely through the EU's CSDP (Common Security and Defence Policy) missions. Regardless of the missions, an imbalance was growing between the deployments and the efforts needed pre- and post-deployment. Even though various consecutive ministers of Defence warned that the organisation's limits had been reached, it took until 2014 to actually (slowly) reverse the trend. The trigger for change was a set of external events and pressure from international partners. First, the Russian annexation of Crimea triggered the Wales pledge at the level of NATO to grow military budgets in ten years' time to 2% of GDP. But the urgency was still not seriously felt and Dutch defence spending remained significantly below that pledge and below the European average.

Six years after the Wales pledge, in 2020, then-Minister Ank Bijleveld-Schouten decided to try to paint a dire, but truthful picture. The Defence Vision 2035, published during her term, concluded that if the defence organisation was really to live up to fulfilling its constitutional tasks in the years to come, the organisation would need an additional €13–17 billion extra per annum on top of the yearly €11 billion budget. The stated solution Bijleveld brought forward was that, to become sustainable, the defence organisation needed to make difficult choices.

But making choices had been exactly what the Dutch defence organisation had never excelled at. Already back in the early 1990s, the Priority Note mentioned earlier was widely criticised for having failed in one fundamental aspect: actually setting priorities. Similarly, in 2010, a large study was conducted on the most appropriate focus of a future defence organisation.[23] The interdepartmental study was a thorough analysis of future scenarios and strategic shocks, Dutch interests and objectives and the possible role of the armed forces. The outcome was that the best option was to keep all the armed forces' strategic functions, be it in a smaller form. The armed forces were to be a 'Swiss pocket knife', and when in 2011 the government decided to cut €1 billion from an overall budget of €8.5 billion, the cutbacks mainly led to salami tactics for the entire capability portfolio (e.g. fewer F-16s, transport aircraft and mine hunters), but also the divestment of a complete capability such as the remaining two tank battalions. One of the measures was also to reduce the number of staff officers at the Ministry of Defence and the services with a third. Ironically, one of the capacities to be removed was the Operational

Policy branch, precisely the branch that was considered a key unit in the thinking about the future of the armed forces. The cuts came at the price of decreasing readiness, as combat support and combat service support, stocks of ammunition and at the worst moments also safety measures, were gradually hollowed out.

The cuts lasted until 2018, when finally the Dutch defence budget increased. The first €1.5 billion that was released with the 2018 Defence White Paper, was mainly used to repair the existing armed forces. The real increase in the defence budget came in 2022 with the coalition agreement of the new Rutte cabinet (€3 billion extra) and another €2 billion extra by a motion or vote from a member of the House of Representatives related to the war in Ukraine. This time, the organisation could spend not just on filling holes and solving long-standing problems such as the ICT environment and real estate, but also improving the readiness and employability of the armed forces by investing in combat support and combat service support and stocks, as well as investing in cyber and specific combat units, while €500 million was allocated to improvements of the working conditions of personnel.

## Approaches, processes, methods and techniques

### *Structuring defence planning*

Over the past 15 years, the Ministry of Defence has made a number of attempts to structure the future force development process. Of these, the 2010 project mentioned earlier was the most thorough and later projects more or less built on its outcomes. The method used in the first project in 2010 can be best described as a mix of portfolio-based planning (or capability-based planning)[24] and task-based planning. In the policy options, different configurations to address multiple risks were developed. However, the focus was on being able to continue to carry out all strategic functions of the armed forces in some shape or form, with a key focus on the protection of the international legal order and promoting stability, through relatively small deployments such as in the Balkans and in Eritrea, Iraq and Afghanistan.[25]

Every new project was started with an ad hoc working group within the central organisation but with interference of the services. The absence of continuity stemmed from, among other things, the dissolution of the Operational Policy Department of the Defence Staff, as the task to provide future guidance from a military perspective was not taken over by any other department.

As described previously, defence planning was mainly financially driven. In order to retain the Swiss pocket knife as much as possible, planners were looking for cost-efficiency. The necessary cuts were found in business operations, stocks, combat support units and combat service support units. The remaining money that was available for investments was mainly spent on the mid-life update of existing capacities or their replacement (like for like). Investments in truly new and innovative capabilities were only made sparsely. The planning was mainly done by the services, who drew up individual studies for their future force without much

consideration for the armed forces jointly. While the MoD took notice of such studies, they were mostly regarded as attempts to influence future investments.

### Organisational reform and a new future force development process

A change in thinking emerged in 2020. In the government's earlier mentioned
'Defence Vision 2035', a green paper, attempts were made towards a threat-
based analysis. Perhaps not as thoroughly as its predecessor in 2010, the green
paper was a large project based on analysis by the ministry itself, with significant inputs from think tanks and seminars with Dutch industry, politicians and
officials from other ministries. While the Defence Vision was being developed, a
large organisational reform was being carried out. This led to the establishment of
a new Directorate-General for Policy, which merged all the smaller policy-focused
units that previously held responsibility for drafting the main strategic policy documents. Additionally, it made room again for a Directorate for Operational Policy
and Plans, which was rather well resourced and saw an increase in the number of
military officers in policymaking.

Within the reformed organisation, a new Defence planning process was developed and efforts were made to institutionalise a future force development process
at the joint level. The process was loosely based on the capability-based approach
used by the US, the UK, Canada, Australia and New Zealand in their Technical
Cooperation Programme.[26] In addition to reincarnating the Operational Policy
Department, the MoD posted an officer in the UK MoD's Development, Concepts
and Doctrine Centre (DCDC). The ministry also kept commissioning strategic analytical work by the two main Dutch think tanks dealing with international security,
Clingendael and the Hague Centre for Security Studies (HCSS). Plans were also
drafted to create a cyclical strategic process, with a green paper coming out every
four years to inform the regular government white papers and ensure better alignment with the budgetary review cycles in the ministry.

The future force development process that was being developed aimed to generate policy that aligns the strategic interests and goals of the Kingdom of the Netherlands ('ends'), the strategic and operational concepts ('ways') and the necessary
assets ('means'). In this process, the aforementioned green paper provides guidance
based on trend analysis and a view on how to mitigate them for the next 15 years.
Other important documents are the Strategic Concept, which is a non-political
appreciation of how the MoD should prepare for the future security environment;
the Future Operational Environment, which depicts the operational environment
in which the armed forces must operate; the Future Operational Concept, which
translates strategic ends into ways; and the Capability Plan, which finally translates
all of this into policy options for future expenditures. As this is a new process, at
the time of writing, the first Strategic Concept was being drafted.

This new process is based on scenario thinking and modelling, to identify
which capabilities are needed in the future. A gap analysis will identify the gaps
between the current and planned (or funded) force and the required capabilities of
the future force. These gaps are translated into several courses of action about how

the MoD can close the gap as efficiently or effectively as possible, given the available amount of money as a constraint. Given the challenges of rising uncertainty about the future security environment—and the limited resources to address this uncertainty—the Ministry of Defence will need to make choices.

The goal of this future force development process is to equip and prepare the armed forces for the most relevant future threats vis-à-vis the national security interests of the Netherlands and its allies. Consequently, the Dutch MoD concentrates on upholding a balanced portfolio of military capabilities, which considers the ongoing developments in technology, warfare and other trends.

It is necessary to make two comments on this process. First, force development is a continuous and cyclical process. Having learned from the previous attempts between 2010 and 2020 with its start and stop character and the planning groups as adhocracies, the MoD opted for the establishment of a strategic planning cycle and the reincarnation of the Operational Policy department to arrange a continuous and cyclical process. Secondly, while this description might suggest that the process is a rational and linear one, in reality it is not. Already at the start of its first cycle, it was foreseen that it would be full of inter-service rivalry, bureau-politics and zero-sum games. Nonetheless, due to this process, the MoD is able to discuss the right issues at the right level at the right time, instead of limiting the discussion to the number of tanks, frigates and planes.

### Implementing the new process

With this new organisational structure and defence planning process established, the MoD started the development of the new Defence White Paper of 2022 as the first litmus test for the new way of working and thinking. With a change at the ministerial level, the Defence Vision that had been drafted only two years earlier was pushed into the drawer. Reflecting on the implementation of the process so far, it can be concluded that the MoD already did not comply with the newly established approach to the future force development. Although the green paper showed a broad threat inventory (mainly to spur an increase in the budget), during the development of the white paper it proved difficult to make a thorough analysis and to have a good discussion at the strategic level about the consequences of this threat analysis for the armed forces. Discussions about the character of future conflict, the consequences for the armed forces, the necessary choices and the implications for strategic and operational concepts as well as capabilities, were difficult and mostly remained stuck at the level of capacities.

The Defence White Paper 2022 got mixed reviews. On the one hand, there was recognition of the extra budget for the MoD by the vast majority of the Dutch parliament and society (especially in light of the war in Ukraine) and also for certain investments. But there was also criticism, for example, about the lack of a good security analysis, the lack of insight into choices and goals and the lack of coherence with the goals of the EU and NATO.[27] NATO itself, through its Defence Planning Capability Review (DPCR), criticised in particular the lack of investments in the land domain. Although the Defence Vision 2035 has made being a 'reliable partner' one of three core characteristics of the future defence organisation, the NATO and EU capability objectives are still not automatically the starting points for defence planning. 'The Netherlands makes its own choices', is the MoD's explanation.[28]

While the process showed that the sense of urgency to review the MoD's capability portfolio has grown, the outcome does not entirely reflect this. Certainly, because of the Russian aggression, the rise of China and the perceived dependence on the US, there is a growing emphasis on warfighting (rather than on stabilisation operations). Also, the MoD recognises that, due to rising costs of defence equipment, priorities must be set. However, in practice, the process of force development and defence planning, decision-making dynamics and the actual composition of the portfolio have barely changed.

One ongoing feature of these dynamics is that, based on the input of the services, the MoD tries to use opportunistic considerations to choose capabilities that resonate with politics or for which one thinks one can receive money, while trying to maintain the broadest set of capabilities as possible. Initiatives to disinvest from certain capabilities in order to make room for new ones remain scarce. Most of the budget is still earmarked for mid-life updates, replacements of existing capacities, or more of the same capacities. In the Dutch armed forces, the impact of the legacy force and especially of the services is large, as they tend to do most of the conceptual thinking. Given the zero-sum nature of investments, the inter-services rivalry is significant. To cope with this rivalry, decision-making at the top level is consensus-oriented, with the ultimate decision-making powers vested in the Minister and the Assistant Minister of Defence (and ultimately the Cabinet, in which the Minister of Foreign Affairs and the Minister of Finance are highly influential as well).

With regard to the ultimate choices made in the defence planning process, three constants can be recognised. First, the Netherlands professes NATO as the cornerstone of its security architecture, but mostly in words only. In practice, the Netherlands wants to retain freedom for its own choices. In recent years, the Netherlands has increasingly been seen as a free rider on the implementation of NATO's 'three c's': cash, capabilities and commitments. Recently, the situation has improved slightly when it comes to adhering to the 2% of GDP pledge, but it still deviates substantially from NATO's requirements, particularly in the land domain.

The second constant is the strategic choice the Netherlands makes with regard to the design of the armed forces: it holds on to the Swiss pocketknife model, retaining a broad set of armed forces. As depicted in Table 10.1, the current force does not differ substantially from to the force in 2009 and even in 1990. Of course, some capabilities are cut away and some new capabilities are introduced and undeniably, the current force is more modern, but most of the capabilities existed already in the nineties. To some degree this is understandable. Strategic surveys of the past decades underline 'uncertainty' as the most constant and important conclusion. In this light, it makes sense to apply a 'broad toolbox approach' in order to maintain a heterogeneous portfolio of capabilities.[29] However, given the consensus-oriented decision-making processes, it is debatable whether this is really a deliberate choice.

In the context of inter-service rivalry, it could be argued that the Swiss pocketknife model is first and foremost the result of avoiding stark (political) choices with regard to the composition of the future force. This dynamic is haunting the organisation, arguably until the current day.

The third constant is that the defence planning process remains financially driven. Of course, finances and budgets are common denominators and important constraints for planning,[30] but in this case, it is often considered more of a driver than a constraint. Choices made in previous white papers, the implementation of budget cuts and the harmonisation of the Defence Lifecycle Plan underline this observation.[31] It seems that the Netherlands is most comfortable with making choices based on financial pressure, rather than a strategic appreciation of its security environment.

## Military innovation and adaptation

### *The Netherlands as a top-tier global innovator*

Over the past five years, the Netherlands has averaged as a top-five global innovator. Particularly with respect to human capital and research, infrastructure and market sophistication, the Netherlands is considered an attractive and competitive environment for innovation.[32] The best-performing science and technology cluster is centred around the city of Eindhoven,[33] which is well known as the hometown of Philips. Nowadays, this region is dominated by the success of ASML, which builds the world's most sophisticated and sought-after chip-manufacturing machines. Additionally, the Netherlands is renowned for its performance in agricultural science, water management and high-tech sectors, including optics and radar. At first glance, these circumstances seem very promising for the quality and quantity of military innovation in the Netherlands. In practice, however, the country plays a relatively modest role, particularly when looking at the volume of its national defence industry.

### *A modest role for military innovation*

Military innovation can be defined as 'a change in operational praxis that produces a significant increase in military effectiveness'.[34] In order to get a better understanding of the Netherlands' performance in military innovation, it is useful to refer to Michael Raska's broadly accepted framework for analysing this phenomenon. Within this framework, we would position the Netherlands in the upper region of 'Tier 2b' (Figure 10.2). For *conceptual paths* (x-axis), we argue that the Dutch armed forces are by and large situated as a smart follower of other military organisations. This neatly fits into Raska's description of '*adaptation*', since the Dutch MoD is very adept at adjusting existing military means and methods, but mostly within the confines of existing tactics, concepts and structures.[35] A good example of this is the introduction of the F-35 fighter aircraft, which operates within the US tactical and organisational structure that was developed in tandem with the

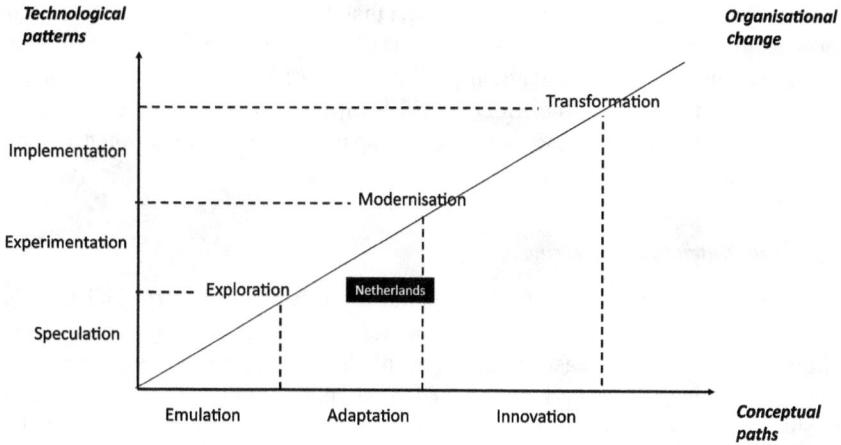

*Figure 10.2* The Netherlands' military innovation trajectories.

*Source:* Figure created by editors based on Michael Raska, Military Innovation of Small States—Creating a Reverse Asymmetry (Routledge, 2016)[37]

platform. Similarly, and for good reason, the Dutch MoD uses NATO standards and structures for most of its military means, instead of developing its own methods. As a result, while the MoD is a useful contributor to military improvements, the actual heavy lifting is mostly done by its larger peers in countries like the US, the UK, France and Germany.

Also with reference to Figure 10.2, for *technological patterns* (y-axis), it is fair to argue that the Netherlands is occasionally rising above the level of *speculation*, which can be credited to a broad range of efforts in the field of concept development and experimentation (CD&E). These activities are widespread throughout the MoD and are supported by strong knowledge and experience within Dutch national research institutes such as MARIN, TNO and NLR, as well as by Dutch companies such as Thales Nederland, Defenture and ISISPACE. Although some elements of *implementation* are visible within the MoD, they are not so mature and institutionalised to warrant a classification at this level. For instance, the Netherlands has a unique proposition with the advanced SMART-L radar built by Thales, but this technology has not led to the transformation of the navy or to a new naval doctrine. To be sure, the technology has the potential to facilitate such changes, but it would have to be accompanied by a much broader (and costlier) innovation programme. Consequently, there is a lot of room for improvement to qualify for this level of innovation, particularly on the part of leadership and the development of a more daring transformation strategy.[36] As explained in the previous pages, such a strategy is challenging to develop in the Dutch case.

When triangulating the scores for *conceptual paths* and *technological patterns*, it is possible to ascertain the magnitude of *organisational change*. As stated earlier, the Netherlands falls into Tier 2b, which represents the lower end of *modernisation*.

This assessment can be supported by the fact that the Netherlands is not a frontrunner when it comes to the modernisation of its armed forces, which is confirmed by public statements and documents, including the 2022 White Paper. Even though exceptions can be found, the MoD is still hampered by restricted resources for innovation, a relatively broad portfolio of capabilities and very limited combat (service) support capabilities.

### *Explaining Dutch underperformance*

Based on its stellar ranking in global innovation, it could be expected that the Dutch defence industry would equally be among the top-ranking military innovators. For a variety of reasons, however, this is not the case. To begin with, the European defence industry is dominated by a small number of countries, most notably Germany, France, the UK, Italy and Sweden. These countries are home to conglomerates like Airbus, BAE Systems, Rheinmetall, Leonardo and Saab, which all produce complete weapon systems, including military vehicles and aircraft. To be sure, companies such as Airbus and KNDS have their headquarters in the Netherlands (presumably for fiscal reasons), but the main R&D and production sites are not on Dutch soil. Consequently, most of the Dutch defence industry is tailored to develop and produce parts of weapon systems, in areas where it enjoys a relative advantage. Examples can be found in radar technology, aircraft components, system integration and the design and final assembly of ships. As a result, (potential) technical innovations and adaptations with regard to weapon systems used by the Dutch armed forces, are limited in scope and scale. Moreover, the potential scale of European defence projects is already dampened by a fragmented European defence industry (Europe has six times more types of major weapon systems than the US),[38] and by fierce competition from non-European companies (mostly from the US).

Secondly, the Dutch government spends a relatively modest amount of money on innovation programmes, particularly with respect to the military industry.[39] Although the Netherlands does have a standing policy for spurring innovation within its so-called top sectors, the amount of capital that is invested in defence and security is small and not very effective.[40] One of the reasons for this lack of effectiveness is that state funding is mostly allocated to fundamental research (low Technology Readiness Level [TRL]), which is hard to valorise for existing (or new) companies.[41] Additionally, Dutch companies have few incentives to co-finance innovation when the only projected customer is their own government.

As a third element, an often-heard critique is that Dutch policy for military innovation lacks focus, which makes it harder to effectively allocate the already limited resources to projects that need additional funding to succeed.[42] As a result, many initiatives remain stuck at the CD&E level, with little to no impact on enhancing military performance. A fourth argument is that the Dutch MoD is mostly concerned with replacing existing weapon systems with more advanced versions, rather than looking for more innovative ways to increase military effectiveness. It is a well-known phenomenon that the traditional armed services (maritime, land, air) want to retain their current weapon platforms and organisational structure, born from fear of losing their relevance and budget allocation.[43] Obviously, such clinging to the past is detrimental to innovation, which is not to say that all forms of conservatism are at odds with future military effectiveness. It should be noted, however, that military surprise (by an opponent) becomes more likely, when military organisations are only interested in incremental changes, while methods of warfare and potential adversaries are evolving rapidly. A fifth and final explanation is that, over the past decades, the Dutch government has not experienced much pressure to innovate and adapt in the field of defence and security, which was in line with a general lack of urgency, search for relevance and a thinking in budget cuts, which has been the case for many decades and only recently started to change.

In sum, there are multiple drivers that explain Dutch underperformance in military innovation, but an underlying tenet appears to be that a lack of urgency has prevented a change in the status quo. In theory, a major change in the threat environment could be an imperative for change. Therefore, it is interesting to explore whether the presumed lack of urgency is affected by the Ukraine War of 2022, which is discussed next.

### *A changing threat environment*

At times of rapid change, it could be expected that innovation and adaptation are regarded as more urgent than before. Obviously, such change does not only pertain to the rising threat that emanates from Russia's aggression in Ukraine. It also stems from the rise of China, uncertainty about future US military support to Europe, climate change, lingering effects of COVID-19 and the rapid pace of technological change. It is fair to argue, however, that the Ukraine War serves as the strongest argument to up the ante on military innovation.

Judging from the 2022 White Paper, it is obvious that the threat perception has increased significantly. In response, innovation is now regarded as a key theme. Given the large increase in defence spending over the coming years—approximately 40%—it should be expected that the funding for innovation will also grow substantially. Nonetheless, it remains to be seen whether the stated policy goals will translate into an actual increase in innovation funding.

Having said that, even when assuming that the resources for innovation will grow over the coming years, this does not automatically translate into more innovation. After all, the Dutch MoD still has to cope with the challenges that were described earlier: lack of scale, mostly low TRL, lack of focus, replacing old with new and a lack of urgency. With the latter being 'solved' by the Ukraine War, the other four remain. While the first two are more structural and will require more time to improve upon, the lack of focus is something that can be mitigated in the short term. Interestingly, the 2022 White Paper appears to recognise this, by describing that 'the defence organisation is working on an implementation plan to bring more focus and concrete substance to the 2021–2025 Strategic Knowledge and Innovation Agenda'.[44] While this statement seems to acknowledge the critique that the innovation programme of the MoD lacks focus, it is encouraging that this is going to be dealt with.

Interestingly, the 2022 White Paper already elaborates on where such focus can be found: new domains, labour extensive working methods and sustainability.[45] Remarkably, however, these topics are merely additional areas of innovation, on top of existing programmes. There is no mention of divestments or shifting priorities. Consequently, rather than creating focus, these extra tasks will further dilute the scarce resources that are made available for innovation efforts. Arguably, this could stifle national ambitions to improve military innovation, despite the favourable structural conditions that exist.

In the same vein, it should be relatively easy to depart from the practice of replacing old with new systems. This, however, is easier said than done, which is also reflected in the 2022 White Paper. Rather than making radical changes, the policy settles for continuing the trusted method, while making few investments in emergent areas such as (Counter) UAS (Unmanned Aircraft Systems), Electronic Warfare and Smart Munitions. In the short term, this will keep the military services content, since they can continue doing what they are used to. In the long term, however, this may result in military surprise and shortcomings, which are not easily reversible—if at all—in times of military confrontation with a more resourceful adversary.

Overall, the rapidly changing threat environment has not yet translated into a more apparent focus on and within military innovation. Even though there are positive signs that point to future improvements, such as increased military spending, it is too soon to conclude that military innovation is likely to pick up pace in the Netherlands.

## Conclusion

With reference to Figure 10.3, we can conclude that in the case of the Netherlands, its national ambitions with regard to being an international defence and security actor are nowadays predominantly shaped by cultural and historical constraints. After all, the country is endowed with a relatively large economy and in historical times, it has proven that its small population was not a serious constraint for global power projection, while its geographical location was more of an enabler than a constraint. Instead, it has been the course of history and the related cultural aspects that have made the Netherlands into the defence actor that it is and have subsequently shaped its defence planning processes, including the budgetary constraints that are often driving these processes.

The Netherlands understands that it can by no means act alone and is much better off allying and aligning with a Great Power to receive a credible security guarantee. While 'balancing' with the US has remained the main priority, the country has put efforts into teaming up with its fellow small-state neighbours to generate cross-border collaborations and increasingly, force integration. The defence organisation's strategic choices have often been driven by external events and financial pressures and especially since the end of the Cold War, defence planning has been approached through the primary lens of cost-efficiency while keeping a broad set of forces. Until very recently, threats were hardly a serious driver of

| | FINANCIAL | DEMOGRAPHIC | CULTURAL | GEOGRAPHICAL | OTHER |
|---|---|---|---|---|---|
| **SMPS ADJUST THEIR APPROACHES BASED ON SYSTEMIC CHANGE OUT OF THEIR CONTROL** | **SMPS REACT TO CHANGES IN THE RELEVANT SECURITY COMPLEX[ES] THE SMP IS PART OF** | | | | |
| *Alliances, Dependencies and National Ambitions* | | | | | |
| SMPS FACE CONSTRAINTS IN SAFEGUARDING NATIONAL SECURITY | FINANCIAL | DEMOGRAPHIC | CULTURAL | GEOGRAPHICAL | OTHER |
| SMPS NEED TO POSITION THEMSELVES VIS-À-VIS GLOBAL AND REGIONAL POWER CONSTELLATIONS | BALANCING | HEDGING | BANDWAGONING | NON-ALIGNING | OTHER |
| SMPS HAVE TO ENGAGE WITH OTHER STATES TO INCREASE THEIR DEFENCE CAPABILITIES | SECURITY GUARANTEES | TRANSACTIONAL | FORCE INTEGRATION | SPECIALISATION | OTHER |
| SMPS HAVE ARMED FORCES FOR DIFFERENT PURPOSES | INSTRUMENT OF DIPLOMACY | PRESTIGE PROJECT | REDUCE STRATEGIC RISK (DEFEND) | DOMESTIC STABILITY | OTHER |
| *Approaches, Processes, Methods and Techniques* | | | | | |
| SMPS TAKE A MORE CONSTRAINED APPROACH TO DEFENCE PLANNING | THREAT-NET ASSESMENT BASED | PORTFOLIO-BASED | TASK-BASED | MOBILISATION | OTHER |
| SMPS HAVE LIMITED RESOURCES TO DESIGN AND IMPLEMENT DEFENCE PLANNING PROCESSES | TAILORED ANALYTICAL PROCESS | OUTSOURCING STRATEGY | ACCEPT DILUTION / FAIR SHARE | EXTERNAL TEMPLATE | OTHER |
| SMP EMPLOY MULTIPLE METHODS AND TECHNIQUES | COPY GREAT POWERS | INFORMAL CONSENSUS | TAILORED METHODS | AD HOC | OTHER |
| SMPS RELY ON DIFFERENT PROCESSES TO DECIDE ON FORCE POSTURE | ANALYTICAL RESULTS | POLITICS DECIDES | MILITARY DECIDES | GREAT POWER DECIDES | OTHER |
| *Military Innovation* | | | | | |
| SMPS FACE CHALLENGES IN KEEPING PACE WITH TECHNOLOGICAL ADVANCES | BUILD DOMESTIC TECH/INDUSTRY | TECH TRANSFER AGREEMENTS | ACQUIRE READY INNOVATIONS | PARTNER WITH GREAT POWER | OTHER |
| SMPS PURSUE VARIOUS MILITARY INNOVATION PATHS AND PATTERNS | EMULATION / SPECULATION | ADAPTATION / EXPERIMENTATION | INNOVATION / IMPLEMENTATION | OTHER | OTHER |
| SMPS FIND THEMSELVES AT DIFFERENT STAGES OF ORGANISATIONAL CHANGE | EXPLORATION | MODERNISATION | TRANSFORMATION | OTHER | OTHER |

*Left axis labels: GENERAL COMMONALITIES; VARIATIONS WITHIN THESE COMMONALITIES*

*Figure 10.3* Structured focused comparison framework for the Netherlands.

*Source:* Authors

defence planning, as most of the threats remained abstract and were considered too far away or unlikely to materialise. As a result, in the past couple of decades, the deployment of the Dutch armed forces was mostly led by pragmatism and political opportunism and the portfolio-based approach made the armed forces into an instrument of diplomacy rather than a necessary entity for the (direct) protection of national interests.

During times of relative peace, defence planning efforts often remained ad hoc and changeable; coherence between ends, ways and means was not a central consideration. While consensus-building is the overarching driver in Dutch bureaucratic culture, the influence of intra-service rivalry, as well as civil-military rivalry should not be underestimated. At the same time, politicians have also been able to push their own top-down decisions in various instances, particularly in favouring national priorities over international capability requirements. The methods and techniques used for decision-making can be characterised as 'by informal consensus' and potentially 'ad hoc' as there is no clearly institutionalised process. The outcome of the various techniques applied throughout time seems to be consistently pointing towards 'keeping all options open'. Based on the inputs of the services, the MoD settles for opportunistic considerations to choose capabilities that resonate with politics, while trying to maintain the broadest set of capabilities as possible. Initiatives to disinvest from certain capabilities in order to make room for new ones, remain scarce. In combination with the budget cuts, this led to a significantly hollowed-out force, which only recently has started to see some long overdue repairs and new investments. Since the Ukraine War, urgency has risen, threat-based thinking is making a comeback and budgets are expanding. But as these developments generate different questions than before, the way the MoD is conducting defence planning has not changed much.

When it comes to incorporating military innovation as part of the defence planning process, the Dutch innovation profile would suggest it could play a prominent and agenda-setting role. In practice, however, the Netherlands is a relatively modest player in this area. The Dutch MoD tends to cope with a set of challenges, which withholds it from claiming a more active role: lack of scale, research on low TRL, a lack of focus and prioritisation, hesitance to replace old ways with new ways and a lack of urgency. As it is facing challenges with keeping pace with technological advances, it tends to partner with the US and other larger innovators, as well as at times acquire ready innovations. While here as well, the Ukraine War has raised the level of urgency and the most recent White Paper actually notes innovation as a core theme, so far, this has not (yet) translated into more apparent military innovation. Even though there are positive signs that point to future improvements, such as increased military spending, it is too soon to conclude that military innovation is likely to pick up pace in the Netherlands.

## Notes

1 David Ormrod, *The rise of commercial empires: England and the Netherlands in the age of mercantilism, 1650–1770* (Cambridge: Cambridge University Press, 2003).
2 Charles Wilson, *Profit and power: A study of England and the Dutch Wars* (The Hague: Springer, 1978); Koen Stapelbroek, 'The Dutch Debate on Commercial Neutrality (1713–1830)', in: idem (ed.), Trade and War: The Neutrality of Commerce in the Inter-State System (Helsinki 2011) 114–142: 114–142; Marc Frey, 'Trade, ships, and the neutrality of the Netherlands in the First World War', *The International History Review* 19, no. 3 (1997): 541–562.
3 Joris J.C. Voorhoeve, *Peace, profits and principles: A study of Dutch Foreign Policy* (The Hague: Springer, 1979).
4 Maartje Abbenhuis, *The art of staying neutral: the Netherlands in the First World War, 1914–1918* (Amsterdam: Amsterdam University Press, 2006).
5 Tim Sweijs et al., 'Why Are Pivot States So Pivotal? The Role of Pivot States in Regional and Global Security".
6 Margriet Drent, Dick Zandee and Lo Casteleijn, *Defence Cooperation in Clusters. Identifying the Next Steps* (The Hague: Clingendael, October 2014).
7 Cagan Koc, 'Dutch Army to Merge Land Combat Units with Germany This Year', *Bloomberg*, 1 February 2023, www.bloomberg.com/news/articles/2023-02-01/dutch-army-to-merge-land-combat-units-with-germany-this-year?sref=q1A0pZuo.
8 'Joint Declaration, Netherlands-France', *Government of the Netherlands*, 12 April 2023, www.government.nl/documents/publications/2023/04/12/franco-dutch-pact-for-innovation-and-sustainable-growth#:~:text=On%2012%20April%202023%20the,food%20transition%20and%20sustainable%20mobility.
9 A.L. ter Beek, H. van den Broek and B.-J.M. baron van Voorst tot Voorst, *Defensienota 1991*, https://repository.overheid.nl/frbr/sgd/19901991/0000033229/1/pdf/SGD_19901991_0006658.pdf; *Prioriteitennota*, 1993, https://repository.overheid.nl/frbr/sgd/19921993/0000291535/1/pdf/SGD_19921993_0007821.pdf.
10 'Aantal personeel', *Ministerie van Defensie*, 1 January 2023, www.defensie.nl/onderwerpen/overdefensie/het-verhaal-van-defensie/aantallen-personeel.
11 The first number is the number of operational units or pieces and the second number is the number of units or pieces in depot.

12  In the numbers below are the measures of the 2022 White Paper included. The first number is the actual number (current force), the second number is the planned number (funded force).

13  CV-90, Boxer, Bushmaster and Fennek

14  The Netherlands are in the midst of replacing their F-16's into F-35 fighters. The replacement programme's completion is planned by the end of 2024.

15  The Multinational Multi Role Tanker Transport (MRTT) Unit is a cooperation between Belgium, Germany, Luxemburg, the Netherlands, Norway and the Czech Republic and consists of ten A-330 Airbus planes.

16  In 2023 the Netherlands Ministry of Defence announced the procurement of the Caracal helicopter to replace the Cougar helicopter.

17  The numbers in Table 10.1 are derived from Ministry of Defence, *Eindrapport Verkenningen: Houvast voor de krijgsmacht van de toekomst* (Ministerie van Defensie, 29 March 2010), https://zoek.officielebekendmakingen.nl/blg-68805.pdf and www.defensie.nl/.

18  See figures published at the Veteranen Institute: 'Materieel', *nederlands veteranen instituut,* 19 July 2021, www.nlveteraneninstituut.nl/nieuws/checkpoint-5-2021-materieel-facts-figures/#:~:text=18,geleased%20van%20de%20Duitse%20krijgsmacht.

19  J.J.W.C. van Dinther, 'Parlementaire instemming bij de inzet van de krijgsmacht', *Tijdschrift voor Constitutioneel Recht* 2, no. 1 (2011): 4. And see Willem Heringa and Melissa Wevers, 'Militaire missies en de grondwet', *Montesquieu Instituut,* www.montesquieu-instituut.nl/id/vjoymy9105n4/militaire_missies_en_de_grondwet, last accessed 3 August 2023.

20  Frans Osinga, 'Netherlands defence and security policy: Coping with the "new normal"?', in *Security in Northern Europe: Deterrence, Defence and Dialogue,* ed. John Andreas Olsen (Abingdon: Routledge, 2019), 75–87.

21  A.R. Korteweg, 'The superpower, the bridge-builder and the hesitant ally: how defense transformation divided NATO (1991–2008)', (PhD diss., Leiden University, 2011).

22  Ministry of Defence, *Eindrapport Verkenningen: Houvast voor de krijgsmacht van de toekomst* (Ministerie van Defensie, 29 March 2010), https://zoek.officielebekendmakingen.nl/blg-68805.pdf.

23  Ministry of Defence, *Eindrapport Verkenning*.

24  S. Fruhling, *Defence planning and uncertainty. Preparing for the Next Asia-Pacific War* (Abingdon: Routledge, 2014).

25  So-called wars of choice, see, for instance, in Richard N. Haass, *War of Necessity, War of Choice. A Memoir of Two Iraq Wars* (New York: Simon & Schuster, 2009).

26  Within the framework of the Technical Cooperation Program (TTCP), the US, UK, Australia, Canada and New Zealand developed a capability based approach to defence planning. See, for instance, the TTCP Technical Report by Ben Taylor, *Analysis Support to Strategic Planning* (The Technical Cooperation Program, June 2013), https://cradpdf.drdc-rddc.gc.ca/PDFS/unc194/p801995_A1b.pdf.

27  See, for instance, Arnout Brouwers, 'Kamer is uiterst kritisch over Defensienota, maar steunt wel investeringen in krijgsmacht', *Volkskrant* (Dutch newspaper), 14 September 2022.

28  Spokesman of the Ministry of defence in RTL-4 news of 15 November 2022. 'Minister Ollongren deelde kritiek van NAVO niet met de Kamer', *RTLnieuws,* 15 November 2022, www.rtlnieuws.nl/nieuws/nederland/artikel/5346877/minister-ollongren-deelde-kritiek-van-navo-niet-met-de-kamer.

29  Osinga, 'Netherlands Defence and Security Policy'.

30  Colin Gray, *Strategy and Defence Planning. Meeting the Challenge of Uncertainty* (Oxford: Oxford University Press, 2016).

31  The Defence Lifecycle Plan is the overview of the allocation of budgets to planned investments of the Defence organisation for the next 15 years. It consists of planned

investments in large materiel such as armoured vehicles, frigates, real estate and IT systems.

32 'GII at a glance 2022', in *Global Innovation Index 2022: What Is the Future of Innovation-Driven Growth?*, eds. Soumitra Dutta, Bruno Lanvin, Lorena Rivera León and Sacha Wunsch-Vincent (Geneva: WIPO, 2022), www.wipo.int/edocs/pubdocs/en/wipo-pub-2000-2022-section1-en-gii-2022-at-a-glance-global-innovation-index-2022–15th-edition.pdf.

33 Dutta, Lanvin, Rivera León and Wunsch-Vincent, eds., *Global Innovation Index 2022*.

34 Adam Grissom, 'The future of military innovation studies', *Journal of Strategic Studies* 29, no. 5 (2006): 907.

35 Frans Osinga and Rob de Wijk, 'Innovating on a shrinking playing field', in *A Transformation Gap? American Innovations and European Military Change*, eds. Terry Terriff, Frans Osinga and Theo Farrell (Stanford, CA: Stanford University Press, 2010), 108–143.

36 H.G. Geveke, 'Klaar zijn voor hyperwar', *Militaire Spectator* 188, no. 12 (2019): 580–595.

37 Michael Raska, 'A structured-phased evolution: The third generation force transformation of the Singapore armed forces', in *Military Innovation in Small States: Creating a Reverse Asymmetry*, ed. M. Raska (New York: Routledge, 2016), 130–162.

38 Munich Security Conference Foundation, 'Post-truth, post-west, post-order?', *Munich Security Conference Foundation*, 8 February 2017, https://espas.secure.europarl.europa.eu/orbis/document/munich-security-report-2017-post-truth-post-west-post-order.

39 'Kamerbrief over Missiegedreven Topsectoren- en Innovatiebeleid', *Rijksoverheid*, 15 October 2021, www.rijksoverheid.nl/documenten/kamerstukken/2021/10/15/kamerbrief-over-missiegedreven-topsectoren-en-innovatiebeleid.

40 'Kamerbrief over Missiegedreven Topsectoren'.

41 Ibid.

42 H.G. Geveke, 'Technologische Revoluties en Defensie', *Militaire Spectator* 185, no. 7/8 (2016): 288–300.

43 Osinga and de Wijk, 'Innovating on a shrinking playing field'.

44 Ministry of Defence, *A stronger Netherlands, a safer Europe: Investing in a robust NATO and EU (Summary), Defence White Paper 2022* (2022 June): 9, file://vuw/Personal$/Homes/25/s2514133/Downloads/WEB_Engels_Defensienota+samenvatting.pdf.

45 Ministry of Defence, *A stronger Netherlands*, 36.

## Bibliography

### *Books and articles*

Abbenhuis, M., The Art of Staying Neutral: The Netherlands in the First World War, 1914–1918 (Amsterdam: Amsterdam University Press, 2006).

Drent, M., D. Zandee and L. Casteleijn, Defence Cooperation in Clusters. Identifying the Next Steps (The Hague: Clingendael, October 2014).

Dinther, J. van, 'Parlementaire instemming bij de inzet van de krijgsmacht', Tijdschrift voor Constitutioneel Recht 2, no. 1 (2011).

Frey, M., 'Trade, ships, and the neutrality of the Netherlands in the First World War', The International History Review 19, no. 3 (1997): 541–562.

Fruhling, S., Defence Planning and Uncertainty. Preparing for the Next Asia-Pacific War (Abingdon: Routledge, 2014).

Geveke, H., 'Technologische Revoluties en Defensie', Militaire Spectator 185, no. 7/8 (2016): 288–300.

———, 'Klaar zijn voor hyperwar', Militaire Spectator 188, no. 12 (2019): 580–595.

Gray, C., Strategy and Defence Planning. Meeting the Challenge of Uncertainty (Oxford: Oxford University Press, 2016).

Grissom, A., 'The future of military innovation studies', Journal of Strategic Studies 29, no. 5 (2006): 907.

Haass, R., War of Necessity, War of Choice. A Memoir of Two Iraq Wars (New York: Simon & Schuster, 2009).

Korteweg, A., 'The superpower, the bridge-builder and the hesitant ally: how defense transformation divided NATO (1991–2008)', (PhD diss., Leiden University, 2011).

Ormrod, D., The Rise of Commercial Empires: England and the Netherlands in the Age of Mercantilism, 1650–1770 (Cambridge: Cambridge University Press, 2003).

Osinga, F., 'Netherlands defence and security policy: Coping with the "new normal"?', in Security in Northern Europe: Deterrence, Defence and Dialogue, ed. John Andreas Olsen (Abingdon: Routledge, 2019), 75–87.

——— and R. de Wijk, 'Innovating on a shrinking playing field', in A Transformation Gap? American Innovations and European Military Change, eds. Terry Terriff, Frans Osinga and Theo Farrell (Stanford: Stanford University Press 2010), 108–143.

Stapelbroek, K. 'The Dutch debate on commercial neutrality (1713–1830)', in: idem (ed.), Trade and War: The Neutrality of Commerce in the Inter-State System (Helsinki: Helsinki Collegium for Advanced Studies, 2011), 114–142.

Voorhoeve, J., Peace, Profits and Principles: A Study of Dutch Foreign Policy (The Hague: Springer, 1979).

Wilson, C., Profit and Power: A Study of England and the Dutch Wars (The Hague: Springer Netherlands, 1978).

## Websites

Beek, A. ter, H. van den Broek and B. baron van Voorst tot Voorst, Defensienota 1991, at: https://repository.overheid.nl/frbr/sgd/19901991/0000033229/1/pdf/SGD_19901991_0006658.pdf.

Budget MoD 1990, at www.orbat85.nl/documnents/HTK/1989-1990-21300-X-48.pdf.

Budget MoD 2000, at https://zoek.officielebekendmakingen.nl/kst-26800-X-1.pdf and Budget MoD 2010.

Budget MoD 2010, at www.parlementairemonitor.nl/9353000/1/j4nvgs5kjg27kof_j9vvij5epmj1ey0/vi8i7srxgyx9/f=/kst800012a.pdf.

Defensie Website, www.defensie.nl/onderwerpen/overdefensie/het-verhaal-van-defensie/aantallen-personeel.

GII at a glance 2022, in Global Innovation Index 2022: What is the future of innovation-driven growth?, eds. Soumitra Dutta, Bruno Lanvin, Lorena Rivera León and Sacha Wunsch-Vincent (Geneva: WIPO, 2022), at: www.wipo.int/edocs/pubdocs/en/wipo-pub-2000-2022-section1-en-gii-2022-at-a-glance-global-innovation-index-2022–15th-edition.pdf.

Heringa, W. and M. Wevers, 'Militaire missies en de grondwet', Montesquieu Instituut, at www.montesquieu-instituut.nl/id/vjoymy9105n4/militaire_missies_en_de_grondwet.

Joint Declaration, Netherlands-France, Government of the Netherlands, 12 April 2023, at: www.government.nl/documents/publications/2023/04/12/franco-dutch-pact-for-innovation-and-sustainable-growth#:~:text=On%2012%20April%202023%20the,food%20transition%20and%20sustainable%20mobility.

Kamerbrief over Missiegedreven Topsectoren- en Innovatiebeleid, Rijksoverheid, 15 October 2021, www.rijksoverheid.nl/documenten/kamerstukken/2021/10/15/kamerbrief-over-missiegedreven-topsectoren-en-innovatiebeleid.

Koc, C., 'Dutch Army to Merge Land Combat Units With Germany This Year', Bloomberg, 1 February 2023, www.bloomberg.com/news/articles/2023-02-01/dutch-army-to-merge-land-combat-units-with-germany-this-year?sref=q1A0pZuo.

Ministry of Defence, A stronger Netherlands, a safer Europe: Investing in a robust NATO and EU (Summary), Defence White Paper 2022 (2022 June), https://english.defensie.nl/downloads/publications/2022/07/19/defence-white-paper-2022.

Ministry of Defense, Eindrapport Verkenningen: Houvast voor de krijgsmacht van de toekomst (Ministerie van Defensie, 29 March 2010), at: https://zoek.officielebekendmakingen.nl/blg-68805.pdf.

Munich Security Conference Foundation, 'Post-Truth, Post-West, Post-Order?', *Munich Security Conference Foundation*, 8 February 2017, at https://espas.secure.europarl.europa.eu/orbis/document/munich-security-report-2017-post-truth-post-west-post-order.

Prioriteitennota, 1993, at: https://repository.overheid.nl/frbr/sgd/19921993/0000291535/1/pdf/SGD_19921993_0007821.pdf.

Taylor, B., Analysis Support to Strategic Planning (The Technical Cooperation Program, June 2013), at https://cradpdf.drdc-rddc.gc.ca/PDFS/unc194/p801995_A1b.pdf.

Veteranen Institute: 'Materieel', nederlands veteranen instituut, 19 July 2021, www.nlveteraneninstituut.nl/nieuws/checkpoint-5-2021-materieel-facts-figures/#:~:text=18,geleased%20van%20de%20Duitse%20krijgsmacht.

**Media**

Brouwers, A., 'Kamer is uiterst kritisch over Defensienota, maar steunt wel investeringen in krijgsmacht', *Volkskrant (Dutch Newspaper)*, 14 September 2022, at https://www.volkskrant.nl/nieuws-achtergrond/kamer-is-uiterst-kritisch-over-defensienota-maar-steunt-wel-investeringen-in-krijgsmacht~bdcb9c59/?referrer=https://www.google.com/.

'Minister Ollongren deelde kritiek van NAVO niet met de Kamer', *RTLnieuws*, 15 November 2022, at www.rtlnieuws.nl/nieuws/nederland/artikel/5346877/minister-ollongren-deelde-kritiek-van-navo-niet-met-de-kamer.

# 11 Becoming a good ally

## Slovak defence planning since independence

*Michal Onderco*

## Introduction

Central European countries all underwent a major transformation of their defence policy and defence planning since the end of the Cold War. At the end of the Cold War, these countries were in somewhat similar situations—their militaries were structured similarly (based on top-heavy conscript armies), their military plans were obsolete (no one was planning the attack on Western Europe anymore), they shared identical military equipment and their former patron was disappearing fast from geopolitical reality.[1]

Slovakia was in a situation which was in some ways even more dire than other post-communist countries. It did share the predicament of a small state—the inability to shape the systemic forces that influence it.[2] It also very early on developed an interest in integration into what the local experts call 'Western institutions', chiefly the European Union and NATO.[3] The desire to become a member of these institutions translated into a series of reforms which determined the shape of defence policy and force posture in these countries up until the present day.

However, the transformation of the defence policy in these countries goes beyond achieving NATO membership. Basics of security and defence policymaking needed to be established. Armed forces needed to be 'depoliticized' where the link between the Communist Party and the armed forces had to be cut after the fall of communism. In 1992, the Czechoslovak Federal Assembly passed a law which banned membership of political parties for soldiers.[4] The domestic military industry, which during the Cold War was well connected to the Soviet military-industrial complex, was decimated by the decline in interest in Soviet-era goods, and by the time Slovakia gained independence in 1993, it had basically disintegrated.[5]

In this chapter, I will track the development of defence planning in Slovakia since its independence in 1993, with a focus on three aspects—the choice of allies and how it affects national military ambitions; the approach to military innovation and adaptation; and the method for doing so. I will demonstrate that up until 2016 (and arguably until 2021), Slovakia was enjoying the geopolitical honeymoon, and its security policy was mainly focused on joining NATO and demonstrating its value as an ally. Only more recently has Slovakia restarted taking security more seriously and started to invest accordingly.

DOI: 10.4324/9781003398158-11

## Alliances and ambitions

### Choice of allies

After the collapse of communism in Central Europe, Czechoslovakia was freed from its bond with the Soviets and could develop its own, independent foreign policy. In these early days, Czechoslovak dissidents who were elevated to the corridors of power developed different ideas about a European security architecture *without* blocks.[6] However, faced with the clear message from Washington that NATO would remain a centrepiece of the European security architecture,[7] and increasingly aware that other Central and Eastern European (CEE) countries were eyeing closer security cooperation with the United States (and possibly NATO membership), Czechoslovak officials also quickly recalibrated. Although Slovakia became independent only in 1993, its Ministry of Foreign Relations has been functional since 1990. As a long-time Slovak diplomat Miroslav Mojžita writes, Slovakia's foreign policy was at that time indistinguishable from the foreign policy of 'any other country in Europe'.[8] Slovakia at that time saw the strong involvement of the United States in European security as not only positive but also as desirable. Entering NATO had become a key task in Slovakia's foreign policy, and the country's security policy started to reflect this goal too.

Interestingly enough, the desire to integrate with NATO was not driven by threat perception at the time. The fear of any territorial attack on Slovakia was at the time not salient, and Slovakia did not consider that it needed NATO protection to cover its direct security gaps. By contrast, the desire to join NATO stemmed from the ambition to belong to the West. It is important to realise that Slovakia was at the time in a rather peculiar situation. Russia had recently withdrawn its troops from Slovakia and the rest of Central Europe. It was weak and mired primarily in domestic problems. Wars in the Balkans were ongoing, but not seen as a direct security risk. Therefore, Slovakia was enjoying a geopolitical honeymoon. Its desire to join NATO is to be seen as part of the goal of rejoining the Western institutions rather than a reaction to any concrete threat.

Slovakia's early independence years were marked by the 'struggle against Mečiar'.[9] Vladimír Mečiar became Slovakia's prime minister after independence and ruled the country in an increasingly autocratic fashion between December 1994 and October 1998. Under his leadership, the country came to be seen as the 'black hole of Europe' (using the [in]famous words of US Secretary of State Madeleine Albright). As Slovakia's constitution makes the President the Commander-in-Chief of the Armed Forces, Mečiar's battles with President Kováč also translated into battles about the authority over defence.[10] The struggles culminated in a referendum in 1998, which was supposed to include a question on entry into NATO that was however removed by the Minister of Interior Gustáv Krajči. Krajči's step led to the invalidation of the whole referendum, and the whole experience was seen as the lowest point of Mečiar's rule.

Mečiar was voted out of office in 1998, and the incoming Dzurinda government doubled down on the efforts to enter both the European Union (EU) and NATO. After a series of reforms, Slovakia entered NATO in 2004. Although the desire to enter NATO was contested domestically under Mečiar, it became broadly shared afterwards and by the time of the accession, there was no serious political force questioning Slovakia's accession.

Already the 1994 Defence Doctrine, the first strategic document adopted after Slovakia's independence, was tailored toward the country's participation in NATO's Partnership for Peace (PfP) programme and underscores the goal of the integration of the country with the West. This document is extremely short—it is eight double-spaced pages long, and opens with a statement which captures the spirit of the times very well: 'The Slovak Republic does not consider any state to be its enemy and does not feel threatened by any country'.[11] At the same time, the document argues that 'the historical experience shows that unequivocal security guarantees are necessary for the stability of the Central European area'.[12] Therefore, it argues for becoming a member of the North Atlantic alliance and participation in the PfP as a tool thereof.

The 2005 Security Strategy clearly spells out that international institutions of which Slovakia is a member are 'tools for realisation of security policy', specifically citing NATO.[13] The same strategy spells out that '[t]he relations between the Slovak Republic and the United States of America will have a special position for promoting the security interests of the Slovak Republic. The USA is a strategic ally of the Slovak Republic'.[14] The Defence Strategy, published in the same year, doubles down on this commitment, arguing that 'the basic objective of its defence policy will be pursued by the Slovak Republic from the position of Euro-Atlantic orientation. [Slovakia] considers membership of NATO and the EU as a decisive guarantee of its security and defence capability. [Slovakia] shares the basic objectives and tasks arising from the NATO Strategic Concept and the European Security Strategy'.[15]

The 2005 Security Strategy reflects the security situation in the post-9/11 world. It discusses terrorism and non-state actors as the major sources of instability. There is no discussion about any direct security or military threats to Slovakia. It lists 'international terrorism' and the risk of WMD use by non-state actors as the biggest threats to the country. Curiously, the section devoted to Russia is very short and argues that Slovakia 'will build relations with Russia based on mutual benefits of economic cooperation'.[16] The Defence Strategy, adopted in the same year, reached a similar conclusion: 'a political-military conclusion can be made that the Slovak Republic is not threatened by an immediate large-scale conventional military conflict in the long term and the severity and scope of other threats of a military nature are decreasing. However, the probability and danger of non-military threats is increasing, especially international terrorism'.[17]

The fundamental role of NATO and the EU is also underlined in the 2016 White Paper on Defence.[18] This White Paper is also the first one to note that the security environment in Europe shifted fundamentally in the aftermath of the Russian invasion of Ukraine.[19] This document demonstrates a much stronger concern about Russia and its actions in the region, while continuing to be concerned about the instability emanating from the Middle East. Curiously, the document does not mention China. The document also mentions the concerns about 'hybrid warfare' from Russia.

Active participation in NATO activities also reappears in the 2017 Proposal for a long-term defence development plan.[20] The 2021 Security Strategy reiterates this point, arguing that '[the USA] is a strategic transatlantic ally of the Slovak Republic in NATO, with whom we share common values, and with whom we are bound by a strong historical alliance and commitment to collective defence'.[21] The same document also links NATO membership clearly to the security situation of Slovakia, arguing that 'the risk of a direct threat to the Slovak Republic from an armed attack is low thanks to its membership in NATO and the EU'.[22]

However, these documents paint a very different picture of the country's security situation. Of course, the concerns about Russia are strongly present. The 2021 Security Strategy also adds the continuing high risk of international terrorism and instability originating in the Middle East and the Sahel region. China is labelled as a 'systemic rival of the European Union', but is not linked to security or military threats.[23] Both hybrid warfare and disinformation are mentioned as key threats to national security.

The desire to strengthen relations with the United States is a controversial policy in Slovakia. The support for NATO membership increased since 2018 but is among the lowest in Central and Eastern Europe.[24] Slovakia, together with Bulgaria, is one of the most anti-American countries in NATO.[25] In 2021, Slovakia was the country in the CEE with the lowest share of population with a positive view of the United States; in 2022 it was the second last.[26] It is therefore no surprise that when the government moved to sign and ratify a bilateral defence cooperation agreement with the United States in 2021, it led to massive societal polarisation. Nevertheless, the agreement was ratified in early 2022, amid protests and after a very contentious parliamentary vote.

The 2022 Russian invasion of Ukraine fundamentally transformed the perception of security in Slovakia. There is a universally shared perception among the Slovak elites that Russia poses a fundamental threat to Slovakia's security. This translated into two behaviours—first, strong support for Ukraine's defence; and second, speeding up the country's own defence modernisation plans. It has become commonplace for Slovak officials to say that if Ukraine does not succeed in its defence, Slovakia could be next on Russia's list; as well as that Russia 'must lose' in Ukraine.[27] At the same time, the perception of Russia as a fundamental threat to Slovakia's security is not broadly shared among the public. The government's aid to Ukraine is subject to fierce contestation at home and in the parliament. The future of Slovakia's support for Ukraine is one of the key elements in the upcoming elections in September 2023.

To conclude, Slovakia's desire to join NATO and the desire to enter into an alliance with the United States is a fairly straightforward demonstration of bandwagoning behaviour by small states.[28] Given the limited resources (material and otherwise) and given the strategic culture which stimulated distrust towards European powers, commitment to the entry into NATO and strengthening the alliance with the United States remains the only possible logical choice in the eyes of Slovakia's defence policy establishment. In more recent years, this perception has been rationalised by references to America's unique military capabilities and the relative military weakness of Western European countries vis-à-vis the United States. Altogether, this makes Slovak officials and intellectuals much less interested in entertaining ideas about developing such capabilities without the relationship with the United States. With the exception of the parliamentary opposition, NATO membership and the alliance with the United States are unquestioned and unquestionable in Slovakia in 2023.

## Innovation and adaptation

### *Innovating the hardware*

The process of integrating Slovakia into NATO has also meant that Slovakia has needed to transform its armed forces from the Cold War communist army into a smaller, leaner force. This required adaptation in multiple ways. Slovakia inherited a military which was focused on fighting a conventional war in Europe, long considered an unlikely scenario in Central and Eastern Europe. It consisted primarily of Soviet legacy mechanised infantry, with an oversized number of personnel heavily reliant on conscripts. Slovakia drastically decreased the size of its armed forces. Between 1993 and 2021, Slovakia decreased the number of its uniformed personnel from approximately 47,000 to a little less than 16,000.[29] At the end of 2004, conscription was abolished. The number of tanks was decreased by 97 per cent, and the number of fighter jets by 92 per cent.[30] As reported in Figure 11.1, defence expenditure was in steady decline until 2014 and did not increase significantly until

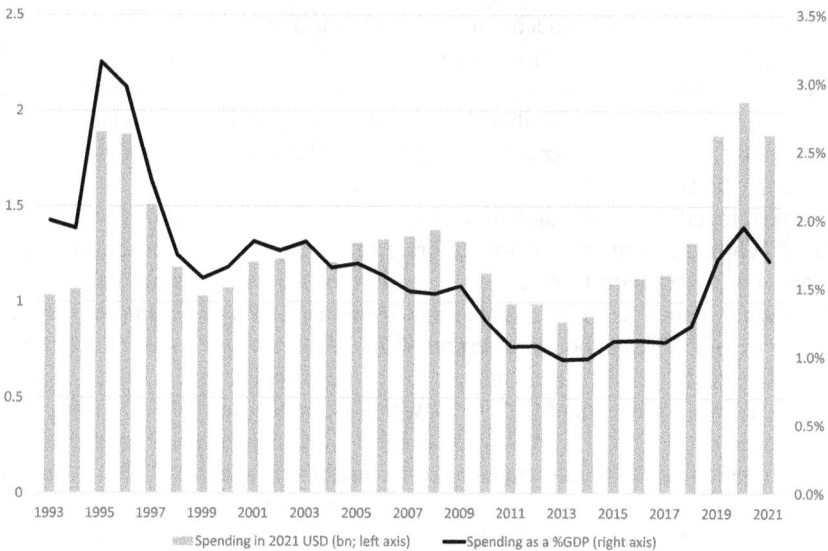

*Figure 11.1* Defence spending in Slovakia since 1993.

*Source:* SIPRI[31]

2019 (which was largely driven by the decision to procure F-16 fighter jets). Following the 2008 global financial crisis, even larger defence expenditure cuts were introduced.

The existing threat perceptions, discussed in the previous section, also influenced how Slovakia structured its armed forces. While nominally the protection of the territory was listed as a primary task of armed forces in all strategic documents, in practice the armed forces faced significant shortfalls. To compensate for these shortfalls, Slovakia intensified its participation in multinational forces abroad. It has contributed troops to UN peacekeeping operations (although on a fairly moderate scale), to the multinational forces in Iraq, the NATO mission in Afghanistan, enhanced forward presence (EFP) in Lithuania, as well as to a number of EU-deployed missions. Slovakia often deployed forces to missions where no high-intensity engagement was expected. In settings where high-intensity warfare was foreseeable (such as Iraq or Afghanistan), Slovakia committed troops to relatively low-risk tasks such as base protection. This was a reflection of the risk-averse attitudes of the political leaders rather than an expression of the armed forces' preferences.

A major step towards regional coordination was the desire to establish a Visegrad battlegroup, in cooperation with Poland, Czechia and Hungary. Although the ambition stems from early 2011, it has been mired in difficulties for a long time. It has been set up twice thus far, in 2016 and 2019, on both occasions under Polish leadership for a period of half a year. In recent years, successive Visegrad presidencies (including Slovakia in 2022–2023) reiterated their commitment to committing a Visegrad battlegroup to EU operations.[32]

In the aftermath of the Russian invasion of Ukraine, Slovakia became a host nation for the NATO Enhanced Forward Presence Battalion.[33] The deployment of the NATO troops to Slovakia was seen as a reaction to the shortage of Slovakia's existing equipment, especially when it comes to air defence. At the same time, Slovakia signed an agreement with Visegrad countries on air policing, which allowed the country to bridge the period between retiring the old Mig-29 jets and delivery of the new F-16-II jets.[34]

Taken together, this meant that over time, much of the equipment became outdated. While the ministry and armed forces often touted ambitious modernisation goals, these were hard to square with the limited (and decreasing) budgets. The 2013 White Paper on Defence stated that 70 per cent of land equipment was past its service life.[35] Throughout a large part of the 2010s, a major domestic debate was whether to continue the supersonic air force. Slovakia was, until 2021, reliant on the ageing fleet of Russian Mig-29 jets, which ultimately resulted in a fleet of four functional jets (in 2016, the number was still 12, but due to sanctions and deteriorating relations with Russia, the number declined over time). In 2018, the government of Peter Pellegrini decided to invest in procuring 14 F-16-II fighter jets, to be delivered in 2024. While the decision to purchase the F-16-II fighter jets was lauded as a bold investment and a strategic step in developing the alliance ties with the United States, observers remarked that it was eating up scarce resources while Slovakia fell short on other investment goals.[36] Acquisition of F-16 jets tripled

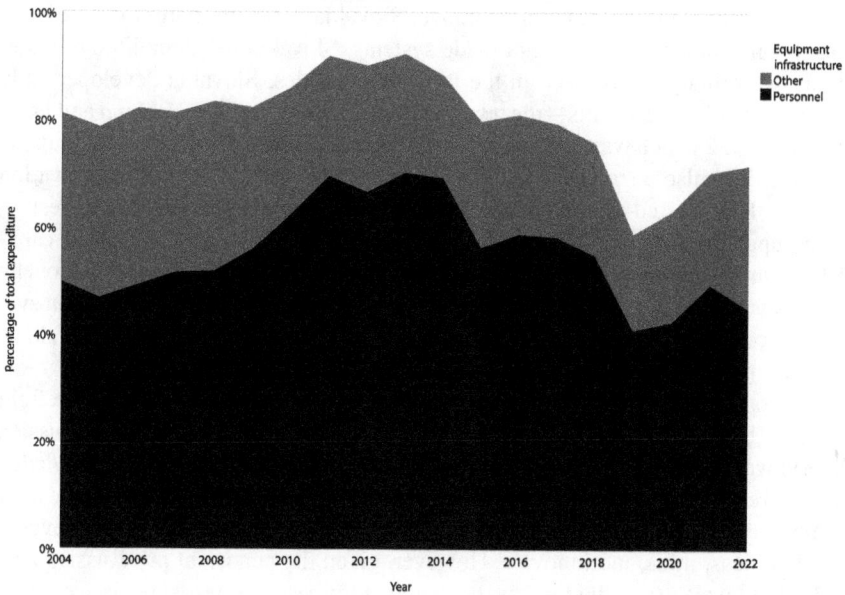

*Figure 11.2* Structure of defence expenditure in Slovakia.

*Source:* Author, based on NATO data[39]

the past spending on equipment. Slovakia continues to overspend, compared to its peers, on personnel and to underspend on equipment and operations, as can be seen in Figure 11.2. The NATO Defence Planning Capability Review, released to the public for the first time in October 2022, has criticised Slovakia for not being on track to develop a heavy infantry brigade, a goal which the allied defence ministers agreed to in 2021. This report also criticised Slovakia's land capabilities for being 'obsolete with substantive qualitative shortfalls in its combat . . . elements'.[37] The report criticised the readiness of Slovak armed forces in almost all areas, including those which make up an important part of Slovakia's contribution to NATO, such as chemical, biological, radiological and nuclear defence (CBRN).

Russia's invasion of Ukraine turbocharged Slovakia's military modernisation. In April 2022, Slovakia announced the purchase of Patria armoured modular vehicles from Finland for 1.2 billion euros. In June 2022, Slovakia announced the decision to procure 152 Swedish-produced armoured combat vehicles for 1.6 billion euros. It has also announced the intention to procure 15 Leopard-II tanks from Germany. In March 2023, the US announced that it authorised the sale of 12 combat helicopters to Slovakia.[38] At the same time, Slovakia transferred large parts of its ageing military equipment to Ukraine, including the S-300 air defence system and remaining Mig-29 fighter jets. Overall, the war meant a reversal of long-term trends in Slovakia and a step into a completely new era when it comes to military procurement.

As is clear from the description above, Slovakia's transformation is based primarily on acquisitions of foreign-made systems. Slovakia has done little to foster domestic military innovation. In the last three decades, Slovakia developed only two major military systems—the teleoperated demining machine *Božena* and howitzer *Zuzana 2* which was for the first time deployed in the EFP mission in Lithuania and has since also been sold for battle use in Ukraine in 2022. Military innovation has mainly consisted of purchasing major technological systems from the Western allies, upgrading the existing armed forces but not developing any new, niche capability. One example of military innovation in cooperation with Western European firms was the development of the MOKYS military encrypted communication system, in cooperation with BAE. Very little investment was reserved for Slovakia's own military producers.

The only area where Slovakia was relatively in the avant-garde was the fight against 'hybrid warfare'. Given Slovakia's position at the forefront of Russian hybrid warfare in Europe, early on Slovakia's experts developed niche knowledge in this area.[40] While the government's interest in this area varied over time, the fight against hybrid threats is one of the few areas where Slovakia is able to develop its own ideas, think and innovate. However, given the persistent pro-Russian attitudes in Slovakia (and the fact that the Russian Embassy in Bratislava is according to a Beacon Project study the most active of Russian and Chinese embassies in Europe),[41] the success in the fight against hybrid threats has its clear limits.

### Innovating the software

Perhaps even more important than updating and upgrading the military hardware was updating and upgrading the defence policy 'software'—organisational structures, human resources, doctrine and planning. Many of these changes and updates were driven by the demands of Slovakia's NATO Membership Action Plan. Slovakia's MAP asked for the development of stronger civilian control and the upgrading of its planning and staffing policies.[42]

The initial reform was outlined in the 1999 *Concept for Reform of the Ministry of Defence Until 2022*, which was heavily criticised by the expert community as being insufficient.[43] These findings were echoed in a major study conducted by US Army Major-General Joseph Garett (hence unofficially known as the Garrett study). The Garrett study found not only that the reform was insufficient but also that the armed forces would face jurisdictional difficulties should they actually need to be used. The report also concluded that the military personnel was not prepared for NATO membership.[44]

The criticism in the Garrett study, combined with the highest political priority given to the NATO membership, meant that further reforms were pushed. This meant that the MoD was re-staffed with new senior civilian staff who replaced soldiers or 'undressed soldiers' (soldiers who left active duty but re-entered as civilians). Having said this, the top-level leadership at the Ministry has been consistently composed of civilians, often with a diplomatic background. At the end of the process, a number of strategic documents were adopted, many of which were in place until very recently.

The new models introduced firm civilian control over the armed forces. This control has organisational and institutional elements. Organisationally, civilian control was fostered through a firm control over the budget (during the communist era, the armed forces were in charge of their own budget). Through the introduction of managerial practices and practices of evaluation according to stringent standards, civilians exercised control over the armed forces.[45]

Institutionally, Slovakia did not adopt major innovations either. The only major institutional innovation introduced since independence was the introduction of the Command for Special Operations Forces in 2019. In 2018, the Analytical Unit of the Ministry of Defence, earlier mainly tasked with policy analysis, also became responsible for the execution of the Value for Money analyses, which are presently required by law for all procurements above one million euros. These analyses explicitly compare Slovakia to Western European NATO members of similar size, including Denmark and the Netherlands.[46]

Institutional innovation within the Slovak armed forces therefore mainly comes in the form of emulation of structures which are known in other allied countries. Slovakia is therefore a good example of emulation, which is a result of the desire to square the NATO membership requirements with limited material and human resources.[47] In such a situation, Slovakia's officials have largely emulated foreign trends. The period between 1994 and 2004 was a period of 'catching up' when Slovakia tried to be a good student and executed transformations in reaction to the demands from the alliance and the United States as the leading ally. The period between 2005 and 2016 was a period when Slovakia emulated other small- and medium-sized countries in the alliance, by contributing to missions abroad and redirecting its foreign policy to out-of-area missions. The post-2016 (and especially post-2021) period is one where emulation happens primarily through acquisition of equipment, which is procured off the shelf and primarily from a single pool of allies and friendly countries. Low efforts to diversify the suppliers of military equipment and low efforts to bring genuine intellectual innovation in strategic thinking are key elements of military innovation in Slovakia since independence.

**Methods and techniques of transformation**

Defence planning in Slovakia has similarly undergone changes since independence. One of the main intellectual changes was the switch from the execution of war plans to the practice of defence planning, which starts with defining interests.[48]

The defence and strategy planning in Slovakia has been done in a haphazard fashion and often in response to major external demands (or shocks). The desire to produce these documents is fully driven by the political leadership, even if the penholding resides with the civil servants in different parts of the administration. The presence—or absence—of the desire to update strategic documents also translates into the haphazard sporadicity with which they are being produced. The initial defence and strategic documents were produced in expectation of Slovakia's NATO membership. They were updated after critical feedback from abroad (as described previously). In 2005, after entering NATO, Slovakia produced a new set of strategic documents, which were meant to reflect the reality of a country which was now a full member of NATO. These documents reflected the feelings of the times, when the main source of insecurity was meant to come from terrorism, political instability abroad, failed and/or rogue states.[49] These documents indicated the risk of conventional war in Europe as low.

These documents were produced in a process, which was led by civilian officials, who often spent training stints either in the United States, the United Kingdom, at the NATO School in Oberammergau or at the George Marshall Center in Garmisch-Partenkirchen, Germany. Comparatively fewer civilians had exposure to continental military and defence thinking (for instance, fluency in German or French is very rare among the defence experts in Slovakia). Military leaders too were often trained in military schools in the United States or the United Kingdom, which often meant a growing familiarity with the trends in these countries (which in the early 2000s meant a focus on terrorism and irregular warfare). While armed forces were involved in the process, there was no doubt that the process was led by the civilian leaders. Civil society, especially representatives of various think tanks, are often invited to participate in the process. Slovak civil society is marked by a high level of circulation between the government and expert community and almost all civilian analysts of any repute have sooner or later ended up working in mid- and senior-level positions at the ministry.[50] These processes hence reflect both technocratic expert-level knowledge and political realities (reflected in the fact that governments need to provide a permissive environment for planning to happen in the first place). It might also be added that the whole field is male-dominated, and few female experts have been included. Even a cursory look at the references in this piece testifies how very male-dominated the field is.

Slovakia's strategic documents were not updated until 2021. In 2016, the Ministry of Defence led a process to produce a White Paper on Defence, which did not however hold any formal status. Security and defence strategies were produced in the late 2010s, but due to domestic political debates, they were never formally adopted (though they served as a basis for the 2021 documents).

In Slovakia's defence planning and force structure, the demands stemming from the desire to join NATO and resource constraints were binding. The feeling that defence planning could be done in a more systematic manner is palpable. In 2022, the Analytical Unit of the Ministry of Defence launched a study which compared defence planning in Belgium, the Czech Republic, Denmark, Lithuania and Hungary, with the aim of deriving lessons for Slovakia's defence policy planning.[51] This review also divided Slovakia's defence policy planning into 'strategic/long-term', 'medium-term' and 'short-term planning', where various strategies are listed under 'strategic/long-term phase', internal guidelines are listed under 'medium-term' and the annual budget under 'short-term' planning.

## Conclusion

In terms of the structured focused comparison (see Figure 11.3), it is clear that the national defence goals have been, until recently, not driven by concrete immediate threats, but by the desire to act as a good ally. This makes Slovakia similar to

| | Option 1 | Option 2 | Option 3 | Option 4 | Option 5 |
|---|---|---|---|---|---|
| SMPS ADJUST THEIR APPROACHES BASED ON SYSTEMIC CHANGE OUT OF THEIR CONTROL | SMPS REACT TO CHANGES IN THE RELEVANT SECURITY COMPLEX[ES] THE SMP IS PART OF | | | | |
| *Alliances, Dependencies and National Ambitions* | | | | | |
| SMPS FACE CONSTRAINTS IN SAFEGUARDING NATIONAL SECURITY | FINANCIAL | DEMOGRAPHIC | CULTURAL | GEOGRAPHICAL | OTHER |
| SMPS NEED TO POSITION THEMSELVES VIS-À-VIS GLOBAL AND REGIONAL POWER CONSTELLATIONS | BALANCING | HEDGING | BANDWAGONING | NON-ALIGNING | OTHER |
| SMPS HAVE TO ENGAGE WITH OTHER STATES TO INCREASE THEIR DEFENCE CAPABILITIES | SECURITY GUARANTEES | TRANSACTIONAL | FORCE INTEGRATION | SPECIALISATION | OTHER |
| SMPS HAVE ARMED FORCES FOR DIFFERENT PURPOSES | INSTRUMENT OF DIPLOMACY | PRESTIGE PROJECT | REDUCE STRATEGIC RISK (DEFEND) | DOMESTIC STABILITY | OTHER |
| *Approaches, Processes, Methods and Techniques* | | | | | |
| SMPS TAKE A MORE CONSTRAINED APPROACH TO DEFENCE PLANNING | THREAT-NET ASSESMENT BASED | PORTFOLIO-BASED | TASK-BASED | MOBILISATION | OTHER |
| SMPS HAVE LIMITED RESOURCES TO DESIGN AND IMPLEMENT DEFENCE PLANNING PROCESSES | TAILORED ANALYTICAL PROCESS | OUTSOURCING STRATEGY | ACCEPT DILUTION / FAIR SHARE | EXTERNAL TEMPLATE | OTHER |
| SMP EMPLOY MULTIPLE METHODS AND TECHNIQUES | COPY GREAT POWERS | INFORMAL CONSENSUS | TAILORED METHODS | AD HOC | OTHER |
| SMPS RELY ON DIFFERENT PROCESSES TO DECIDE ON FORCE POSTURE | ANALYTICAL RESULTS | POLITICS DECIDES | MILITARY DECIDES | GREAT POWER DECIDES | OTHER |
| *Military Innovation* | | | | | |
| SMPS FACE CHALLENGES IN KEEPING PACE WITH TECHNOLOGICAL ADVANCES | BUILD DOMESTIC TECH/INDUSTRY | TECH TRANSFER AGREEMENTS | ACQUIRE READY INNOVATIONS | PARTNER WITH GREAT POWER | OTHER |
| SMPS PURSUE VARIOUS MILITARY INNOVATION PATHS AND PATTERNS | EMULATION | ADAPTATION | INNOVATION | OTHER | OTHER |
| | SPECULATION | EXPERIMENTATION | IMPLEMENTATION | OTHER | |
| SMPS FIND THEMSELVES AT DIFFERENT STAGES OF ORGANISATIONAL CHANGE | EXPLORATION | MODERNISATION | TRANSFORMATION | OTHER | OTHER |

*Figure 11.3* Structured focused comparison framework for Slovakia.

*Source:* Author

other European powers, such as the Netherlands, described in other chapters of this book. However, as opposed to the Netherlands, Slovakia's posture was not driven by cultural or historical experiences, but by the relatively benign security environment and absence of immediate threats, combined with the broadly shared political goal to integrate into the EU and NATO.

The desire to become allied with the United States and join NATO has determined all the steps in Slovakia's defence planning. The transformations made in the process were in response to NATO's demands and recommendations. Slovakia's defence has ever since been driven by the goal to become a member of NATO, and after entry into NATO, by the goal of becoming a 'good ally'. Given its relatively small size and limited resources—which were even further cut up to 2015—Slovakia's priority was to comply with the goals determined by the desire to join NATO while modernising its armed forces by procuring the new equipment produced by its allies or other friendly countries (such as Israel or South Korea). The scholars of the European Union have for a long time discussed the process known as 'Europeanization'—the desire of countries to adopt European rules to integrate into the EU.[52] The case of Slovakia shows that a similar process can happen also in the case of defence planning, where the desire to join an international organisation—and to be a good ally—can lead to a total transformation of national defence policy.

This also meant that defence planning was not a priority for the successive governments, but something that had to be done. The relatively little attention given to it is demonstrated by the fact that between 2005 and 2021, no official security strategy was published. This little attention to defence planning and low prioritisation also translated into defence procurement, where Slovakia did not prioritise

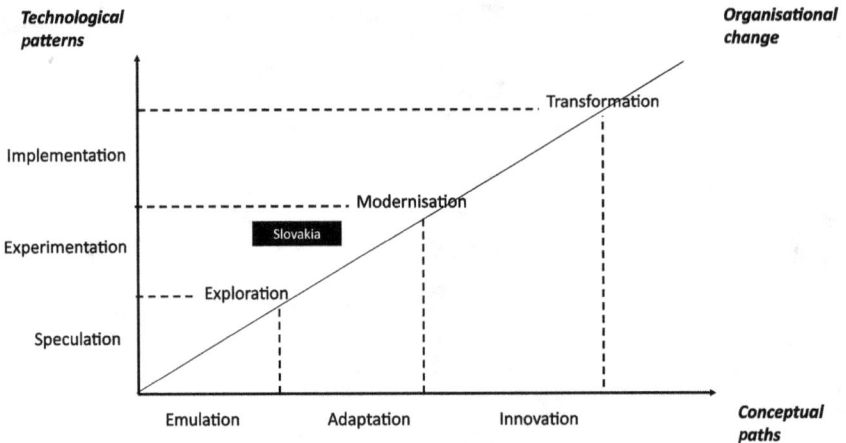

*Figure 11.4* Slovakia's military innovation trajectories.

*Source:* Figure created by author based on Michael Raska, Military Innovation of Small States—Creating a Reverse Asymmetry (Routledge, 2016)[53]

domestic innovation, but instead focused on acquisition from abroad. One possible explanation for this pattern in Slovakia is the combination of the desire to build an alliance with the United States and the lack of perception of any unique needs. Figure 11.4 indicates how this pattern of defence innovation compares with the broader framework.

In terms of Raska's typology, Slovakia sits somewhere between the 'first' and the 'second' band.[54] When it comes to conceptual paths, Slovakia sits squarely in the 'emulation' camp. Slovakia emulated the security policy of other NATO powers and particularly its threat perception reflected the mainstream of the NATO countries. Slovakia's threat perception until 2016 did not include any regional threats. Therefore, for Slovakia, the threats were primarily outside the region. This sets Slovakia apart from other countries in the CEE region. When it comes to technological patterns, Slovakia sat between speculation and experimentation. The only area where Slovakia developed genuinely new units and force structures (i.e. not those emulating NATO countries) is represented by the inroads to the hybrid warfare and growing awareness of disinformation. Slovakia's desire to purchase 'off-the-shelf' technological solutions from Western countries with little technological innovation fits Raska's definition of modernisation in terms of organisational change. This move, however, happened only in the recent year. If we were to prepare this figure before 2020, Slovakia would sit much more closely to 'Exploration' in this chart.

To allow a comparison of Slovakia with other small and middle powers in this book, let us now place Slovakia into the analytical matrix. This matrix and the placement are available in Figure 11.3. Starting with the constraints, it is clear that Slovakia's challenge was mainly linked to financial constraints (years of budget cuts) and cultural/historical constraints (linked to historical threat perceptions). In the decision-making process, the government is the leading actor. When it comes to relationships with the great powers, Slovakia is a clear bandwagoner. As for the other nations, Slovakia seeks primarily security guarantees (via NATO), but there are nascent attempts at force integration within the framework of the NATO battle-group. Armed forces were mainly meant to be an instrument of diplomacy and then primarily focused on bolstering Slovakia's desire to be a 'good ally'. Defence planning was primarily threat-based, with the proviso that until very recently, the most relevant threats were emanating from non-state actors and the Middle East. As was described earlier, Slovakia's decision-making is strongly top-down, with the Ministry of Defence in the driving seat. The defence planning follows an analytical (even if ad hoc) process, but as mentioned previously, there is a strong influence from the Western (especially American and British) defence thinking. The resulting force posture choices are a result of political decisions. The main approach to keeping pace with the technological advances is by acquiring ready military innovations by purchasing goods from allies.

The resulting pattern shows Slovakia sitting uncomfortably when it comes to defence innovation. Whereas defence policy in countries such as Estonia was built around an existential angst which drove much of the innovation, this was not the case in Slovakia. Absent any immediate security threats, Slovakia had seen, for a long time, the threats to be primarily originating in the Middle East (and Sahel) and be mainly composed of non-state and terrorist actors. While the country was able to contribute to addressing those through limited contributions to multinational forces, it did not create perceptions of urgent security needs at home. Hence, the perceived need for military innovation was much smaller. Once a member of NATO, military innovation—regardless of manner—has not been a priority, and defence force planning has been haphazard at best. A 2023 expert assessment, for instance, evaluated that Slovakia does not have any strategy or deployment of artificial intelligence tools for the military.[55] This lack of military innovation was also remarked by NATO, which argued that NATO's and Slovakia's defence planning priorities 'remain, in practice, not particularly well aligned'.[56]

The ongoing Russian invasion of Ukraine somewhat changed this lethargic state. Since shortly before the war, Slovakia has updated its strategic plans, but also started a modernisation spree with a number of major procurement processes with high price tags. At the same time, almost the whole technocratic scene has been united in supporting Ukrainian defence with whatever means necessary. This enthusiasm is not broadly shared, and certainly faces domestic contestation, by both the general public and the parliamentary opposition. Given the upcoming elections in Slovakia in September 2023, it remains to be seen whether the reinvigoration of defence spending and security thinking in Slovakia persist over time.

### Acknowledgements

This chapter was completed in 2023 and reflects the situation as of that time. I thank the participants in the Amsterdam workshop, a Slovak defence official who wishes to remain anonymous, as well as the editors for their helpful suggestions. I am also thankful to Terézia Vagačová and Sandra Fudorová for their diligent research assistance.

## Notes

1 For more historical detail, see Michal Onderco, 'Czech Republic, Hungary, Slovakia', in *The Handbook of European Defence Policies & Armed Forces*, eds. Hugo Meijer and Marco Wyss (Oxford: Oxford University Press, 2018), 279–296.
2 Robert O Keohane, 'Lilliputians' Dilemmas: Small States in International Politics', *International Organization* 23, no. 2 (1969): 291–310; Michael Raska, *Military Innovation in Small States* (London: Routledge, 2015).
3 On EU integration, see Sharon Fisher, John Gould and Tim Haughton, 'Slovakia's Neoliberal Turn', *Europe-Asia Studies* 59, no. 6 (2007): 977–998; on NATO integration, see Jeffrey Simon, *NATO and the Czech and Slovak Republics: A Comparative Study in Civil-Military Relations* (Lanham, MD: Rowman & Littlefield, 2004).
4 Zdeněk Kříž, 'Czech Military Transformation: Towards Military Typical of Consolidated Democracy?', *The Journal of Slavic Military Studies* 23, no. 4 (2010): 617–629; Marie Vlachová and Štefan Sarvaš, 'Democratic Control of Armed Forces in the Czech Republic', in *Democratic Control of Military in Postcommunist Europe: Guarding the Guards*, eds. Andrew Cottey, Timothy Edmunds and Anthony Forster (Basingstoke: Palgrave Macmillan, 2002), 44–63.
5 Miroslav Wlachovský, 'Obrana a armáda', in *Slovensko 1996. Súhrnná správa o stave spoločnosti a trendoch na rok 1997*, ed. Martin Bútora (Bratislava: Inštitút pre verejné otázky, 1996), 101–111.
6 Dissident, advisor, diplomat and later Czech foreign minister Jaroslav Šedivý describes these ideas in his different writings. See Jaroslav Šedivý, *Černínsky palác v roce nula. Ze zákulisí polistopadové zahraniční politiky* (Praha: Ivo Železný, 1997); Jaroslav Šedivý, *Dissident-Led Foreign Policy* (unpublished manuscript, 2019).
7 Liviu Horovitz and Elias Götz, 'The overlooked importance of economics: Why the Bush Administration wanted NATO enlargement', *Journal of Strategic Studies* 43, no. 6–7 (2020): 847–868.
8 Miroslav Mojžita, *Kňažko / Demeš / Kňažko: Formovanie slovenskej diplomacie v rokoch 1990 až 1993* (Bratislava: Veda, 2019).
9 Elisabeth Bakke and Nick Sitter, 'Patterns of stability: Party competition and strategy in Central Europe since 1989', *Party Politics* 11, no. 2 (2005): 243–263.
10 Simon, *NATO and the Czech and Slovak Republics*.
11 'Obranná doktrína SR 1994', *Ministerstvo Obrany Slovenskej Republiky*, 1994, accessed 1 March, 2023, www.mosr.sk/data/files/4968_1994-obranna-doktrina-sr.PDF, 1.
12 'Obranná doktrína SR 1994', 2.
13 'Bezpečnostná stratégia SR 2005', *Ministerstvo Obrany Slovenskej Republiky*, 2005, accessed 1 March, 2023, www.mosr.sk/data/files/4264_bezpecnostna-strategia-sr-2005.pdf, 14.
14 'Bezpečnostná stratégia SR 2005', 14.
15 'Obranná stratégia SR 2005', *Ministerstvo Obrany Slovenskej Republiky*, 2005, accessed 1 March, 2023, www.mosr.sk/data/files/4265_obranna-strategia-sr-2005.pdf, 4.
16 'Bezpečnostná stratégia SR 2005', 15.
17 'Obranná stratégia SR 2005', 4. Interestingly enough, WMD proliferation is often discussed in the context of terrorism. Slovakia is quite likely one of the few countries which has a single directorate in its foreign ministry which combines the WMD and terrorism portfolios.
18 'White Paper on Defence of the Slovak Republic', *Ministerstvo Obrany Slovenskej Republiky*, 2016, accessed 1 March, 2023, www.mosr.sk/data/WPDSR2016_LQ.pdf, 67.
19 'White Paper on Defence of the Slovak Republic', 6, 20, 33.
20 Ministerstvo obrany Slovenskej republiky, 'Návrh Dlhodobého plánu rozvoja obrany s dôrazom na výstavbu a rozvoj ozbrojených síl Slovenskej republiky s výhľadom do roku 2030', *Ministerstvo Obrany Slovenskej Republiky*, 2021, accessed 1 March 2023, www.mosr.sk/data/files/3869_3561_dlhodoby-plan-2017.pdf.

21 'Bezpečnostná stratégia SR 2021', *Ministerstvo Obrany Slovenskej Republiky*, 2021, accessed 1 March 2023, www.mosr.sk/data/files/4263_210128-bezpecnostna-strategia-sr-2021.pdf, 14.

22 'Bezpečnostná stratégia SR 2021', 6.

23 'Bezpečnostná stratégia SR 2021', 17.

24 Globsec, 'Globsec Trends 2022', 2022, *Globsec,* accessed 11 May 2023, www.globsec.org/what-we-do/publications/globsec-trends-2022-central-and-eastern-europe-amid-war-ukraine.

25 Michal Onderco, 'European Crises and Foreign Policy Attitudes in Europe', in *Towards a Segmented European Political Order: The European Union's Post-Crises Conundrum*, eds. Jozef Bátora and John Erik Fossum (London: Routledge, 2020), 225–242.

26 Globsec, 'Globsec Trends 2022'.

27 The definitions of Ukraine's victory and Russia's loss differ and are vastly different when officials speak on- and off-record.

28 Stephen M. Walt, *The Origins of Alliances* (Ithaca, NY: Cornell University Press, 1987).

29 International Institute for Strategic Studies, *The Military Balance* (Abingdon: Routledge, 1993); International Institute for Strategic Studies, *The Military Balance* (Abingdon: Routledge, 2021).

30 Numbers from Onderco, 'Czech Republic, Hungary, Slovakia'.

31 Source: SIPRI, 'SIPRI Military Expenditure Database', 2022, *SIPRI,* accessed 1 March 2022, www.sipri.org/databases/milex/.

32 International Institute for Strategic Studies, *The Military Balance* (Abingdon: Routledge, 2023).

33 International Institute for Strategic Studies, *The Military Balance* (2023).

34 For the official press release, see www.mosr.sk/52469-sk/madarsko-sa-pripoji-k-cesku-a-polsku-a-zapoji-sa-do-ochrany-vzdusneho-priestoru-slovenska/.

35 International Institute for Strategic Studies, *The Military Balance* (Abingdon: Routledge, 2014).

36 Ministry of Finance and Ministry of Defence, 'Defence Spending Review (Final Report)', 2020, accessed 2 March, 2023, www.mosr.sk/data/files/4527_slovakia-defence-spending-review-final-report.pdf.

37 North Atlantic Council, 'Defence Planning Capability Review 2021/2022. The Slovak Republic', (2022).

38 See also International Institute for Strategic Studies, *The Military Balance* (2023).

39 *Compilation and visualisation based on NATO Website*, accessed 27 September 2023, www.nato.int/cps/en/natolive/topics_49198.htm.

40 Globsec, 'Slovak Republic Hybrid Threat Vulnerability Study', 2018, *Globsec,* accessed 4 September 2019, www.globsec.org/wp-content/uploads/2018/12/Slovak-Republic-Hybrid-Threats-Vulnerability-Study_Executive-Summary.pdf.

41 Una Hajdari, 'Russian Embassy in Slovakia Uses Facebook to Push Propaganda. Why Are So Many Slovaks Buying It?', *Euronews*, 29 March 2023, accessed 3 August 2023, www.euronews.com/2023/03/29/russian-embassy-in-slovakia-uses-facebook-to-push-propaganda-why-are-so-many-slovaks-buyin.

42 Matej Navrátil and Michal Onderco, 'Slovakia: Creating and Transforming the Civil-Military Relations', in *Oxford Research Encyclopedia of Military and Politics*, ed. William R. Thompson (Oxford: Oxford University Press, 2020).

43 Zoltan D. Barany, *The Future of NATO Expansion: Four Case Studies* (Cambridge: Cambridge University Press, 2003).

44 Barany, *The Future of NATO Expansion: Four Case Studies.*

45 Mário Nicolini, 'Slovensko v NATO, EÚ a BR OSN: zmena a kontinuita', in *Panoráma globálneho bezpečnostného prostredia 2006–2007*, eds. Lubomír Lupták, Robert Ondrejcsák and Vladimír Tarasovič (Bratislava: Odbor bezpečnostnej a obrannej politiky Ministerstva obrany Slovenskej republiky, 2007), 115–132.

46 Ministry of Finance and Ministry of Defence, 'Defence Spending Review (Final Report)'.
47 Michael Raska, 'Strategic Competition for Emerging Military Technologies: Comparative Paths and Patterns', *PRISM* 8, no. 3 (2019): 64–81.
48 This point was underlined by an interviewee for Navrátil and Onderco, 'Slovakia: Creating and transforming the civil-military relations'.
49 'Bezpečnostná stratégia SR 2005'.
50 For a critical look at this process and pathologies it produces, see Lubomír Lupták, *(Ne) bezpečnosť ako povolanie* (Brno: Doplněk, 2017).
51 Maroš Cuník and Barbora Hrozenská, 'Obranné plánovanie. Komparatívna analýza piatich štátov', 2022, accessed 1 March 2023, www.mosr.sk/data/files/4891_2022-a-01-obranne-planovanie-komparativna-analyza-piatich-statov.pdf.
52 See, for instance, Frank Schimmelfennig and Ulrich Sedelmeier, eds., *The Europeanization of Central and Eastern Europe* (Ithaca, NY: Cornell University Press, 2005).
53 Michael Raska, "A Structured-Phased Evolution: The Third Generation Force Transformation of the Singapore Armed Forces", in *Military Innovation in Small States: Creating a Reverse Asymmetry*, ed. M. Raska (New York: Routledge, 2016), 130–162.
54 Raska, 'Strategic Competition for Emerging Military Technologies'.
55 Maggie Gray and Amy Ertan, 'Artificial Intelligence and Autonomy in the Military: An Overview of NATO Member States' Strategies and Deployment', *The NATO Cooperative Cyber Defence Centre of Excellence*, 2023, accessed 16 August 2023, https://ccdcoe.org/library/publications/artificial-intelligence-and-autonomy-in-the-military-an-overview-of-nato-member-states-strategies-and-deployment/.
56 North Atlantic Council, 'Defence Planning Capability Review 2021/2022. The Slovak Republic'.

## Bibliography

Bakke, Elisabeth and Nick Sitter. 'Patterns of stability: Party competition and strategy in Central Europe since 1989'. *Party Politics* 11, no. 2 (2005): 243–63.
Barany, Zoltan D. *The future of NATO expansion: Four case studies*. Cambridge: Cambridge University Press, 2003.
Cuník, Maroš and Barbora Hrozenská. 'Obranné plánovanie. Komparatívna analýza piatich štátov'. 2022, accessed 1 March 2023. www.mosr.sk/data/files/4891_2022-a-01-obranne-planovanie-komparativna-analyza-piatich-statov.pdf.
Fisher, Sharon, John Gould, and Tim Haughton. 'Slovakia's Neoliberal Turn'. *Europe-Asia Studies* 59, no. 6 (2007): 977–98.
Globsec 'Slovak Republic Hybrid Threat Vulnerability Study'. 2018, accessed 4 September 2019. www.globsec.org/wp-content/uploads/2018/12/Slovak-Republic-Hybrid-Threats-Vulnerability-Study_Executive-Summary.pdf.
———. 'Globsec Trends 2022'. 2022, accessed 11 May 2023. www.globsec.org/what-we-do/publications/globsec-trends-2022-central-and-eastern-europe-amid-war-ukraine.
Hajdari, Una, 'Russian Embassy in Slovakia uses Facebook to push propaganda. Why are so many Slovaks buying it?'. *Euronews*, 2023, accessed 3 August 2023. www.euronews.com/2023/03/29/russian-embassy-in-slovakia-uses-facebook-to-push-propaganda-why-are-so-many-slovaks-buyin.
Horovitz, Liviu and Elias Götz. 'The overlooked importance of economics: why the bush administration wanted NATO enlargement'. *Journal of Strategic Studies* 43, no. 6–7 (2020): 847–68.
International Institute for Strategic Studies. *The Military Balance*. Abingdon: Routledge, 1993.
———. *The Military Balance*. Abingdon: Routledge, 2014.
———. *The Military Balance*. Abingdon: Routledge, 2021.

————. *The Military Balance*. Abingdon: Routledge, 2023.

Keohane, Robert O. 'Lilliputians' dilemmas: small states in international politics'. *International Organization* 23, no. 2 (1969): 291–310.

Kříž, Zdeněk. 'Czech military transformation: towards military typical of consolidated democracy?'. *The Journal of Slavic Military Studies* 23, no. 4 (2010): 617–29.

Lupták, Lubomír. *(Ne)bezpečnosť ako povolanie*. Brno: Doplněk, 2017.

Ministerstvo Obrany Slovenskej Republiky. 'Obranná doktrína SR 1994'. 1994, accessed 1 March 2023. www.mosr.sk/data/files/4968_1994-obranna-doktrina-sr.PDF.

————. 'Bezpečnostná stratégia SR 2005'. 2005a, accessed 1 March 2023. www.mosr.sk/data/files/4264_bezpecnostna-strategia-sr-2005.pdf.

————. 'Obranná stratégia SR 2005'. 2005b, accessed 1 March 2023. www.mosr.sk/data/files/4265_obranna-strategia-sr-2005.pdf.

————. 'White paper on defence of the Slovak Republic'. 2016, accessed 1 March 2023. www.mosr.sk/data/WPDSR2016_LQ.pdf.

————. 'Bezpečnostná stratégia SR 2021'. 2021a, accessed 1 March 2023. www.mosr.sk/data/files/4263_210128-bezpecnostna-strategia-sr-2021.pdf, 14.

————. 'Návrh Dlhodobého plánu rozvoja obrany s dôrazom na výstavbu a rozvoj ozbrojených síl Slovenskej republiky s výhľadom do roku 2030'. 2021b, accessed 1 March 2023. www.mosr.sk/data/files/3869_3561_dlhodoby-plan-2017.pdf.

Ministry of Finance and Ministry of Defence. 'Defence spending review (final report)'. 2020, accessed 2 March 2023. www.mosr.sk/data/files/4527_slovakia-defence-spending-review-final-report.pdf.

Mojžita, Miroslav. *Kňažko/Demeš/Kňažko: Formovanie slovenskej diplomacie v rokoch 1990 až 1993*. Bratislava: Veda, 2019.

Navrátil, Matej and Michal Onderco. 'Slovakia: Creating and transforming the civil-military relations'. In *Oxford Research Encyclopedia of Military and Politics*, edited by William R. Thompson. Oxford: Oxford University Press, 2020.

Nicolini, Mário. 'Slovensko v NATO, EÚ a BR OSN: zmena a kontinuita'. In *Panoráma globálneho bezpečnostného prostredia 2006–2007*, edited by Lubomír Lupták, Robert Ondrejcsák, and Vladimír Tarasovič, 115–32. Bratislava: Odbor bezpečnostnej a obrannej politiky Ministerstva obrany Slovenskej republiky, 2007.

North Atlantic Council. *Defence Planning Capability Review 2021/2022. The Slovak Republic*. Bruxelles: NATO, 2022.

Onderco, Michal. 'Czech Republic, Hungary, Slovakia'. In *The Handbook of European Defence Policies & Armed Forces*, edited by Hugo Meijer and Marco Wyss, 279–96. Oxford: Oxford University Press, 2018.

————. 'European crises and foreign policy attitudes in Europe'. In *Towards a Segmented European Political Order: The European Union's Post-crises Conundrum*, edited by Jozef Bátora and John Erik Fossum, 225–42. London: Routledge, 2020.

Raska, Michael. *Military Innovation in Small States*. London: Routledge, 2015.

————. 'Strategic competition for emerging military technologies: Comparative paths and patterns'. *PRISM* 8, no. 3 (2019): 64–81.

Schimmelfennig, Frank and Ulrich Sedelmeier, eds. *The Europeanization of Central and Eastern Europe*. Ithaca: Cornell University Press, 2005.

Šedivý, Jaroslav. *Černínsky palác v roce nula. Ze zákulisí polistopadové zahraniční politiky*. Praha: Ivo Železný, 1997.

————. *Dissident-led foreign policy*. Unpublished manuscript, 2019.

Simon, Jeffrey. *NATO and the Czech and Slovak Republics: A Comparative Study in Civil-Military Relations*. Lanham: Rowman & Littlefield, 2004.

SIPRI. 'SIPRI Military Expenditure Database'. 2022, accessed 1 March 2022. www.sipri.org/databases/milex/.

Vlachová, Marie and Štefan Sarvaš. 'Democratic Control of Armed Forces in the Czech Republic'. In *Democratic Control of Military in Postcommunist Europe: Guarding the*

*Guards*, edited by Andrew Cottey, Timothy Edmunds and Anthony Forster, 44–63. Basingstoke: Palgrave Macmillan, 2002.

Walt, Stephen M. *The Origins of Alliances*. Cornell studies in security affairs. Ithaca: Cornell University Press, 1987.

Wlachovský, Miroslav. 'Obrana a armáda'. In *Slovensko 1996. Súhrnná správa o stave spoločnosti a trendoch na rok 1997*, edited by Martin Bútora, 101–11. Bratislava: Inštitút pre verejné otázky, 1996.

# 12 Canada and defence planning

## From making a virtue of necessity to the necessity of virtue

*Paul Dickson*

## Introduction

Canada's defence planning in the twenty-first century has had to address a series of significant challenges, some external, many self-imposed; all a function of a Canadian strategic culture that since the end of the Cold War balances two historical realities: Canada rallies militarily only when faced with a strategic threat; and its defence planning and preparation must balance a desire to maintain relevance based on a credible commitment to bilateral and multilateral defence cooperation with the reality that defence is rarely viewed as a national or political imperative.[1] C.P. Stacey characterised the Canadian approach to defence preparations in a classic study of Canada's military problems, written on the eve of war in 1939: 'long periods when the national defences are almost utterly neglected' interrupted by 'short violent interludes, arising out of sudden foreign complications, when the country awakes to the inadequacy of those defences and tries to make up for earlier inactivity'.[2]

The end of the Cold War collapsed the post-war threat foundations of Canadian defence policy, ushering in an era in which defence threats were either too abstract to compel defence planning decisions or not direct enough to provide the level of shock historically required to forge a national consensus to rearm and seriously rethink force levels and commitments. Canada's commitments during the Cold War, at least from 1950 until the 1970s, might have surprised Stacey, but not so the fundamentals of the debates or the re-assertion of the traditional indifference to defence preparations, unless forced otherwise, that characterised the period from the end of the Cold War to present day.

The generalisation that Canadians—that is the Canadian government and body politic—rarely take defence planning seriously except when confronted with an existential level external shock has limited value in understanding the impact of the dynamics of defence planning in the Canadian context in general, and the defence establishment specifically.[3] It does, however, reflect the strategic culture and environment in which defence planners must operate. Applying the analytical framework of this comparative study and focusing on the period from 2001 to the present, this chapter examines the strategic environment in which defence planning takes place as constant factors that inform Canada's thinking about defence. It will

DOI: 10.4324/9781003398158-12

then explore the primary drivers of that planning as well as the impact of planning processes and defence organisation. The chapter concludes with an assessment of the interaction of these elements and the resulting Canadian approach to thinking about future conflict and innovating to address the challenges of preparing for those conflicts.

### Part 1: national ambitions, alliances, dependencies: from the imperial hinterland to the continental strategic perimeter

Geography and history shape Canada's national military ambitions, inform its alliances as well as determine some, but not all, of its dependencies. History has resulted in a region defined by a continent of three countries but from 1945 dominated by a global superpower. The North American history of conflict and conquest bequeathed to Canada a tradition of its defence being intimately linked to successive global powers: France, Great Britain and the United States. Geography has resulted in a country where scale, distance and austere environments determine domestic defence considerations, even if geography has proven both a defence blessing and curse. These enduring conditions—the 'permanent elements' or 'invariants'—of Canada's strategic position shape but do not singularly determine early defence planning choices.[4] In response, Canada made a virtue of necessity, adopting a policy of being non-threatening to its neighbour while maintaining sufficient capability to encourage support from its main partner.

Prior to Canada's establishment as an independent dominion in 1867, the struggle for control of North America between France and Great Britain, and later between Great Britain and the United States, created the political and social dynamic within which Canadian defence and defence issues would be conducted, debated and settled. The United States was determined to dominate the continent and remove any British presence. US invasions of the loyal British colonies during the Rebellion and again in 1812–1814 were thwarted, but only just.[5] The division of the northern half of the continent was formalised over the next century, with Canada's economic development and political maturation occurring under the shadow of the ongoing, if muted, US-British antagonism. Canada's official statehood dates from 1 July 1867 through the confederation of four British colonies. It was prompted in no small part by the combination of fears generated by the American military potential demonstrated during the US Civil War. Combined with the increasing disinclination of the British to pay for the defence of indefensible parts of its empire of self-governing and increasingly autonomous dominions, this contributed to Canada's statehood. For Canada, it would maintain forces just sufficient to claim it was at least contributing to its own defence.

There was truth to both perspectives, and they infused Canada's strategic culture. Then, as now, the Canadians, and British, charged with thinking about defence faced the paradox of Canada being at once invulnerable and indefensible.[6] Great Britain determined Canada's defence and foreign policy, but arguably both post-1867 Canadian and British leadership made choices about contingencies and posture in service of maintaining the peace with the main threat to Canada: the

United States. Canada's first post-1867 Confederation Prime Minister—Sir John A. MacDonald—arguably pursued a defence policy predicated on a posture designed to avoid any pretext for an American invasion, and thus allowed the states' limited resources to be devoted to developing the economic, transportation and political sinews of the new and increasingly vast state.

As Canada's land mass expanded west, north and, finally in 1949, east to New-foundland, adding territories and provinces, and as the threat from the US diminished, geography, to paraphrase the words of one prominent defence strategist, evolved to become a 'sustainer' of Canadian security. But this was only once the main threats no longer emanated from within North America.[7] Geography was Canada's defence for nearly the first four decades of the 20th century, until a Second World War was on the horizon.[8]

A second feature of Canada's national approach to defence appeared during this period: its' commitments took the form of expeditions, and the scale and scope were at the discretion of the political leadership. These decisions were shaped by the political-cultural reality of, first, Canada's French-English duality, and then, increasingly, its diverse immigrant, and emerging Canadian character, which effectively worked with geography to act as a brake on achieving a domestic consensus on the degree, if not the nature, of the military threat facing Canada, and thus on commitments, planning and preparation. While, until 1933, Canada had no constitutional authority to declare war—when Britain was at war, Canada was at war, there was no distinction—political leadership navigated public sentiment to determine the nature of its commitment.

This balancing act introduced many strands of thought into Canada's strategic culture that would continue into the twenty-first century. Canada would assume a dependence on a major ally or alliance but be driven in equal parts by the indifference of invulnerability and the angst of indefensibility. The complacent attitude towards defence—what some characterised as an isolationist element—was reflected in the oft-quoted politician's 1924 characterisation that in international affairs Canada was 'a fireproof house, far from inflammable materials'.[9] Canada's reliance on others was a by-product of the second concern. The two combined to create a third feature, reinforced by the distances imposed on Canada's defenders, that shaped approaches to twenty-first-century defence: defence as a primarily expeditionary undertaking, even in order to reach the vast expanses of the Canadian north.

The Second World War and its aftermath marked a major shift in Canada's defence posture, notably its view of domestic and continental defence, its international ambitions and, ultimately, its primary alliance partner. First, driven by US concerns, Canada became a pivot state for North America with the 1938 enunciation of the principle of the indivisibility of the defence of North America.[10] The need to manage and balance the subsequent continental security and sovereignty dilemmas introduced a new form of the geography defence condition through to the end of the Cold War. The scale of Canada's commitment to the Second World War and its emergence from the war as one of two nations left largely unscathed and relatively prosperous shaped Canada's emerging image of itself as a 'middle power', aligned with the US but committed to an active role in the new international order and multilateral institutions it helped to shape.[11] One consequence was the internal and external pressures to develop and maintain 'multi-purpose, combat capable' forces in order for Canada to purchase just enough to 'punch above its weight', a rhetoric and level of ambition that often created a challenge to maintain the resources required to interoperate and align with US and NATO forces. Less evident is the degree to which this is considered a controversial challenge, traditionally balancing the reality of indivisibility of North American defence, the pragmatic logic of cooperating with the US to address the persistent 'crisis of means', with a Canadian internationalist establishment that leads the resistance to excessive American influence and avoidance of closer military integration as a 'third rail' of Canadian defence discussions.[12]

The main streams of Canada's post–Second World War international and strategic perspectives emerged as the Cold War dawned: internationalism, continentalism and nationalism. They were reflected in the enduring three defence policy priorities of this period: domestic, North American and contributing to international security. There was a periodic re-balancing between the three as reflected in force posture and overseas commitments, but they were mutually reinforcing, with a general assumption that international commitments like the United Nations' peacekeeping and the forward positioning of forces under NATO command in Europe were foundational to the defence of the homeland as well as hedging or counterweights to US influence. Continental defence created more angst amongst politicians and defence observers. The establishment of the bi-national command, the North American Air (later aerospace) Defense Command (NORAD), in 1958 was interpreted either as evidence of Canada's sovereignty and defence commitment, or as the necessary 'defence against help' to ward off American threats to Canada's sovereignty.[13] The main driver for domestic defence—usually focused on the Arctic—was to assert sovereignty and autonomy. Canada's image of itself as a 'middle power' was cemented during this 'Golden Age' of Canadian diplomacy and defence commitments. By 1964, Canada's navy maintained a fleet of 50 warships, including an aircraft carrier, crewed by 21,500 personnel. The army peaked at 52,000, maintaining a permanent rotational presence of a NATO brigade stationed in western Europe, six divisions of reserves for home defence and, from 1956, a growing commitment to peacekeeping. The Canadian government also committed an air division of 12 front-line fighter squadrons to Europe. In 1958, Canada and the US joined in the formation of the NORAD, with a Canadian as deputy commander. Under NORAD, the Royal Canadian Air Force (RCAF) expanded, operating interceptor squadrons and early warning radar sites across Canada.

Through the Cold War, Canada practised 'hedged balancing' but as the Cold War ended, and a new era of international threats loomed, Canada's defence commitments were increasingly characterised as 'bandwagoning'.[14] Whether the threats to Canada's sovereignty were real, the dependencies and drivers remained consistent through the Cold War and the post–Cold War period. In the absence of a US or NATO (or Five Eyes) proscribed existential or immediate military threat to Canada or North America, there were dramatic cuts to the Canadian Armed Forces

(CAF), cuts which neither appeals to the maintenance of Canada's self-image as a 'helpful' fixer, reflected in the peacekeeping tradition, or new force planning methods, based on abstracted capability requirements, could slow.[15] The Canadian Armed Forces establishment declined from the peak of the golden years to a post–Cold War low of a total establishment of 61,600 in 2002. The defence budget declined from 4.19% of GDP in 1960 to 1.13% of GDP in 2006.[16] The decline in numbers and budget was uneven, dictated by policy decisions in the 1970s to minimise permanent commitments in Europe and then to restore commitments during the 1980s, peaking with the government's ambitious 1987 White Paper, *A Defence Policy for Canada* and ending abruptly with the 1994 *White Paper on Defence*, which reduced the budget by a third and severely cut the forces.[17]

The debates and logic behind the various stances all indicated that defence remained a hard sell to the wider Canadian public. Peacekeeping, even as it evolved into more complex and riskier peace-making and the number of Canadians on peacekeeping missions declined precipitously during the twenty-first century, was insufficient to drive modernisation or even re-capitalisation. Its hold on the Canadian public's view of Canada in the world remained firm but acted as a constraint on force planning. The tension between our national ambitions, as driven by our post-war role and self-image and the complacency in the absence of an immediate state-based existential threat and geographic challenges emerged as the primary defence planning shaper from the post-Cold War to the present day.

The twenty-first century can be divided into two periods, although there were multiple defence constants that ran through both periods, even as they manifested in different forms. The first period from 2001 to 2011 was characterised by a focus on the war against terror, specifically the combat and training missions in Afghanistan.[18] The evolution and nature of the operations Canada conducted in and to support Afghanistan became a key driver for defence and force planning but also reinforced long-standing themes and trends in Canada's approach to both. Most obvious was the important role of external shocks to force planning, particularly in driving capability acquisition and focused re-capitalisation. While the attacks of 9/11 prompted Canada to immediately provide limited support to overthrow the Taliban government of Afghanistan, the government of the day and the majority of Canadians did not support the US-led invasion of Iraq, remaining instead focused on the NATO-led International Security Assistance Force (ISAF) multilateral mission to rebuild Afghanistan. As that mission evolved into a more complex mix of nation-building and increasingly robust counter-insurgency, its impact on force planning was significant, both for the focus on regenerating the army, in particular and re-capitalisation of some major fleets, notably the air force, but also because the evolved mission provided an opportunity to restore a post-Cold War (and peacekeeping scandal) gutted military and defence department by using a combat mission and Canadian's support for its military in combat to start to re-equip the Canadian Armed Forces and re-focus them on preparing for operations.[19] The major equipment acquisitions—including strategic lift (CC-177 Globemaster III), the first of its Unmanned Aerial Vehicles, light armoured vehicles (LAV) and replacement tanks (Leopard 2 Tank), including restoring armoured capabilities deemed both legacy and unsuited to modern battlefields—were fast-tracked and driven by operational requirements.

Arguably more significant was that while the profile of the military and defence in general, was elevated in public discourse and respect, the missions in Afghanistan reinforced the traditional contours of Canadian defence planning drivers. The scale of Canada's mission in Afghanistan was a result, in part, of re-balancing Canada's defence relationship with the US after Iraq as it concurrently resisted contributing more to North America, reflected primarily in resisting participation in Ballistic Missile Defence. For others among the Canadian public, the combat mission was at odds with its sense of Canada as helpful fixers and peacekeepers.[20] Even as defence leaders argued for an expansion of the mission to include combat in Kandahar in 2005, the Prime Minister favoured humanitarian missions in Darfur and Haiti. The Afghanistan conflict also reinforced for defence leaders and planners the necessity of crisis to drive forward force planning and development. The RCAF was, for example, critiqued by some for not recognising the opportunity to rethink its fleet mixes to acquire more UAVs.[21] Finally, the reality of defence as the single largest non-statutory government expenditure combined with the limited defence constituency continued to make it an inevitable target during periods of fiscal restraint. Policy promises to increase expenditure and force sizes in 2005 and 2008 were reversed by 2014.

As one policy observer asserts, 'Cutting the defence budget in austere times is in fact one of the few areas of Canadian public policy where there has been a rock solid bi-partisan consensus for a generation'.[22]

Canada's combat mission in Afghanistan ended in 2011. Its training mission continued to 2014, and a limited naval mission continues to the present. Its main legacy for defence and force planning was threefold: some new equipment and capabilities, including a new Defence Procurement Strategy; a hope that the expanded constituency for defence might translate into a more comprehensive renewal; and a new cadre of senior leaders who believed the future of conflict was intra-state conflict—the 'messy middle' of the conflict spectrum.[23] The consequence was a new variation of the old themes. Ambitious plans to rebuild the military post Afghanistan went nowhere, in part due to internal constraints discussed in the next sections, but also because of the Canadian strategic dilemma: what capabilities and capital equipment were required for the 'messy middle'. And, even after Russia invaded Ukraine in 2014 or as China became increasingly aggressive in its ambitions, its rhetoric and its actions there was no agreement on the nature of the threat to Canada. As a former Chief of Defence Staff reflected on the challenges of the period, 'Canadian Armed Forces do not procure capabilities unless they're absolutely necessary to the attainment of our mandate'.[24] In 2016, Canada had only approximately 1,100 personnel deployed on 14 missions.

The development and promulgation of the new defence strategy *Strong, Secure, Engaged* (SSE) in 2017, and a pending update, reflected an emerging recognition that the threat environment had changed, introduced a level of ambition for Canadian military commitments and stable, predictable funding to provide better guidance to force planning decisions and promised more financial and human resources.[25]

However, it could not overcome the traditional complacency of the public (and many political leaders) and the lack of consensus on how the changes in the environment threatened Canada, as well as the institutional constraints on change with the department and government. The defence policy's muted characterisation of the threat—threats to the international order and rule of law rather than Canada directly—was a reflection of the traditional Canadian defence paradox: invulnerable and indefensible. The Arctic was viewed as vulnerable to new technologies but the driver for the NORAD modernisation was alignment and contribution to US efforts. And despite the growing consensus that the threat had changed, particularly since February 2022, and concerns that Canada's spending on the military is diminishing its international influence,[26] those concerns are only just translating into a commitment to an updated defence policy, which has yet to be released. As of the writing of this chapter, it is unclear how the concerns will translate into resource commitments, but the government announced in an autumn 2023 cross-government review of expenditures to identify areas to cut.[27]

As of 2023, the 2017 defence policy commitments have not been fully realised. The CAF has eight core missions under the traditional three geographic focuses: Canada, North America and international. CAF has a significant role in aid to the civil power—including emergency aid during crises like the COVID-19 pandemic and wildfire—and is responsible for the air component of search and rescue (as well as coordinating air and maritime). Canada has committed to significant investments in North America's early warning systems and Arctic capabilities. Internationally, the latest defence policy stated a level of ambition that introduced ambitions to be prepared to lead and sustain a medium-size deployment (500–1500) as well as concurrently deploying a similar-sized force and a mix of up to five smaller (100–500) limited time and sustained deployments. These were in addition to maintaining its NORAD obligations, NATO Article 5 commitments and the Disaster Assistance Response Team (DART).[28] To execute these missions, the 2017 strategy committed to an expansion of 3,500 (to 71,500 total) regular forces and 20,000 Primary Reserve Force members. That expansion of the establishment has not occurred, and the current Chief of Defence Staff (CDS) estimates the CAF is 15% short of 68,000.[29] There was no commitment to increasing the percentage of GDP allocated to defence.

Major fleet re-capitalisation was also announced in 2017 but will take a decade or more to be delivered, for reasons discussed later. In the air, Canada has finally committed to a fifth-generation fighter capability, purchasing 88 F35s to replace the 76 remaining CF-188 Hornets. The RCAF maintains its strategic and tactical lift capabilities, with the first of nine new strategic tankers transport (Airbus 330–200) arriving in 2023. Tactical airlift also supplements the RCAF's fixed and rotary wing capability tasked with an expansive domestic search and rescue role and provides it with the expeditionary and domestic reach central to Canada's defence commitments, at home and abroad. Delays in replacing its utility helicopter fleet suggest that the replacement may mirror the US Future Vertical Lift programme of crewed and uncrewed platforms. Canada has committed to replace its existing CP-140 Aurora anti-submarine warfare and surveillance aircraft but has not made a decision on how or when.

Canada has also re-committed to a blue water navy, with an increasing focus on Arctic capable vessels, a concern that is also driving a debate about renewing Canada's submarine capabilities. The National Shipbuilding Strategy (NSS) is an almost complete renewal of Canada's surface combat and non-combat fleets as well as its shipbuilding infrastructure. Major investments include 15 surface combatants based on the Type 26 frigates design, six offshore patrol vessels, two joint support ships and a range of icebreakers, search and rescue and research vessels for the Canadian Coast Guard.[30] The Royal Canadian Navy (RCN) also places a premium on being able, if necessary, to lead allied fleets, as it has when contributing to NATO reassurance missions. The RCN's ability for surveillance and control with rotary-wing capabilities on the frigates will be increasingly complemented by UAVs and autonomous underwater vehicles (AUVs). Challenges in the procurement process have delayed delivery and created significant cost overruns, notably for the NSS, where the first delivery is expected in the early 2030s.

The Canadian Army was also targeted for re-capitalisation of 'much of its land combat capabilities and its aging vehicle fleets while modernizing its command-and-control systems'. This was to include replacing its light-armoured and armoured combat support fleets and, notably, to include all-terrain vehicles, snowmobiles and other utility vehicles for use in the Arctic. Command and control, intelligence, surveillance and reconnaissance systems were to be modernised as was much of the individual soldiers' kit. New investments included ground-based air-defence systems and associated munitions. The re-capitalisation has been slow, and was arguably insufficient, a conclusion suggested by the war in Ukraine, which has highlighted ongoing deficiencies such as the lack of ground-based anti-aircraft systems and weapons to counter drones as well as the state of Canada's tank fleet. Deployments have also highlighted the pressing need for accelerating basic equipment purchases for soldiers, as projected under the Soldier Combat System project, due to be completed in 2035.[31] It also maintains a niche auxiliary for the northern region, the Canadian Rangers, a localised capability generated for employment by the CAF to assist in sparsely settled remote, northern, coastal and isolated areas. The army reserves also bear the brunt of emergency response aid to civil power. It deployed 1400 personnel in 2021–2022 across the country, including the northern territories.

Canada's Special Forces, expanded during the counterterror missions, has also grown. It has its own command reporting to the CDS, and the 2017 policy boosted the establishment of 1900 by 605 personnel and promised $1.5 billion for new equipment. Cyber and space capabilities also continue to be enhanced, notably through the expansion of the missions, in collaboration with Canadian government cyber actors, to include defensive and offensive cyber operations. Major re-capitalisation is slow. Securing and operationalising many key enablers for joint force integration, sustainment and digitalisation are glacial, hindering the effectiveness of the attempts to modernise.

The promises in 2017 reflected Canada's recognition of the return of great power conflict and competition, and the consequent need for collective security at home and in Europe after 2014. The slow pace of reconstitution and increased investment from 2017 to the present reflected the more traditional constraints, not least of which was the apathy of the Canadian public. The commitments are consistent across Canada's three main defence priorities and reflect, to a degree, its level of ambition. The naked Russian aggression in 2022 shows some signs of accelerating the need to rebuild and modernise. Canada has committed to enhancing its contribution to NATO deterrence and to restoring NORAD, and hemispheric, defence. In the former, Canada's primary mission in Europe remains the NATO collective defence deterrence mission. In July 2023, Canada announced plans to augment its existing 800-person contribution to the enhanced Forward Presence Battlegroup Latvia with an additional 1200 personnel as well as new equipment by 2026, turning it into a Brigade. It also continues to train Ukrainian forces and participate in the maritime and air policing mission. In Canada and North America, in addition to the investments in fighters, space, radar and naval assets to survey and respond in the Arctic as well as Pacific and Atlantic maritime approaches, in June 2022, Canada announced a 38.6 billion 20-year funding commitment to modernize NORAD.[32] The navy programme is driven by increasing concerns for the security of North America and the lines of communication with Europe. Canada also contributes to other hemispheric maritime efforts, supporting US-led efforts to counter the flow of illegal narcotics into North America by interdicting drug trafficking activities in the Caribbean and Central America. Canada also contributes to land and maritime missions in the Middle East and Africa. The most notable shift in emphasis is Canada's increase in military and security-based engagement, including an enhanced posture, in the Indo-Pacific, guided by a new Indo-Pacific Strategy to assert Canada as a Pacific nation. Equally notable is the limited movement on the policy promise of reinvigorating Canada's contribution to UN peacekeeping, a challenge given an already stressed CAF. In July 2023, 57 personnel were contributing to seven UN missions, from a high point in 1993 of 3,336 personnel on deployments that included Bosnia and Somalia.[33]

## Part 2: approaches, processes, methods and techniques: too far from inflammable materials

Policy guidance for defence planning is, in theory, driven by periodic foreign and defence policy reviews, informed by assessments of the current geopolitical environment, fiscal situation and allied evaluations and changes. The policy should be informed by anticipated changes in the future security environment (including policy changes by allies), assessments of the implications of new technologies or changes to the economic environment. In the Westminster system of government, major issues and changes to address them are provided by a White Paper or a similar document. In Canadian practice, the issue of major policy in this form is not systematised and indeed is increasingly rare. External shocks, including internal economic challenges such as structural deficits, as well as what could be described as internal stress—domestic political or even the pent-up pressure resulting from trying to address the impending consequences of too many defence deficiencies—drive the development and issue of new policy documents. Canada issues no periodic national foreign policy and no guidance for what its military should look like beyond the historical sense of itself in the world as a supporter of multilateralism and the rule of law.

In the last four decades, Canada has issued six defence policy papers, in the form of White papers or under another title. While most purported to forecast ongoing challenges, they were driven by existing changes to the threat environment.[34] Indeed, arguably, new governments were reacting to shifts in US and allied perceptions of the threat environment as well as public sentiment as much as by internal analysis and advice. There was, for example, almost a decade and a half between the 1994 *White Paper on Defence*, focused on the post-Cold War peace dividend, and the 2008 *Canada First Defence Strategy*, issued in recognition of the war on terror and the belief that the intra-state threat environment was the norm. Even after the issue of policy direction, the level of ambition remained vulnerable to new fiscal realities. Canada's most recent policy document—*Strong, Secure and Engaged*—remains both underfunded and, as will be explained, challenging to execute given other constraints imposed by glacial and ponderous Canadian procurement processes.[35]

Translating policy direction and guidance into defence strategy and planning is not fully institutionalised. There is no consistent development process or regular issue of a departmental strategy document to discipline or drive a process. There are sections of the department devoted to policy and strategy but, when not involved in a policy event, they focus on specific environments, issues— cyber, for example—or responses to emerging challenges. The latter focus on the political and operational levels consumes staff effort at the expense of the strategy. The de-centralised stove-piped efforts come at the expense of a department and joint approach to defence and force planning. This has three consequences. First, strategy and force development is often slowed or stopped as its one-off nature weighs it down with expectations that it needs to solve everything at once or concerns that it isn't being undertaken correctly. Second, subordinate service or functional strategies are misaligned or attempt to fill in gaps due to the absence or slow production of departmental strategy. And finally, strategy development does not reconcile current operational requirements with future operating requirements.

A centralised force development process has promoted future security and operating environment studies to achieve those ends but with mixed results. The strategic level is not habituated to regularised strategy development, nor is there any legislative or political requirement to do so. There are means for senior leadership to develop and communicate service, function or domain-specific strategic guidance and direction, much of the latter experience being intuited rather than institutionalised, and a result, at least in part, of the vacuum left by the absence of departmental strategies.[36] Across the department and Canadian Armed Forces (CAF), strategy development and direction remains an 'event' rather than a snapshot of existing outputs of a process, subject to ongoing review and scheduled revision. The status as an event places substantial pressure to address most ills on any attempts to develop and publish department-level strategy.

The inconsistent, reactive policy and strategy direction shape the perception of the utility of strategy development and discourage a culture of strategic thinking. There are also additional structural challenges which reinforce service cultures and are an impediment to systematic and broad defence and force planning which, when combined with the scale and level of Canada's military ambition, combine to further inhibit the development of a cross-service, joint force, perceptive and consequent department-wide direction.

The structural, or institutional, constraints begin with the organisation of the department, the legacy of the search for efficiencies of scale and to both limit inter-service rivalries and better coordinate the activities of the three traditional services that began in earnest after the Second World War. This search begat the policies of Integration and Unification. By the 1960s, the main result had been to create an imbalance between political control and the independence of the services. A 1964 White Paper, acting on recommendations from a Royal Commission on Government Organization, enacted, first, an integrated headquarters structure with a single CDS and then in 1968, the Canadian Forces Reorganization Act which abolished the Royal Canadian Navy, Canadian Army and Royal Canadian Air Force as separate entities to be replaced by the Canadian Forces (later the CAF) under the CDS. This was also an attempt to eliminate service cultures and their identification with them. Instead, the 1964 White Paper saw it as the first step towards a 'unified [an interoperable] defence force for Canada'.[37] Traditional ranks and uniforms were discarded in favour of one single uniform and a common rank structure. Further departmental integration created a diarchy of a military chief—the CDS—and a Deputy Minister, to better balance civilian control of the military National Defence Headquarters (NDHQ) with the department and CAF and to distinguish between defence and military advice.[38]

The integrated Department of National Defence (DND) and NDHQ structure has evolved considerably. Many administrative functions, such as recruiting, were integrated. Enabling functions remained more challenging to separate, particularly when they were or are critical service enablers. Services had the capacity to forge ahead where the department of CAF joint staffs could not. The future-oriented capability-based planning method used for force development and the organisations and processes established to implement it has promoted centralisation through new structures and processes, particularly in the first decade and a half of the twenty-first century. Unification, however, was never completed. The tension between the unified character of the CAF and the re-assertion of service identity has been a constant theme since the policies were enacted and became politicised as a metric of support for the armed forces. New governments reintroduced service uniforms, ranks and even redesignated them by their historic names (Royal Canadian Navy, for example) to emphasise historic ties as well as an indicator of general support for defence in the military.

The tension between a CAF identity and the re-assertion of service identity impacts strategic direction, and thus defence and force management and planning. Canada has no standing joint forces or joint command.[39] Its operational headquarters—Canadian Joint Operations Command (CJOC)—functions as a joint champion. Force development is centralised and coordinated by a committee but not synchronised or mediated by top-down strategic-level direction. Canada's operational commitments—its military ambitions—are shaped by this to a degree. Services generate service-oriented task forces: tailored force packages, sustainable as expeditionary operations, designed to integrate and be interoperable with US and NATO allies. This approach is also a function of scale which constrains the maintenance of large standing operational structures.

Within the DND, the future-oriented capability-based planning method used for force development and the organisations and processes established to implement it would look—in outline—familiar to any NATO or western-influenced defence and military practitioner.[40] Since 2000, DND has used Capability-based Planning (CBP) as the DND/CAF force development methodology. CBP was adopted because it was perceived as the most appropriate and effective force development methodology to deal with an uncertain strategic environment and to understand defence outputs from a system's perspective.[41] CBP was designed to address the uncertainty facing defence planners at the end of the Cold War when the main question was defence against whom? Flexibility and adaptability were posited as the means to drive preparedness in the absence of a main adversary and a focus on the future would pull defence planning towards the next wars, not the last ones.[42] A focus on capability requirements would create efficiencies by abstracting the requirement and making decisions service neutral.[43]

DND's CBP process continues to undergo refinement.[44] It is increasingly threat-informed and retooling in anticipation of the threat of state-on-state high-intensity warfare, integrated into a whole-of-government approach to address threatening, below-the-threshold activities. Canada's threat lens is focused primarily on two regions, navigated through NORAD (continental) and NATO but with increasing attention to the Five-Eyes Indo-Pacific efforts.[45] The implementation of its defence and force planning methods is also shaped by the departments' structural and cultural idiosyncrasies. The institutionalisation—and the reach and influence—of centralised and joint force planning and development is limited. Centralised force development is hindered by the absence of a joint champion and departmental direction. Canada makes incremental changes through its political direction, but most capability and capital equipment acquisition, when it occurs, is driven by external events, usually changes in the threat environment and those that affect our allies, in particular our primary ally, the US.

The challenges within the department both reflect and are also products of cross-government challenges and the historical themes that contribute to Canada's place in the world as well as its own sense of that place. Interoperability with the US and NATO is both necessary and expensive. That requirement when combined with the size of the CAF as well as a desire to retain a certain international standing by deploying a multi-purpose, combat-capable force (versus, for example, a constabulary one) results in a one-platform or fleet reality where each must be multi-roled, capable of contributing to the defence of North America and elsewhere in the globe as an element—and ideally a leading element—of an integrated Allied force. Large-scale purchases or fleet acquisitions must also meet industrial and regional development requirements, effectively politicising defence economics. In the post–Cold War era, Canada's Afghan commitments only underlined how an abstract capability-based approach to force planning further hindered the overwhelming consensus necessary to reconcile the tensions between political and social imperatives and defence spending. Canada, despite these clear—and traditional—financial and political constraints on defence commitments relative to social and economic development remains fixated on maintaining options as if it were operating in the post–Second World War environment. All these secondary factors interact with a glacial procurement system to create decision paralysis and a feast or famine approach to major re-capitalisation.

Applying Fruhling's four conceptual frameworks within which to address this interaction between planning and uncertainty, Canada's strategic circumstances have created a tradition of what could be identified as a blend of mobilisation and task planning, which focuses on a risk 'that is uncertain because the threat is only prospective, and may arise at an unknown point in the future' and where the scale of effort to contribute to an alliance effort is limited except in the event of a full-scale conflict perceived by Canadian leaders as an existential threat. In practice, the uncertainty consequent upon the multiple constraints hinders planning.[46] Notably, there is little dialogue between policymakers and defence planners. The iterative requirement of the relationship is absent. Defence and force planners should provide their analysis and advice to allow policymakers to assess the risks of the strategic environment, weigh options and evaluate recommended capability investments and divestments.[47] In practice, the threat environment of the post–February 2022 world has prompted internal departmental calls for planning based on net assessments focused on identified and dominant threats where, even if in different regions, there is a 'well-understood and immediate risk'.

## Part 3: military innovation and adaptation: the North American transformation gap

Institutional dynamics and strategic culture suggest that Canada does not decide on transformative changes for defence, but has it forced upon it by events. Although Canada expends institutional effort to identify the elements required to innovate in defence, it struggles to modernise—that is keep pace with allies, primarily the US—to the degree that in the 2020s, modernisation would be transformational (see Figure 12.1).

Unsurprisingly, Canada measures its modernisation and transformation ambitions and progress against the constantly evolving yardstick of interoperability with the US. The US, and to a lesser extent the Five Eyes and NATO, perspective on the threat environment provides the 'pacing' assessment. This establishes aspirations and much anguish within the defence and security stakeholder communities, primarily because major equipment re-capitalisation, investments in technology, or changes to force design or posture are only evident when Canada rallies in response to strategic events. One caveat is necessary. The crisis sense of urgency required to prompt a fulsome modernisation, and possibly a transformational change, may

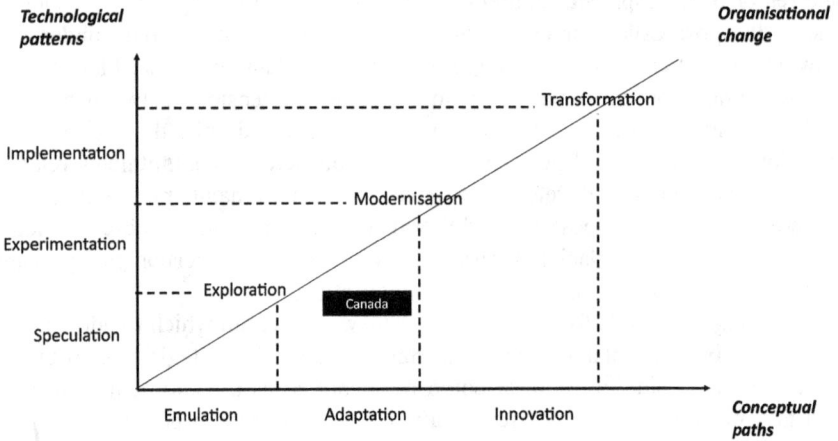

*Figure 12.1* Military innovation trajectories.

*Source:* Figure created by authors based on Michael Raska, Military Innovation of Small States—Creating a Reverse Asymmetry (Routledge, 2016)[48]

increasingly result from a growing and increasingly public belief that the Canadian armed forces and Canada's approach to defence and international commitments are converging to create a real institutional crisis.

Canada does not traditionally innovate in defence, although it will emulate innovations.[49] This assessment may need to be qualified if newer domains such as space are considered in isolation but taken as a whole it remains valid.[50] As the preceding sections suggest, Canada has adopted a force planning process which promotes innovation and experimentation to drive transformation. The department funds internal and external research to solicit innovative concepts and technological developments, for example, the Innovation for Defence Excellence and Security (IDEaS) programme.[51] It has initiated a digitalisation strategy. In practice, however, and generally, the consequence or indeed the legacy of its geography, history and defence choices—its strategic culture of reaction—forces it to, at best, emulate and implement in peacetime. Canada is challenged to get past exploration.

The same institutional, political and cultural roadblocks that hinder force planning act as a brake on innovation. Innovation is, or has been, largely driven from the top down, and technology rather than concept driven. While initiatives like IDEaS are meant to drive what Adamsky describes as the speculation phase, in practice, the department rarely has the time, scale or resources—financial or analytical—to pursue major innovations, or to execute and integrate those which it identifies as important.[52] Applying new concepts of forced employment on its own would also be a risky venture for a country that will only operate with the US and NATO allies.[53] Canada's continental commitments make operating with the US the primary driver for innovation, and, as suggested, addressing that gap is its primary defence fixation, even if a fulsome public discussion of that fact is politically challenging.[54]

Defence is thus challenged to get past the exploration phase, where speculation meets real-world pressures to reinforce the reality of emulation. The other primary drivers of innovation, deriving from Canada's desire to use defence acquisition as an industrial policy to maintain key traditional sectors, equitably distributed across Canada's regions, notably the aerospace and ship-building sectors, but also to promote new cyber and space-related technologies, also results in a highly politicised and bureaucratised procurement system.[55] These new technologies are often dual-use, driven by the requirements to surveil and control Canada's diverse, and austere, geography.[56] In Canada, defence procurement is a complex process involving several federal departments and agencies, notably the DND, Public Services and Procurement Canada (PSPC), Innovation, Science and Economic Development Canada (ISED) and the Treasury Board of Canada Secretariat. This multidepartmental approach to defence procurement, whereby each department and agency is responsible for a particular stage in the process, is unique to Canada.[57] The structural and cultural obstacles to defence and force planning also contribute to a risk-averse culture.

Barry Posen posits that military organisations 'place a premium on predictability, stability and certainty'. Modern western military and defence organisations are thus highly structured, hierarchical, risk averse and are subject to the same bureaucracy as government. These characteristics are viewed as optimal for the conduct of operations and accountability, but, if an innovative culture requires thinking disruptively and embracing failure, they are obstacles to innovation.[58] Internally, achieving consensus among the services is challenging given the abstract capability-based planning system and the procurement system. In addition, the capital costs of major fleets to address a range of high-intensity scenarios also risks procurement being perceived as a zero-sum game, particularly when re-capitalisation is overdue for all fleets. The culture of innovation may also reflect a more general Canadian trend. While the direct connection is unclear, the Canadian commercial sector in general has a low propensity to innovate, spending less on research and development than in comparable countries.[59]

Modernisation, while often conflated with transformation, is recognised as necessary but not sufficient to maintain Canada's ability to fight alongside allies and its credibility with its own military personnel. The current tenor of the debate is the need to address concurrently a capability gap crisis, a culture and leadership crisis and a reconstitution crisis (recruiting and retention) crisis, all of which are interrelated. They are collectively driving modernisation and institutional efforts that will be, if effected, potentially transformative by modernising Canada's military.

The digitalisation planning is a useful illustration of both the approach to innovation and transformative efforts, as well as ongoing challenges to move to modernisation posed by the institutional dysfunction. The department self-assesses that it is currently 'digitally aware', the first stage on a four-stage model, which is 'reflected by legacy analog systems and processes, stove-piped capability development, and generally low levels of digital literacy'. The work underway currently will move the CAF and the department along the path from digitally enabled to digitally transformed. That state will provide the baseline for a culture change, and thus the final stage of continuous digital innovation.[60]

The plan qualifies its ambitions, noting that at the strategic level, 'work continues on the development of a broader Department of National Defence (DND)/CAF Digital Strategy', as well as structures and processes—the infrastructure—necessary to 'accelerate our enterprise-wide digital transformation'. It also notes the concurrent efforts to modernise its approach to data—the DND/CAF Data Strategy and 'the Artificial Intelligence Strategy' are currently under development—as well as the 'other strategic policy guidance are shaping our organizational response to the digital challenge'.[61] The de-centralised initiative, while addressing the generally low levels of digital literacy, is also an ongoing challenge to a more systematic approach and risks maintaining a legacy of multiple, if updated, systems.[62] The Canadian Army's assessment of its own digital strategy is blunter: before digital transformation can take place, the 'first step is to move into the 21st century from a digital perspective',[63] 'changing our entire relationship with technology, bringing the CA [Canadian Army] from an analog organisation with legacy industrial-era force structures, tools, processes, governance models and culture, to a true, data-driven, innovative organisation that maximally leverages the military potential of digital solutions'.[64]

The digitalisation initiative is then, first and foremost, an exercise in modernisation and rationalisation across multiple people, processes, information, technology and institutional domains, including the civilian and military elements of the department.[65] The CAF is issuing strategy and plans before the department, which will be hindered in its infrastructure changes by the procurement system. Meanwhile, Canada's equipment procurement focus is driven by 'urgent operational requirements (UOR)' identified from the lessons of combat in Ukraine as necessary for Canada's commitment to the NATO enhanced forward presence in Latvia, notably anti-tank weapons, air defence and counter-uncrewed aerial systems (UAS).[66] And this reality is viewed as one reason for the CAF's challenges to increase the strength of the CAF to the desired level of 101,500 military personnel.[67] Technological obsolescence and risks are a factor in the recruiting and retention problems. The key discussion points are foundational: people, culture, digitalisation and re-capitalisation.

The increasing systemic focus on the challenges, even if crisis-driven, is heartening. Canada's 2017 defence policy, *Strong, Secure, Engaged*, partly reflects this understanding in its call to adapt the CAF to a changing technological landscape, through innovation and modernisation. Canadian defence planning has incorporated new technology platforms—cybertechnologies and space-based surveillance assets, for example—into the national defence network, but not in a systematic manner. There are also ongoing efforts to harness internal analysis to drive innovation through greater support for gaming, experimentation and forecasting at defence and service warfare centres.

More importantly, the past half-decade has seen accelerated efforts to change the culture of innovation from one of dependence on top-down direction and initiative to one which cultivates ideas from across the organisation and recognises that defence—and government—cannot keep pace with the research and development in the commercial and academic sectors. The CAF is promoting the concept that every soldier is a force developer and agent of change, a concept that is more effective in operational settings than departmental ones. However, there is a more general recognition that accountability must be leavened with an acceptance of failure and the delegation of responsibility to the lowest levels of the organisation—the democratisation of innovation—to promote a less risk-averse culture.[68] Concurrently, initiatives that advance public-private partnerships are attempting to integrate defence processes and planning into the accelerated innovation cycles of industry and academia.[69] Defence recognises that the military and government cannot keep up with the commercial sector, and are establishing mechanisms to leverage innovation from the public sector.[70] The S&T IDEaS programme is an example of a step towards providing funds directly to innovators in academia and industry to accelerate change. The success of these initiatives to move from speculation through to execution and integration will require further changes to decision-making and procurement processes.

## Conclusion

Canada's defence planning continues to be predicated on maintaining the form of an autonomous and activist middle power, parroting the force planning processes and structures deemed necessary for a modern, late twentieth-century, multi-purpose, combat-capable military force with ambitions to be credible partners to the US and NATO while maintaining sufficient autonomy to pursue the strategic choices afforded by geography and proximity to the US[71] (see Figure 12.2).

| GENERAL COMMONALITIES | SMPS ADJUST THEIR APPROACHES BASED ON SYSTEMIC CHANGE OUT OF THEIR CONTROL | SMPS REACT TO CHANGES IN THE RELEVANT SECURITY COMPLEX(ES) THE SMP IS PART OF | | | | |
|---|---|---|---|---|---|---|
| | *Alliances, Dependencies and National Ambitions* | | | | | |
| | SMPS FACE CONSTRAINTS IN SAFEGUARDING NATIONAL SECURITY | FINANCIAL | DEMOGRAPHIC | CULTURAL | GEOGRAPHICAL | OTHER |
| | SMPS NEED TO POSITION THEMSELVES VIS-À-VIS GLOBAL AND REGIONAL POWER CONSTELLATIONS | BALANCING | HEDGING | BANDWAGONING | NON-ALIGNING | OTHER |
| | SMPS HAVE TO ENGAGE WITH OTHER STATES TO INCREASE THEIR DEFENCE CAPABILITIES | SECURITY GUARANTEES | TRANSACTIONAL | FORCE INTEGRATION | SPECIALISATION | OTHER |
| | SMPS HAVE ARMED FORCES FOR DIFFERENT PURPOSES | INSTRUMENT OF DIPLOMACY | PRESTIGE PROJECT | REDUCE STRATEGIC RISK (DEFEND) | DOMESTIC STABILITY | OTHER |
| | *Approaches, Processes, Methods and Techniques* | | | | | |
| | SMPS TAKE A MORE CONSTRAINED APPROACH TO DEFENCE PLANNING | THREAT-NET ASSESSMENT BASED | PORTFOLIO-BASED | TASK-BASED | MOBILISATION | OTHER |
| | SMPS HAVE LIMITED RESOURCES TO DESIGN AND IMPLEMENT DEFENCE PLANNING PROCESSES | TAILORED ANALYTICAL PROCESS | OUTSOURCING STRATEGY | ACCEPT DILUTION / FAIR SHARE | EXTERNAL TEMPLATE | OTHER |
| | SMP EMPLOY MULTIPLE METHODS AND TECHNIQUES | COPY GREAT POWERS | INFORMAL CONSENSUS | TAILORED METHODS | AD HOC | OTHER |
| | SMPS RELY ON DIFFERENT PROCESSES TO DECIDE ON FORCE POSTURE | ANALYTICAL RESULTS | POLITICS DECIDES | MILITARY DECIDES | GREAT POWER DECIDES | OTHER |
| | *Military Innovation* | | | | | |
| | SMPS FACE CHALLENGES IN KEEPING PACE WITH TECHNOLOGICAL ADVANCES | BUILD DOMESTIC TECH/INDUSTRY | TECH TRANSFER AGREEMENTS | ACQUIRE READY INNOVATIONS | PARTNER WITH GREAT POWER | OTHER |
| | SMPS PURSUE VARIOUS MILITARY INNOVATION PATHS AND PATTERNS | EMULATION / SPECULATION | ADAPTATION / EXPERIMENTATION | INNOVATION / IMPLEMENTATION | OTHER / OTHER | OTHER |
| | SMPS FIND THEMSELVES AT DIFFERENT STAGES OF ORGANISATIONAL CHANGE | EXPLORATION | MODERNISATION | TRANSFORMATION | OTHER | OTHER |

*Figure 12.2* Structured focused comparison framework for Canada.

*Source:* Author

Canada's strategic culture—a product of history and geography—and geopolitical position are inherent constraints. Canada's defence conundrum is exacerbated by the public perception that geography has rendered Canada invulnerable. Political priorities, and resources, follow accordingly. Defence spending is forecast to rise in real terms to 51 billion and to 1.59% of GDP by 2026–2027.[72] Still, many of the challenges come from a stagnant policy and strategy development environment and from within the defence structure. Public apathy suggests the requirement for leaders to better articulate the nature of the threats rather than expect Canadians to support imprecise and open-ended aspirations to defend democracy and the current global order. Canada's government—and more specifically the Prime Minister—sets the conditions for the policy and resource environment.

Canada's ability to develop, plan for and deliver the defence and security believed to be commensurate with its vision of itself in the global order is indeed threatened.[73] The reality is that Canadians have not, and do not, perceive defence and security as priorities for spending, relative to economic and social spending, but the threat is not just about money and the artificial GDP targets touted by NATO. Canada remains the sixth biggest spender in NATO, and 14th in the world. But Canada cannot spend the money currently allotted to it or quickly secure re-capitalisation even where political will and financial commitments exist.[74]

Canada partners for security but has also, since the dawn of the Cold War, promoted collective security and multilateralism as principles of defence and security. Interoperability, primarily with the US, is a basic principle. It pursues limited integration with the US, for example, the bi-national NORAD, and has a range of senior posts in US formations and commands, but prefers what was once deemed, in a British context, 'in combination' a modular relationship with more direct lines to Canadian authorities. Canada's defence establishment seeks a cultural disruption to end the endemic challenges and equipment shortfalls that constrain it and wants to move from a model where crisis is a capability gap filler and major re-capitalisation is a feast or famine. There are real risks that we will be unable to operate with allies (read US), and not be invited to international initiatives (read AUKUS). This is reflected in the discernible shift from a deliberate balanced hedging strategy to an unintentional but real possibility of bandwagoning.

As Canada's capabilities decline, its influence has been increasingly viewed as 'value-centric' with a focus on promoting the principle of multilateralism and the 'liberal international order' with the emphasis on values increasing in proportion to a decline in the ability to contribute to multilateral defence efforts.[75]

The threat is less uncertain but abstracted to contributing to efforts to support norms and principles, resulting in an approach to defence planning that blends mobilisation and task planning, which focuses on a risk 'that is uncertain because the threat is only prospective, and may arise at an unknown point in the future'. Historically, when the threat or crisis is concrete, Canada and Canadians respond accordingly. That makes peace-time planning a challenge. The challenges are multiplied by Canada's lack of a joint service culture, as well as the consequent inter-service rivalries, and a bureaucratized, risk-averse military and defence establishment. Canada's adaptation of CBP requires a robust dialogue between and amongst senior political and military leaders.

Politicians decide on procurement, although within a range determined by the need to operate with the US, but posture choices are primarily determined by alliance and great power commitments and concerns. Canada's stretched resources and range of commitments can mean that for smaller missions, providing a contribution is the strategic objective. This resource-constrained environment also affects the capacity to optimise the methods and techniques used for force planning. Small staffs and limited subject matter expertise, as well as the demands of interoperability, can constrain the ability—and desire—to do more than adopt and adapt to the systems and concepts used by Alliance partners. Although Canada maintains a robust intelligence and analytical capability, its efficient use is hampered by the absence of top-down guidance and being blunted through the curation of multiple narrow service and domain-focused stove pipes.

The constraints imposed by the mismatch between scale resulting from inherent constraints and the scope of ambition, exacerbated by Canada's adoption of out-sized force planning methods and structures, drive the way in which defence innovates and interacts with technology. Equipping defence is unsurprisingly an economic development plank. Adding to the complexity, industrialisation is also a means of maintaining regional development equilibrium. And the need to Canadianise platforms to multi-role them—adapt the technology it emulates—adds additional expectations to major fleet and equipment purchases. These factors hinder modernisation—that is, re-equipping Canada's military for mid-twenty-first-century challenges. Canada remains challenged to translate its force planning processes and analytical speculation into ends to transform. Modernisation itself would be transformational.

In sum, the reality is that Canadians have not, and do not, perceive defence and security as priorities for spending, relative to economic and social spending. The resources that are provided are not optimally used and the cumulative effect is telling. In the security environment of the 2020s, Canada is faced with the harsh reality that it may not have the military heft to maintain its sense of self as a middle power in the global order.

## Notes

1  The opinions and assessments expressed in this chapter are the author's alone and do not reflect the views of the Government of Canada or the Department of National Defence.
2  C.P. Stacey, *The Military Problems of Canada. A Survey of Defence Policies and Strategic Conditions Past and Present* (Toronto: Ryerson Press, 1940), 53.
3  For an overview of the challenges of linking defence policy and force development in a Canadian Defence context, see Paul Dickson and Michael Roi, *Government Policy, Military Advice and Force Planning: An Essential Dialogue* (Ottawa: Her Majesty's Printer, 2014); Michael Roi, *Canadian Defence Priorities, CF Force Posture and Strategic Readiness. Linking Government Policy Preferences to Resource Allocations* (Ottawa: Her Majesty's Printer, December 2012).
4  See, for example, Stacey, *The Military Problems of Canada* and R.J. Sutherland, 'Canada's Long-Term Strategic Situation', *International Journal* 17, no. 3 (1962): 199–223.
5  See, for example, Alan Taylor, *The Civil War of 1812* (New York: Alfred Knopf, 2010).
6  Desmond Morton, 'Defending the Indefensible: Some Historical Perspectives on Canadian Defence 1867–1987', *International Journal* 43, no. 4 (1987): 627–644.

7  Robert Kaplan, *The Revenge of Geography* (New York: Random House, 2012), 91.
8  Kaplan, *The Revenge*, 332; see, for example, Richard Preston, *The Defence of the Unde-fended Border: Planning for War in North America, 1867–1939* (Montreal: McGill-Queen's University Press, 1977); Francis M. Carroll, 'The Passionate Canadians: The Historical Debate About the Eastern Canadian-American Boundary', *The New England Quarterly* 70, no. 1 (1997): 83–101.
9  Robert Bothwell, 'The Canadian Isolationist Tradition', *Contemporary International History* 54, no. 1 (1999): 76–87.
10  Dr. P. Whitney Lackenbauer, *Defence Against Help* (Ottawa: Defence and Security Foresight Group, 2020).
11  Geoffrey Hayes, 'Canada as a Middle Power: The Case for Peacekeeping', in *Niche Diplomacy. Studies in Diplomacy*, ed. A.F. Cooper (London: Palgrave Macmillan, 1997), 73–89.
12  See, for example, Joseph T. Jockel and Joel J. Sokolsky, 'Continental Defence: "Like farmers who have a common concession line"', in *Canada's National Security in the Post 9/11 World*, ed. David S. McDonough (Toronto: University of Toronto Press, 2012), 114–140; For pre-11 September 2001 perspective, see Policy Options, 'The Canada-U.S. Defence Relationship: Nostalgia Ain't What It Used to Be', *Policy Options*, 1 April 2002. The article quotes David Rudd's characterisation of Canada's perpetual crisis of means. Available at: https://policyoptions.irpp.org/magazines/continental-defence/the-canada-us-defence-relationship-nostalgia-aint-what-it-used-to-be/.
13  Lackenbauer, *Defence Against Help*.
14  Eric J. Labs, 'Do Weak States Bandwagon?', *Security Studies* 1, no. 3 (1992): 383–416, DOI: 10.1080/09636419209347476.
15  Eric Wagner, 'The Peaceable Kingdom? The National Myth of Canadian Peacekeeping and The Cold War', *The Canadian Military Journal* 12, no. 1 (2007): 45–54.
16  See Macrotrends, 'Canada Military Spending/Defense Budget 1960–2023', *Macrotrends*, www.macrotrends.net/countries/CAN/canada/military-spending-defense-budget; Accessed 21 August 2023.
17  GOC, DND, *White Paper Challenge and Commitment: A Defence Policy for Canada* (Ottawa: Queens' Printer, 1987); GOC, DND, *White Paper on Defence* (Ottawa: Queens' Printer, 1994).
18  Janice Gross Stein and Eugene Lang, *The Unexpected War: Canada in Kandahar* (Toronto: Viking Canada, 2007).
19  Stein and Lang, *The Unexpected War*; See also Sean Maloney, '"Was It Worth It?" Canadian Intervention in Afghanistan and Perceptions of Success and Failure', *Canadian Military Journal* 14, no. 1 (2013): 19–31; and General Walt Natynczyk, 'The Canadian Forces in 2010 and 2011 – Looking Back and Looking Forward', *Canadian Military Journal* 11, no. 2 (2011): 7–11.
20  Maloney, 'Was It Worth It'.
21  Major Matthew Snider, *Targeting Air Power: The Failure of The RCAF to Adapt in the Global War on Terror* (Toronto: Canadian Forces College, 2020).
22  Eugene Lang, 'The Shelf Life of White Papers', *Policy Options*, 23 June 2017, https://policyoptions.irpp.org/magazines/june-2017/shelf-life-defence-white-papers/ (Accessed May 2023).
23  United States Congress House Committee on Armed Services Subcommittee on Intelligence, Emerging Threats and Capabilities, *Perspectives on the Future National Security Environment* (Washington, DC: US Congress, 2013).
24  Murray Brewster, 'More Than a Decade Ago, the Army Had a Plan to Rebuild. It Went Nowhere', *CBC Newsite*, 7 January 2023, www.cbc.ca/news/politics/canadian-armed-forces-equipment-procurement-ukraine-latvia-1.6706444.
25  Government of Canada, *Strong, Secure, Engaged: Canada's Defence Policy* (Ottawa: Queen's Printer, 2017), www.canada.ca/en/department-national-defence/corporate/policies-standards/canada-defence-policy.html.

26 See Christian Leuprecht and Joel J. Sokolsky, 'Defence Policy "Walmart Style": Canadian Lessons in "Not-So-Grand" Grand Strategy', in *Going to War: Trends in Military Intervention*, eds. Stefanie Von Hlatky and H. Christian Breede (Montreal: McGill-Queens University Press), 193–212.

27 Lee Berthiaume, 'Budget 2023: Battle Looming between Defence Officials and Lawmakers, Experts Say', *Global News*, 4 April 2023, https://globalnews.ca/news/9600369/budget-2023-canada-defence-spending/, accessed 23 August 2023.

28 GOC, DND, *Strong, Secure, Engaged*.

29 Murray Brewster, 'Military Personnel Shortage Will Get Worse Before It Gets Better, Top Soldier Says', *CBC News Website*, 6 October 2022. www.cbc.ca/news/politics/eyre-shortage-directive-1.6608107, accessed 23 August 2023.

30 GOC, Public Services and Procurement Canada, *National Shipbuilding Strategy*, www.tpsgc-pwgsc.gc.ca/app-acq/amd-dp/mer-sea/sncn-nss/index-eng.html, accessed 20 August 2023.

31 GOC, DND, *Advancing with Purpose: The Canadian Amy Modernization Strategy* (Ottawa: Queens Printer, 2020).

32 GOC, DND, 'NORAD Modernization Project Timelines', *Government of Canada*, www.canada.ca/en/department-national-defence/services/operations/allies-partners/norad/norad-modernization-project-timelines.html, accessed 22 August 2023.

33 Walter Dorn, 'Tracking the Promises: Canada's Contributions to UN Peacekeeping', *WalterdornNet*, www.walterdorn.net/256, accessed 31 August 2023.

34 Lang, 'The Shelf Like of White Papers', *Policy Options*.

35 GOC, DND, *Strong, Secure and Engaged*.

36 Erica Wiseman, 'The Institutionalization of Organizational Learning: A Neo Institutional Perspective', in *Proceedings of OLKC 2007—"Learning Fusion"* (Warwick University, 2007), 1112–1136.

37 GOC, *White Paper on Defence* (Ottawa: Queen's Printer, March 1964), 19.

38 Doug Bland, *Chiefs of Defence: Government and the Unified Command of the Canadian Armed Forces* (Ottawa: Canadian Institute of Strategic Studies, 1995).

39 Canada's Special Operations Force does operate as a joint force. Lieutenant-General R.R. Crabbe (retired), Vice-Admiral L.G. Mason (retired) and Lieutenant-General F.R. Sutherland (retired), *A Report on the Validation of the Transformed Canadian Forces Command Structure* (Ottawa: DND, 31 January 2007), 24.

40 As the authors note, it soon becomes apparent that the timeframes, methods, processes and institutions involved vary considerably. Magnus Håkenstad and Kristian Knus Larsen, 'Long-Term Defence Planning: A Comparative Study of Seven Countries', *Oslo Files on Defence and Security* 1, no. 5 (2012): 11–12.

41 GOC DND, *Defence Planning Guidance 2001, Chapter 2: Force Planning Guidance* (Ottawa: Queens Printer 2000), www.resdal.org/Archivo/d00000d5.htm, accessed 22 August 2023; Andrew Godefroy, 'Chasing the Silver Bullet: The Evolution of Capability Development in the Canadian Army', *Canadian Military Journal* (2007): 53–66.

42 See Paul K. Davis and Lou Finch, *Defense Planning for the Post-Cold War Era: Giving Meaning to Flexibility, Adaptiveness and Robustness of Capability* (Santa Monica, CA: RAND Corporation, 1993); For some of the challenges that resulted from the execution of CBP, see Stephan De Spiegeleire, 'Ten Trends in Capability Planning for Defence and Security', *The RUSI Journal* 156, no. 5 (2011): 20–28, https://doi.org/10.1080/03071847.2011.626270; Eric A. Hollister, *A shot in the Dark: The Futility of Long—Range Modernization Planning* (Arlington, VA: The Institute of Land Warfare, October 2010). For an overview of the early CBP discussions, see Paul Davis and Zalmay Khalilzad, *A Composite Approach to Air Force Planning* (Santa Monica, CA: RAND Corporation, 1996), 30–35.

43 Paul K. Davis, *Analytic Architecture for Capabilities-Based Planning, Mission System Analysis and Transformation* (Santa Monica, CA: RAND Corporation, 2002), 67–70.

44 Ben Taylor, *Towards and Enhanced Capability Based Planning Approach* (Ottawa: DRDC Reference Document, 2017).

45 Network for Strategic Analysis, *Threat Based Defence Planning: Implications for Canada* (Ottawa: Network for Strategic Analysis, 2021).

46 Stephan Frühling, *Defence Planning and Uncertainty* (New York: Routledge, 2014).

47 Dickson and Roi, *Government Policy*, 4; Mikkel Vedby Rasmussen, *The Risk Society at War* (Cambridge: Cambridge University Press, 2006).

48 Michael Raska, "A Structured-Phased Evolution: The Third Generation Force Transformation of the Singapore Armed Forces", in *Military Innovation in Small States: Creating a Reverse Asymmetry*, M. Raska (New York: Routledge, 2016), 130–162.

49 This section draws on the framework provided by Michael Raska, 'Strategic Competition for Emerging Military Technologies: Comparative Paths and Patterns', *Prism: A Journal of the Center for Complex Operations* 8, no. 3 (2020): 64.

50 Adam Grissom, 'The Future of Military Innovation Studies', *Journal of Strategic Studies* 29, no.5 (2006): 906–908, https://doi.org/10.1080/01402390600901067.

51 For example, DND funds an Innovation for Defence Excellence and Security (IDEas) programme, administered and evaluated by Defence Research and Development Canada, the DND Science and Technology division. 'Featured Opportunities', *Government of Canada*, updated 7 August 2023, www.canada.ca/en/department-national-defence/programs/defence-ideas.html.

52 Dima Adamsky, *The Culture of Innovation* (Stanford, CA: Stanford University Press, 2010).

53 Stephen Biddle, *Military Power: Explaining Victory and Defeat in Modern Battle* (Princeton, NJ, and Oxford: Princeton and Oxford University Press, 2004).

54 Theo Farrell and Terry Terriff, 'Military Transformation in NATO: A Framework for Analysis', in *A Transformation Gap? American Innovations and European Military Change*, eds. Terry Terriff, Frans Osinga and Theo Farrell (Stanford, CA: Stanford University Press, 2010), 1–13.

55 See Douglas Bland, *The Administration of Defence Policy in Canada, 1947 to 1985* (Oakland, CA: University of California, 1987); Daniel Gosselin, 'Unelected, Unarmed Servants of the State: The Changing Role of Senior Civil Servants Inside Canada's National Defence', *Canadian Military Journal* 14, no. 3 (2014): 38–52; Brian Bow, 'Parties and Partisanship in Canadian Defence Policy', *International Journal Special Edition: Electoral Politics and Policy: Annual John W. Holmes Issue on Canadian Foreign Policy* 64, no. 1 (2009): 67–88.

56 For a useful overview of some of the industrial and business drivers (and political constraints) behind Canada's ambitions for aerospace industry see the story of Canada's Cold War fighter jet programme. Palmiro Campagan, *Requiem for a Giant: A.v. Roe and the Avro Arrow* (Ontario: Dundurn Press, 2003).

57 Martin Auger, *Defence Procurement Organizations Worldwide: A Comparison* (Ottawa: Library of Parliament background Paper, 2020), 2–3.

58 Barry R. Posen, *The Sources of Military Doctrine* (Ithaca, NY: Cornell University Press, 1984).

59 GOC, *Budget 2022* (Ottawa: Queens Printer, 2022).

60 GOC, *Canadian Armed Forces Digital Campaign Plan* (Ottawa: Queens Printer, 2022), 6–7.

61 GOC, *Canadian Armed Forces,* 5.

62 GOC, *Canadian Armed Forces Digital*, 6–7.

63 Chris Thatcher, 'Digital Experiments', *Canadian Army Today* 7, no. 1 (2023): 26.

64 Canadian Army, *Modernization Vital Ground: Digital Strategy* (Ottawa: Canadian Army, 2022), 1.

65 Gosselin, 'Unelected, unarmed Servants of the State'.

66 Staff, 'Capabilities Needed-Urgently', *Canadian Army Today* 7, no. 1 (2023): 21.

67 DND, 'CDS/DM Directive for CAF Reconstitution', 6 October 2022, www.canada.ca/en/department-national-defence/corporate/policies-standards/dm-cds-directives/cds-dm-directive-caf-reconstitution.html.

68 Eric von Hippel, *Democratizing Innovation* (Cambridge, MA: The MIT Press, 2005).

69 Daniel Araya, 'Military Tech Is Evolving Fast: It's Time for Canada to Catch Up', *Centre for International Governance Innovation*, 26 May 2022, www.cigionline.org/articles/military-tech-is-evolving-fast-its-time-for-canada-to-catch-up/.

70 Canadian Army, *Modernization*, 2–3; See also, for example: GOC, *Canadian Armed Forces*, 9.

71 Tim Sweijs and Michael J. Mazarr, 'Mind the Middle Powers', *War on the Rocks*, 4 April 2023.

72 Sweijs and Mazarr, 'Mind the Middle Powers'.

73 Bruce Gilley, 'Middle Powers During Great Power Transitions', *International Journal* 66, no. 2 (2011): 245–264.

74 Office of the Parliamentary Budget Officer, *Canada's Military Expenditure and the Nato 2% Spending Target* (Ottawa: Her Majesty's Printer/Office of the Parliamentary Budget Officer, June 2022), 6–8.

75 There is a voluminous body of literature on the question of whether Canada is still a middle power. For example Zachary Paikan, 'Is Canada Still a Middle Power?', *The Institute for Peace and Diplomacy*, 22 April 2021; Eugene Lang, *Searching for a Middle-Power Role in a New World Order* (Calgary: Canadian Global Affairs Institute, June 2019).

## Bibliography

Adamsky, Dima. *The Culture of Innovation*. Stanford, CA: Stanford University Press, 2010.

Araya, Daniel. 'Military tec is evolving fast: it's time for Canada to catch up'. *Centre for International Governance Innovation*, 26 May 2022. www.cigionline.org/articles/military-tech-is-evolving-fast-its-time-for-canada-to-catch-up/. Accessed April 2023.

Auger, Martin. *Defence Procurement Organizations Worldwide: A Comparison, Publication No. 2019–52-E*. Ottawa: Library of Parliament Background Paper, 2020.

Biddle, Stephen. *Military Power: Explaining Victory and Defeat in Modern Battle*. Princeton and Oxford: Princeton and Oxford University Press, 2004.

Bland, Doug. *Chiefs of Defence: Government and the Unified Command of the Canadian Armed Forces*. Ottawa: Canadian Institute of Strategic Studies, 1995.

——— and R.P. Frye. *The Administration of Defence Policy in Canada, 1947 to 1985*. Oakland, CA: University of California, 1987.

Bothwell, Robert. 'The Canadian Isolationist Tradition,' *Contemporary International History* 54:1 (1999), 76–87.

Bow, Brian. 'Parties and Partisanship in Canadian Defence Policy,' *International Journal Special Edition*: Electoral Politics and Policy: Annual John W. Holmes Issue on Canadian Foreign Policy 64:1 (Winter, 2008/2009), 67–88.

Brewster, Murray. 'More than a decade ago, the army had a plan to rebuild. It went nowhere.' *CBC News Site*, 7 January 2023. www.cbc.ca/news/politics/canadian-armed-forces-equipment-procurement-ukraine-latvia-1.6706444. Accessed April 2023.

Campagan, Palmiro. *Requiem for a Giant: A.v. Roe and the Avro Arrow*. Ontario: Dundurn Press, 2003.

Carroll, Francis M. 'The Passionate Canadians: The Historical Debate about the Eastern Canadian-American Boundary,' *The New England Quarterly* 70:1 (March, 1997), 83–101.

Crabbe (retired), Lieutenant-General R. R., Vice-Admiral L. G. Mason (retired) and Lieutenant-General F. R. Sutherland (retired). *A Report on the Validation of the Transformed Canadian Forces Command Structure*. Ottawa: Department of National Defence, 2007.

Davis, Paul K. *Analytic Architecture for Capabilities-Based Planning, Mission System Analysis and Transformation*. Santa Monica, CA: Rand, 2002.

———— and Lou Finch. *Defense Planning for the Post-Cold War Era: Giving Meaning to Flexibility, Adaptiveness and Robustness of Capability*. Santa Monica, CA: RAND Corporation, 1993.

Davis, Paul K. and Zalmay Khalilzad. *A Composite Approach to Air Force Planning*. Santa Monica, CA: Rand, 1996.

De Spiegeleire, Stephan. 'Ten Trends in Capability Planning for Defence and Security,' *The RUSI Journal* 156:5 (October/November 2011).

Dickson, Paul and Michael Roi. *Government Policy, Military Advice and Force Planning: An Essential Dialogue*. Defence Research Development Canada Scientific Letter. Ottawa: Her Majesty's Printer, 2014.

Farrell, Theo and Terry Terriff, 'Military Transformation in NATO: A Framework for Analysis.' In *A Transformation Gap? American Innovations and European Military Change*, edited by Terry Terriff, Frans Osinga, and Theo Farrell, 1–13. Stanford, CA: Stanford University Press, 2010.

Frühling, Stephan. *Defence Planning and Uncertainty*. New York: Routledge 2014.

Gilley, Bruce, 'Middle powers during great power transitions,' *International Journal* 66:1 (Spring 2011), 245–264.

Godefroy, Andrew. 'Chasing the silver bullet: The evolution of capability development in the Canadian army'. *Canadian Military Journal* 8:1 (Spring 2007), 53–66.

Gosselin, Daniel. 'Unelected, unarmed servants of the state: The changing role of senior civil servants inside Canada's national defence,' *Canadian Military Journal* 14:3 (Summer 2014), 38–52.

*Government of Canada (GOC) Budget 2022*. Ottawa: Queens Printer, October 2022.

GOC. *White Paper on Defence*. Ottawa: Queens Printer, March 1964.

GOC. *Strong, Secure and Engaged: Canada's Defence Policy*. Ottawa: Queen's Printer, 2017.

GOC. *Canadian Armed Forces Digital Campaign Plan*. Ottawa: Queens Printer, 2022.

GOC, DND. *Defence Planning Guidance 2001*. Ottawa: Queens Printer, 2001.

GOC, DND. *CDS/DM Directive for CAF Reconstitution*. Ottawa: DND, 6 October 2022.

GOC, DND, Canadian Army. *Modernization Vital Ground: Digital Strategy*. Ottawa: Queens Printer, 2022.

GOC, Office of the Parliamentary Budget Officer. *Canada's Military Expenditure and the NATO 2% Spending Target*. Ottawa: Her Majesty's Printer/Office of the Parliamentary Budget Officer, June 2022.

Grissom, Adam. 'The future of military innovation studies'. *Journal of Strategic Studies* 29:5 (October 2006), 906–908.

Håkenstad, Magnus and Kristian Knus Larsen. 'Long-term Defence Planning: A Comparative Study of Seven Countries'. *Oslo Files on Defence and Security* 5:1 (2012).

Hayes, Geoffrey. 'Canada as a Middle Power: The Case for Peacekeeping.' In *Niche Diplomacy. Studies in Diplomacy*, edited by A. F. Cooper, 73–89. London: Palgrave Macmillan, 1997.

Hippel, eric von. *Democratizing Innovation*. Cambridge, MA: The MIT Press, 2005.

Hollister, Eric A. *A Shot in the Dark: The Futility of Long—Range Modernization Planning. The Land Warfare Papers No. 79*. Arlington, VA: The Institute of Land Warfare, 2010.

Jockel, Joseph T. and Joel J. Sokolsky. 'Continental Defence: "Like Farmers Who Have a Common Concession Line"'. In *Canada's National Security in the Post 9/11 World*, edited by David S. McDonough, 114–140. Toronto: University of Toronto Press, 2012.

Kaplan, Robert. *The Revenge of Geography*. New York: Random House, 2012.

Labs, Eric J. 'Do Weak States Bandwagon?.' *Security Studies* 1:3 (Spring 1992), 383–416.

Lackenbauer, P. Whitney. *Defence Against Help*. Ottawa: Defence and Security Foresight Group, 2020.

Lang, Eugene. 'The Shelf Like of White Papers'. *Policy Options*, 23 June 2017. https://policyoptions.irpp.org/magazines/june-2017/shelf-life-defence-white-papers/. Accessed May 2023.

————. 'Searching for a Middle-Power Role in a New World Order'. *Canadian Global Affairs Institute*, June 2019. https://www.cgai.ca/searching_for_a_middle_power_role_in_a_new_world_order

Leuprecht, Christian and Joel J. Sokolsk. 'Defence Policy "Walmart Style": Canadian Lessons in "Not-So-Grand" Grand Strategy'. In *Going to War: Trends in Military Intervention*, edited by Stefanie Von Hlatky and H. Christian Breede, 193–212. Montreal: McGill-Queens University Press.

Morton, Desmond. 'Defending the Indefensible: Some Historical Perspectives on Canadian Defence 1867–1987,' *International Journal* 43:4 (Autumn 1987), 627–644.

Network for Strategic Analysis. *Threat Based Defence Planning: Implications for Canada.* Ottawa: Network for Strategic Analysis, 2021.

Paikan, Zachary. 'Is Canada Still a Middle Power?', *The Institute for Peace and Diplomacy*, 22 April 2021. https://peacediplomacy.org/2021/04/22/is-canada-still-a-middle-power/

Policy Options. 'The Canada-U.S. Defence Relationship: Nostalgia Ain't What It Used to Be,' *Policy Options*, April 2002. https://policyoptions.irpp.org/magazines/continental-defence/the-canada-us-defence-relationship-nostalgia-aint-what-it-used-to-be/. Accessed April 2023.

Posen, Barry R. *The Sources of Military Doctrine.* Ithaca, NY: Cornell University Press, 1984.

Preston, Richard. *The Defence of the Undefended Border: Planning for War in North America, 1867–1939.* Montreal: McGill-Queen's University Press, 1977.

Raska. Michael. 'Strategic Competition for Emerging Military Technologies: Comparative Paths and Patterns'. *Prism: A Journal of the Center for Complex Operations* 8:3 (2020), 65–81.

Rasmussen, Mikkel Vedby. *The Risk Society at War.* Cambridge: Cambridge University Press, 2006.

Roi, Michael. *Canadian Defence Priorities, CF Force Posture and Strategic Readiness. Linking Government Policy Preferences to Resource Allocations, DRDC CORA Technical Memorandum TM 2012 – 289.* Ottawa: Her Majesty's Printer, December 2012.

Snider, Major Matthew. *Targeting Air Power: The Failure of the RCAF to Adapt in the Global War on Terror.* Toronto: Canadian Forces College, 2020.

Stacey, C. P. *The Military Problems of Canada. A Survey of Defence Policies and Strategic Conditions Past and Present.* Toronto: Ryerson Press, 1940.

Staff. 'Capabilities Needed -Urgently,' *Canadian Army Today* 7:1 (Winter 2023), 21.

Stein, Janice Gross and Eugene Lang. *The Unexpected War: Canada in Kandahar.* Toronto: Viking Canada, 2007.

Sutherland, R. J. 'Canada's Long-Term Strategic Situation.' *International Journal* 17:3 (Summer 1962), 199–223.

Sweijs, Tim. *The Role of Pivot States in Regional and Global Security.* The Hague: HCSS, 2014.

———— and Michael J. Mazarr, 'Mind the Middle Powers.' *War on the Rocks*, 4 April 2023.

Taylor, Alan. *The Civil War of 1812.* New York: Alfred Knopf, 2010.

Taylor, Ben. *Towards and Enhanced Capability Based Planning Approach.* Ottawa: DRDC Reference Document, 2017.

Thatcher, Chris. 'Digital Experiments.' *Canadian Army Today* 7:1 (Winter 2023), 26–30.

United States Congress. House. Committee on Armed Services. Subcommittee on Intelligence, Emerging Threats and Capabilities. *Perspectives on the Future National Security Environment.* Washington: US Congress, 2013.

Wagner, Eric. 'The Peaceable Kingdom? The National Myth of Canadian Peacekeeping and The Cold War.' *The Canadian Military Journal* 12: 1 (Winter 2006–7), 45–54.

Wiseman, Erica. *The Institutionalization of Organizational Learning: A Neo Institutional Perspective.* Proceedings of OLKC 2007—"Learning Fusion". Warwick University, 2007.

# 13 Conclusion

*Tim Sweijs, Saskia van Genugten
and Frans Osinga*

Small and middle powers (SMPs) across the globe are adapting to the challenges
brought about in the current age of geopolitical, technological and climate-related
change. Whether in Europe, the Middle East, North America, Asia or Oceania,
these powers have understood that they operate in a radically different security
environment than they had gotten accustomed to over the past decades. For some,
in particular in Europe, this realisation was triggered by the war in Ukraine; for others in Asia and Oceania, China's shift away from its long-acclaimed policy of non-
interference was a trigger. For those in the Middle East, the so-called Arab Spring,
in combination with a waning US presence, China's advances and continuing
regional turbulence, were drivers of change. SMPs are repositioning themselves in
order to try and protect their national interests as best as possible, thereby using the
often-suboptimal defence capabilities at their disposal. For some, the new situation
translates into increased opportunities and the ability to exert more influence. For
others, these developments leave them vying for security, as they are facing threat
levels not seen for a long time.

Adjustments to defence postures tend to start with such 'big' realisations which
are then translated into defence planning processes, ultimately geared towards
getting the right people, the right equipment, and the right readiness, at the right
time. On paper and according to ideal notions of defence planning, there is a linear
progression from the codification of a threat perception and the formulation of
a strategy to address the threat, all the way down to the procurement of defence
materiel, military training and exercises and the choice of defence partners.[1] As the
collection of chapters in this book shows, for most SMPs, such linearity and cascading logic are far from reality. In contrast, most defence planning stories related
in this book are about SMPs dealing with a geopolitical environment that they
have little control over with limited resources. Their defence planning processes
variably feature a lack of manpower, money, experience and, to outsiders at least,
sometimes even a lack of rationality. Each and every of the 11 cases examined in
this volume is unique, with recurring elements and typical patterns only kicking
in at the more generic level. At the same time, the cases, both individually and
in their combination, illustrate how the defence planning of SMPs fundamentally
differs from great powers, including when it comes to the drivers of their defence
policies, the dynamics of their defence planning and military innovation processes,

DOI: 10.4324/9781003398158-13

and their ability to generate the forces to deal with the challenges in their security environment.

What then are these fundamental commonalities between SMPs that make them different from great powers? First, SMPs face constraints in safeguarding national security. They have limited overall military resources, financial or otherwise, compared to great powers. Thus, while great powers are in a position to lead with regard to all necessary means (financial, industrial, sheer size and numbers, institutionalised influence, human capital and experience), SMPs will not be able to provide that full package. While this may seem a difference in degree, not in kind, these limitations suffuse the entire defence planning process and inform all their choices, including when it comes to alignment strategies vis-à-vis great powers and engagement with other states to increase their defence capabilities. Second, SMPs have a more limited geographical reach and will not be able to sustain a continuous international presence or deploy significant military missions across the globe at the same time. Their defence planning is mostly shaped by local or regional, rather than global threat perceptions and orientations. They take a more constrained approach to defence planning and have more limited resources to design and implement defence planning processes. Third, SMPs face challenges in keeping pace with technological advances. They have considerable difficulties in developing military industries of their own, both because of technical complexities and because of scale and cost. Consequently, they are typically, although not inevitably, dependent on procuring military equipment designed and produced by great powers. While defence planning processes of SMPs may therefore be of a different scale, this does not make the translation of political guidance into strategic concepts and fitting defence capabilities any easier.[2] In fact, the characteristics and constraints associated with SMP defence planning present them with an assortment of trade-offs.

One important trade-off concerns the choice between preserving full independence but not being able to defend oneself when independence is being threatened by a hostile state, and amplifying overall power and resources through alliances and partnerships while giving up on full independence. In other words, SMPs can decide to simply punch at their weight (which often means not punch at all), they can build alliances and partnerships to try and extend their reach, or they can potentially seek shelter under the security umbrella of a great power, thereby fostering a hierarchical, dependency relationship that requires them to keep into consideration the demands of the stronger ally, in the realm of defence but also outside of it.[3] Partnerships can also include private (military) actors, which requires SMPs to deal with large defence industry corporations that themselves have significant heft. SMPs can simply follow great power plans or they can decide to carve out their own path. Most SMPs that are part of an alliance tailor their defence planning to the leading great power, because in effect their security policies have become, at least in the case of NATO, largely aligned with the great power, certainly when it comes to the use of the military instrument at the higher end of the spectrum. In the European context, this has meant that over the past 30 years, European SMPs would generally go where the United States went. Partner choices thus hinge on the important trade-off between the level of autonomy an SMP aspires to keep and the ambitions of the political leadership to expend resources on defence capabilities, in relation to that larger question always looming in the background as to how this generates the required capabilities to provide for national security.

A second set of trade-offs relates to the defence planning approaches that SMPs take and the methods and techniques that they use in the process. Some SMPs, especially those who face specific threats, tailor their entire defence planning to dealing with these threats, whether or not in combination with a strategy of mobilisation. Other SMPs, discerning greater uncertainty in their immediate threat environments, rely on task- or portfolio-based approaches, even if their limited resource base almost precludes effective operations across the entire conflict spectrum. The methods and techniques employed in this process also vary and range from implementing external templates, whether or not they are copy-pasted from great powers, to more tailored analytical processes, and everything in between. Similarly, the ultimate arbiter of the force postures that result from these approaches diverge. Sometimes, the analytical process is leading, but more often decisions are determined by military or political leaders, with varying degrees of parliamentary involvement on the one hand and great power involvement on the other.

A third set of trade-offs concerns dependencies in the sphere of military innovation. A key question for SMPs is whether they want to make use of capabilities based on the most advanced R&D offered by foreign industries or whether they want to build and maintain national R&D infrastructure and the domestic defence industries that support it. Reliance on foreign industries for military equipment can have a significant impact on warfighting abilities, both with regard to how one fights and with whom one fights. The latter becomes increasingly important in an era in which seamless connectivity and synchronisation are key. The existing literature tends to portray SMPs as being weak in terms of military power and unable to generate the necessary capabilities to provide for their own security.[4] The case studies in our volume suggest that the notion that SMPs are neither innovators nor frontrunners, but are instead dependent on great powers to acquire cutting-edge capabilities, is potentially too simplistic.[5] Our volume reveals a much more diverse, less dichotomous picture than frontrunners versus followers. Some SMPs choose to maintain as much of a national industrial defence base as possible, even if that base will as a consequence be less advanced than is internationally available. Other SMPs prefer to rely on available material elsewhere, accepting the associated procurement, delivery and dependency risks. Yet still others specialise in a specific industry or knowledge base that makes them of particular value to allies and partners. Across the board, the acceleration of strategic shocks over the past few years, from the Covid-19 pandemic to Russia's war against Ukraine, in conjunction with other more structural trends, such as the Sino-American competition over leadership in new technologies, has pushed SMPs to reconsider vulnerabilities of supply. In this context, SMPs are striking a balance between spending towards maintaining existing forces and investing in innovation and transformation. The coming decades will bring a host of new military changes; against this background, defence planners of SMPs will need to initiate and manage incremental change in parallel with at least the possibility of revolutionary change.[6] This is rendered more urgent by the fact that many Western SMPs have large and costly legacy forces with force postures tailored to a more benign era in which peace support operations and humanitarian interventions dominated thinking on defence capability development. In the context of techflation, the size of militaries can therefore potentially lead to a dangerous situation, especially in the West: defence organisations need to replenish their arsenal and rebuild their militaries, at the same time as they need to invest in new capabilities.[7]

The sample of 11 SMPs surveyed in this volume may be small but harnesses a large variety. It includes NATO members (initially four countries (Canada, the Netherlands, Slovakia and Turkey), then Finland joined the alliance in April 2023). It also includes countries that are experiencing direct threats in their environment (such as Israel, and increasingly Finland as well as Australia), countries that are tiny in terms of population, but large in terms of spending power (such as the UAE and Singapore), and countries that have long tried to carve out their own path (such as Indonesia) or have a long history of building bridges between different actors (such as Oman). And of course, all 11 SMPs come with differences in governance systems, geographies, histories and cultures.

The objective of the structured focused comparison applied to these 11 cases has been to distil commonalities between them and to look for patterns within those commonalities. As discussed in the introduction, the comparison has been divided along three themes of (1) alliances, dependencies and national ambitions; (2) approaches, processes, methods and techniques and (3) military innovation and adaptation (see Figure 13.1). While the focus has been on finding commonalities, rooted in geographical, cultural, or shared threat perceptions aspects, this volume

| SMPS ADJUST THEIR APPROACHES BASED ON SYSTEMIC CHANGE OUT OF THEIR CONTROL | SMPS REACT TO CHANGES IN THE RELEVANT SECURITY COMPLEX(ES) THE SMP IS PART OF | | | | |
|---|---|---|---|---|---|
| *Alliances, Dependencies and National Ambitions* | | | | | |
| SMPS FACE CONSTRAINTS IN SAFEGUARDING NATIONAL SECURITY | FINANCIAL | DEMOGRAPHIC | CULTURAL | GEOGRAPHICAL | OTHER |
| SMPS NEED TO POSITION THEMSELVES VIS-À-VIS GLOBAL AND REGIONAL POWER CONSTELLATIONS | BALANCING | HEDGING | BANDWAGONING | NON-ALIGNING | OTHER |
| SMPS HAVE TO ENGAGE WITH OTHER STATES TO INCREASE THEIR DEFENCE CAPABILITIES | SECURITY GUARANTEES | TRANSACTIONAL | FORCE INTEGRATION | SPECIALISATION | OTHER |
| SMPS HAVE ARMED FORCES FOR DIFFERENT PURPOSES | INSTRUMENT OF DIPLOMACY | PRESTIGE PROJECT | REDUCE STRATEGIC RISK (DEFEND) | DOMESTIC STABILITY | OTHER |
| *Approaches, Processes, Methods and Techniques* | | | | | |
| SMPS TAKE A MORE CONSTRAINED APPROACH TO DEFENCE PLANNING | THREAT-NET ASSESMENT BASED | PORTFOLIO-BASED | TASK-BASED | MOBILISATION | OTHER |
| SMPS HAVE LIMITED RESOURCES TO DESIGN AND IMPLEMENT DEFENCE PLANNING PROCESSES | TAILORED ANALYTICAL PROCESS | OUTSOURCING STRATEGY | ACCEPT DILUTION / FAIR SHARE | EXTERNAL TEMPLATE | OTHER |
| SMP EMPLOY MULTIPLE METHODS AND TECHNIQUES | COPY GREAT POWERS | INFORMAL CONSENSUS | TAILORED METHODS | AD HOC | OTHER |
| SMPS RELY ON DIFFERENT PROCESSES TO DECIDE ON FORCE POSTURE | ANALYTICAL RESULTS | POLITICS DECIDES | MILITARY DECIDES | GREAT POWER DECIDES | OTHER |
| *Military Innovation* | | | | | |
| SMPS FACE CHALLENGES IN KEEPING PACE WITH TECHNOLOGICAL ADVANCES | BUILD DOMESTIC TECH/INDUSTRY | TECH TRANSFER AGREEMENTS | ACQUIRE READY INNOVATIONS | PARTNER WITH GREAT POWER | OTHER |
| SMPS PURSUE VARIOUS MILITARY INNOVATION PATHS AND PATTERNS | EMULATION / SPECULATION | ADAPTATION / EXPERIMENTATION | INNOVATION / IMPLEMENTATION | OTHER | OTHER |
| SMPS FIND THEMSELVES AT DIFFERENT STAGES OF ORGANISATIONAL CHANGE | EXPLORATION | MODERNISATION | TRANSFORMATION | OTHER | OTHER |

(Left margin label: GENERAL COMMONALITIES; middle label: VARIATIONS WITHIN THESE COMMONALITIES)

*Figure 13.1* Structured focused comparison framework.

*Source:* Authors

has also provided a rich set of case studies on a topic that has so far been under-explored in the literature, as each of the cases shows an (unsurprising) degree of variation and idiosyncrasy. To structure this conclusion, we focus on the common-alities and patterns per theme on the basis of which we have clustered the SMPs and are proposing a three-pronged typology of SMP defence planners.

## Alliances, dependencies and national ambitions

SMPs do not have the bandwidth to provide the full spectrum of defence-related abil-ities that larger powers can afford to secure their environment. The less powerful need to make choices, first and foremost with regard to what the stated role of the defence organisation is, what abilities it should have, what alliances it should engage in, and to what extent it can feel comfortable being dependent—and as such partly compro-mising its sovereignty—on other, more powerful states. Such choices are preceded by how a country perceives its security environment and its own position therein.

### *SMPs face constraints in safeguarding national security*

For starters, SMPs face significant constraints in national security because of their limited overall resource base. Those looking at SMPs' defence planning behav-iour from a purely Western perspective will quickly point out the predominance of financial constraints. But when one takes a wider perspective, the types of con-straints become more varied. For example, some SMPs do not necessarily suffer from financial constraints when it comes to their armed forces, as for example is the case in the UAE, but also in Singapore and Israel. Instead, many of these SMPs suffer from demographic constraints, either in terms of limited numbers or in terms of personnel with the required education and training to operate technologically advanced platforms. They might also face geographical constraints, for example by sharing borders with a great power (such as Finland), being surrounded by enemies while lacking any strategic depth (such as Israel) or governing over too large a ter-ritory to defend with the human resources available (such as Canada). Also primar-ily in the West, the case studies show stronger cultural and historical constraints affecting defence planning than elsewhere. In some SMPs, this has even led, during many decades of peace, to a situation in which the armed forces have been partly placed outside of 'society', with military service being regarded as a thing of the past and politicians that are careful not to put their soldiers into harm's way. What binds all the SMPs together again, is that these limitations will have to be translated into informed defence planning choices, as not everything will be possible.

### *SMPs need to position themselves vis-à-vis global and regional power constellations*

For SMPs, their positioning vis-à-vis the more powerful states is a key considera-tion in the further decision-making processes of defence planning. By the time of publication, five case studies deal with NATO members. They mostly fall clearly in the 'balancing' camp, aligning with the US on most strategic files, tend to procure their military materiel mostly within the alliance, and have common approaches to many parts of the defence planning processes, including readiness requirements and reporting requirements. Canada, similar to the Netherlands and Slovakia, for instance, maintains a desire to remain interoperable with the US and contribute to a wide range of missions, albeit with small contingents. Within this group though, subtle (and not so subtle) differences exist. For most Western powers, the relations with the US, after years of crisis during the Trump Administration, have been tightened again, in particular after the Russian aggression in Ukraine. Finland might momentarily be considered the newest US ally, while Turkey is at times blatantly going against the US, siding with Russia instead and using that contentious relationship to extract benefits as expected in a "hedging" relationship. Similarly, other states in the Middle East, including the UAE and Oman, use different strategies at different levels. When it comes to traditional security ties, they are still relying on the US, while hedging is taking place at many other levels, in particular with regard to defence investments, defence industrial relations and overall relations in the sphere of dual-use materiel. Israel, while clearly in the US camp, has shown its willingness and ability to defy the US line in situations where the repercussions could be high. Those outside of NATO, including Indonesia and Singapore, indeed tend to bring a dose of pragmatism to their positioning vis-à-vis great powers, whereas Australia has established much closer ties to the US in the context of China's military rise.

### *SMPs have to engage with other states to increase their defence capabilities*

International cooperation is for most of the SMPs an important tool to compensate for limited resources. Indonesia might be the outlier here as its considerations with regard to international cooperation are seen not as based on a needs assessment, but rather on a favour-seeking, retail fashion looking for transactional relations. For the overwhelming majority of SMPs, the reason for partnering with other nations, especially stronger ones, is to obtain security guarantees. This is of course the case for NATO members, but also countries such as Oman and the UAE are expecting security guarantees from their international partnerships. The absence of a formal pact in this regard with the long-standing security partner US, and subsequent doubts about the US reliability as a partner in the Middle East region, have been major factors in these powers engaging more intensively with China, and also Russia, in the past decade. Within Europe, SMPs have also teamed up to create scale. While this dynamic has long been mainly driven by financial considerations, since the war in Ukraine, this intra-European collaboration has also gained a greater geopolitical edge. The uptick in cooperation efforts, for example, between the Netherlands and Germany is a case in point and shows that in some cases SMPs will even consider force integration as well as specialisation with trusted partners.

### *SMPs have armed forces for different purposes*

Overall, the armed forces of SMPs serve different primary purposes. In none of the case studies analysed is there a state that truly believes it can be self-sufficient in defending the country against external threats. The three countries that have invested most in trying to be as self-sufficient as possible are Israel, Singapore and Finland. In the case of Finland and Israel, the driver of this ambition is the presence of an actual, direct threat. In the case of Singapore, it is the persistent perception of multiple latent threats and vulnerabilities that drives national security and defence policy. Singapore evolved from a 'poisoned shrimp' island defence, via a porcupine defence strategy, with limited power projection in Singapore's near seas and a brief stint at a network-centric war-based strategy to the current Total Defence strategy that addresses state-level military, hybrid, and terrorist threats. The military is an integral part of a concerted set of diplomatic, deterrence and resilience policies and programmes. Israel, in confronting the threat that manifested itself—be it states such as Syria or Iran, or non-state actors such as Hamas or Hezbollah—responded by taking unilateral action. Finland, in turn, confronted with a looming threat, has shifted gears and decided to compromise on self-reliance in exchange for stronger security guarantees, initially working more closely with Sweden and, since 2023, as part of NATO.

But these 'frontline' states can be considered to be outliers. While most of the SMPs do see their defence organisations as entities that are instrumental in reducing strategic risk, especially for those countries that have been in relatively peaceful spots for the past decades, in reality the armed forces have more often than not been deployed as an instrument of diplomacy. This role is most pronounced in the cases of the Netherlands and Slovakia and is similarly detected in Canada and Oman and to a more limited degree in Australia. Interestingly, these are all countries that have had a very strong Western leaning in their defence relationships. Aware that their armed forces have serious limitations when it comes to engaging in international conflicts, combined with the absence of any serious direct threats to national security, these SMPs have signed up for military missions and operations in ways in which one decisive factor would be following a strategic partner—often, but not always, the US.

In countries with considerable internal tensions, or with governance systems focused on keeping any such tensions low, the armed forces are also tasked with keeping domestic stability, as, for example, in the case in Oman, Turkey and Indonesia. According to the analysis presented in this volume, Indonesia is most focused on that objective, and is described as using its defence planning processes—procurement in particular—as an instrument to maintain such domestic stability. Regional threat developments—such as China's rising aggressive posture in the South Chinese Sea—play a more limited role, which stands in marked contrast to its neighbouring island-state Singapore which sees those developments as key drivers informing its security policy. While in other cases examined, rulers equally care about domestic stability, it is not always the task of the defence organisation to provide it. In the UAE, for example, the defence organisation is maintained mainly to project power outwards, to contribute to regional stability objectives, and to build a strong national identity.

### Approaches, processes, methods and techniques

Based on these different general strategies, the SMPs surveyed in this volume also show considerable variation in the approaches, processes, methods and techniques that they employ in and towards defence planning.

### SMPs take a more constrained approach to defence planning

By and large, all case studies demonstrate SMPs actively observing and analysing threat developments. This can relate to technological developments threatening to make existing modes of operations and/or specific weapon systems in the inventory obsolete, or those assessments point to larger deteriorating tendencies in the global or regional security environment. Depending on the perceived sense of urgency to address these developments by altering for instance the defence posture or range and quality of capabilities, these threat assessments may turn out to be significant drivers of defence policy. However, such linear rationality was often not present in the past three decades that some analysts have described as a strategic pause. Instead of complementing, but often actually dominating, the threat net-assessment approach, most SMPs de facto adhered to portfolio-based planning and task-based planning.

Unsurprisingly, those countries that face a clear and present danger tend to engage in threat net assessment planning. They after all have, or at least think they have, a clear understanding of their threat environment, including principally the adversary/ies they face and the type of conflict their armed forces will engage in. Finland, for instance, has been preparing fairly consistently to defend itself against Russia, also at a time when other states were still expanding energy ties and pursuing the so-called Russia Reset otherwise. In a similar vein, Israel and Turkey have used threat net assessment planning to prepare for the definitely wider spectrum of threats they confront in their respective environments, for Israel in combination with a mobilisation approach, and for Turkey alongside portfolio planning. For Israel, this includes the prospect of an Intifada and engaging in urban combat, as it has, following the Gaza attacks on 7 October 2023, a Third Lebanese War against Hezbollah along its northern front and more persistent operations against Iranian-supported militias in Syria, as well as a larger regional conflagration in a direct conflict with Iran that includes all of the above. For Turkey, this extends to a variety of adversaries along and within its borders, including IS and its offshoots, various Kurdish groups within and along its borders, as well as its long-standing dispute with NATO ally Greece. Australia also relies on threat net assessment planning in conjunction with portfolio-based planning specifically in the context of its deteriorated relationship with China. Slovakia has continued to use threat net assessment planning principles in the context of its contribution to NATO. For Finland and Slovakia, this is the exclusive form of defence planning, whilst Australia, Israel and Turkey combine it with other forms of planning. Singapore employs a sophisticated approach involving long-term strategic foresight, scenario planning, technology studies and operational concept development, all with an eye towards hybrid threats and potential military threats in the region, and all informing capabilities requirements.

The Netherlands and Oman rely predominantly on portfolio-based planning in which they develop a range of capabilities that in theory should allow the armed forces to operate in a variety of contingencies. Canada, too, maintains a portfolio approach, catering to its desire to be a credible partner in NATO and yet also remain an autonomous activist middle-power and retain the ability to participate in both high-intensity and peace operations. For the Netherlands, this approach bridges the defence policy priorities of left- and right-wing political parties, with the left-wing parties favouring the Netherlands to focus on peace operations and centre-right wing parties preferring also to maintain high-end capabilities for Article 5 operations and the first phase of humanitarian operations as part of a coalition. The 2022 Russian invasion of Ukraine has not prompted the abolishment of the portfolio approach, although there is a reorientation to regional planning tailored to the Russian threat in the East and to stabilisation operations in the south. Australia has by now shifted towards a threat-based approach, following a period which centred on coastal defence and stability in the immediate surroundings, in recognition of the increasingly deteriorating security environment in the South Chinese Sea.

The UAE's approach to defence planning is hard to pin down given its often haphazard nature which depends on the whims of the ruler and ideas pitched by external consultants and industry. A common thread that runs through its defence planning, however, is the ability to execute different tasks. Canada employs a combination of task-based and mobilisation planning in the context of considerable uncertainty, which is now, like in the Netherlands' case, supplemented by demands for more threat net assessment-based planning.

Finally, Indonesia pursues a variety of planning approaches described by Laksmana in this book as follows: 'almost like how its people pay for cigarettes—by small retail pieces, rather than wholesale boxes'. The planning is a blend of planning centred on tasks that individual services need to be able to execute, and based on the often very limited available resources that they have at their disposal.

### SMPs have limited resources to design and implement defence planning processes

SMPs find different solutions to deal with the limited human resources that they can allocate to their defence planning processes. Some of them adopt external templates such as Turkey, Slovakia, the UAE and Oman. Others, at least sometimes, partly rely on external consultants in the actual process, such as for instance Australia and Canada. Alternatively, or complementing this, they have used a fair share basis. Countries that have done so include Canada, Turkey and Indonesia. Finland, Israel and to some extent the Netherlands have put in place analytical processes that they adhere to in varying degrees. As in Australia, defence ministries as well as individual services in Canada and the Netherlands have regularly produced strategic vision papers laying out their perspective on international security, the future character of war, and the force implications in terms of capability requirements that derive from them. However, financial constraints have until recently meant that advice derived from analytical processes was effectively put aside; most often financial constraints drove prioritisation regarding investments in new material, modernisation of existing systems and paying for ongoing operations. In particular in the wake of the financial crisis in the late aughts and the early 2010s, this resulted in the prioritisation of ongoing operations, the postponement of modernisation and replacement plans and the cancellation of investments in capabilities such as cyberwarfare capabilities, armed UAVs and cruise missiles. Invariably, making sure to retain a limited number of capabilities of specific types—fighters, submarines, tanks, air defence capabilities—units would be disbanded to free up financial resources in what Hazelbag et al., in their chapter, label the 'cheesegrater' approach. Other strategies involve anchoring service capabilities in international units, a defensive measure adopted by a service to protect its capabilities from future cutbacks as withdrawal from international units was politically expensive.

### SMPs employ multiple methods and techniques

SMPs select from a range of methods and techniques of defence planning often-times importing and tailoring them to their specific needs. In fact, tailor-made processes are quite prevalent: Finland, Indonesia, Israel and Turkey have designed their own tailored processes. Australia, Slovakia and the UAE pick and choose their methods and techniques in an ad hoc fashion, while Indonesia and Israel combine this with tailored processes. Canada, Oman and Turkey copy methods and techniques of great powers, but in Canada this process is not really institu-tionalised, often resulting in mismatches between policy, strategy and operational requirements. The Netherlands, meanwhile, relies on a mish-mash method that is certainly inspired by US templates and tailored to the Dutch context, but also con-tinues to operate through informal consensus in its famous polder model. Here we see each service attempting to influence the process by focusing on the risks of the operational obsolescence of weapon systems and tangible operational problems— missile defence, surviving in a high-threat air defence environment—and subse-quent requirements, but the overall planning process lacks a joint perspective on future warfare. In Singapore, on the other hand, the planning process seems both institutionalised and sophisticated, involving several dedicated organisations— military and non-military—across the government that coordinate diplomatic actions and civil, economic, military and digital defence efforts.

### SMPs rely on different processes to decide on force posture

The governance system of a country appears, perhaps unsurprisingly, as one of the key intervening variables when it comes to considering the role, the ambition and the actual influence of the armed forces on defence planning processes. In the case studies looked at in this volume, the entire spectrum of options can be found. In the Netherlands, parliament holds significant decision-making power with regard to defence procurement and military deployments, while in a governance system such as seen in the UAE and Oman, decision-making power is concentrated at the top, with the ruler ultimately having the last say in everything. In Turkey, the defence planning process is impacted massively by the civil-military rivalry that has always been present in the country, but which saw another spike after the failed military coup attempt in 2016. In a place such as Indonesia, the military is actually dictating the civilian side of defence planning, which creates its own particularities. Many of these governance systems are the product of country-specific histories that have resulted in bureaucratic structures and decision-making cultures which disallow any linear lines of causation within the characteristics of the case studies. Especially in countries where the ambition-setting and role of the military are not clearly defined and where there has long been an absence of a serious, direct threat, the planning processes tend to take place in a more ad hoc, low cost and informal way, without too much top-down control and direction. In contrast, in cases where there has long been a looming, directly felt threat, defence planning seems to have always had more urgency and a clearer direction, tied to concrete threats, as is the case for Finland, Singapore and Israel in particular. In still other cases, those holding decision-making power in defence planning processes use this power partly to provide benefits to domestic industries, entities or individuals that are important stakeholders in their governance systems, such as is the case in Turkey and Indonesia.

Both autocratic and democratic SMPs have put in place top-down processes in which defence planning is conceived and formulated by a higher-level authority—either a supreme leader, as in the case of Oman and the UAE, or a central civilian policy planning directorate, as in the case of Australia, Canada and the Netherlands—that formulates the plans that then guide capability developments. In Australia, for instance, several major ad hoc external reviews have guided the defence policy changes and choices. Canada has published a limited number of white papers on defence even if generally domestic politics—often driven by financial constraints—have produced ad hoc changes in its defence capabilities and orientation. This also applies to the Netherlands where incoming ministers of defence are charged by parliament to develop a vision for the armed forces, and each year debates about spending priorities take place in parliament prior to agreeing on the defence budget. In these latter democratic states, even though the system works top down in theory, interservice rivalry does come into play, when individual services seek to shape the outcomes in the competition for limited resources. In the Netherlands, with a constantly shrinking budget, the main game for each service was to maintain what they had, while fighting for scarce budgets to modernise the existing force and replace ageing systems. This, for instance, resulted in bureaucratic politics to delay or block political decision-making on major system acquisitions which would swallow up large chunks of the available investment budget for years to come, such as the fifth-generation fighter jet. Strategic and political culture also plays an important role. In some SMPs, against the background of competition and rivalry, consensus-seeking efforts are part and parcel of the planning processes, including in Finland, Israel and the Netherlands. Strategic cultures can shape the self-perceived identity of a country as a security actor. In the case of the Netherlands, the self-perception of it being a small country in the international security realm, one that is inclined to soft power instruments and consensus-seeking through international institutions, service interests in investing in offensive weapons such as armed drones and cruise missiles confronted intense political opposition. In other SMPs, civil-military rivalry about who is or rather should be in control of the allocation of resources, is quite intense, as is for instance the case in Turkey and Indonesia.

For NATO countries such as Canada, Slovakia and the Netherlands, although less so for Turkey, alliance considerations feature prominently in the planning processes through which the US has considerable influence. At the same time, while the NATO planning process cycle informs national defence planning in these countries and strives to introduce analytical rigour and logic, absent a direct threat, most countries have found ways to influence that process to suit their own requirements and budgetary limitations. Sometimes, NATO priorities are exploited by governments to sell the national decision to cut on certain capabilities. Alternatively, nations paid lip service to NATO (or EU) defence plans such as the ambitious NATO Transformation programme. While alliance commitments certainly factor into the national defence planning process, in the end, defence policy and expenditures remain a national prerogative. In Israel, alongside the political leadership, the military has a considerable say in the process, as is the case in Turkey and Indonesia. In countries such as the UAE and Oman, the ultimate decision-making power resides at the top, with the ruler. As a result, it is not a given that any analytical products created in the planning process at a lower level are actually used as the basis for reaching decisions within the defence planning process. As decision-making in these countries is highly personalised, much depends on advice provided by trusted advisors as well as personal beliefs and convictions. The ruler's view, or the view of a small, often family-based committee, will in these countries be the decisive factor with regard to the actual outcome and the types of capabilities that are developed and/or acquired.

## Military innovation

Each of the chapters looked at the country's approach to military innovation and the ability to not only produce or acquire but also absorb innovations within the defence organisation, and the implications for the force posture of their respective country. SMPs face additional challenges when it comes to military innovation compared to great powers. As shown previously, their financial resources are more limited, forcing them into trade-offs between innovating and maintaining existing force structures. They can ill afford costly innovation processes involving rapid change and the replacement of existing legacy capabilities, and instead are forced into gradual modernisation processes, accepting unavoidable lower levels of operational effectiveness in future wars. Whereas great powers retain and develop a complete set of capabilities required to win wars, unilaterally if necessary, resource constraints and—by default—limited political ambitions will force SMPs to adopt a different logic in structuring and developing their military capabilities. Often, as is the case in Canada, the Netherlands, Australia and Slovakia, this results in defence policies that promote the development and maintenance of a limited range of general military and niche capabilities—a portfolio approach—which allows a government to deploy relevant contributions in a variety of missions within an international coalition.

Similarly, their industrial base will most likely not be sufficient to generate complete innovative capability packages. Instead, SMPs are dependent on external suppliers, will—with a few notable exceptions—rarely be first movers in military innovation and will seek industrial relations with those external suppliers in the form of co-development, co-production or niche-product development. Innovation may also be inhibited by concerns for the national military industry, resulting in decisions not to acquire state-of-the-art foreign military systems but instead procure what capabilities the national industry can produce. Moreover, their security outlook and defence policy development will be shaped by a regional outlook instead of a global outlook of great powers. Along with alliance politics, and domestic strategic culture, these factors may result in capability developments through acquisition decisions that are often more informed by domestic politics, status in an alliance, the need to remain interoperable with other partner militaries, affordability issues and inter-service rivalry, rather than a deliberate search for operational solutions and an overall strategic perspective on the novel mode of warfare.

### *SMPs face challenges in keeping pace with technological advances*

How does an SMP accommodate technological advances? Does it rely on technology its domestic industry produces, does it seek to establish technology transfer agreements or does it prefer to buy off the shelf, and/or partner with a great power? Almost all SMPs combine different approaches to keep pace with technological advances. Domestic industrial capability and politics are the dominant factors here. Most SMPs investigated tend to acquire off-the-shelf when it comes to major weapon systems in the context of vendor-supplier relationships with a great power. In their approach to keeping pace with technological developments and changes in the threat environment, Australia, the Netherlands, Oman and Israel partner with great powers. Australia, along with Canada, Oman, Slovakia, the Netherlands, Turkey and the UAE, also tend to buy off the shelf. To keep pace with technological advances, in particular in the field of advanced combat aircraft and naval vessels, ISR systems, precision-guided weapons, complex air defence systems, submarines, and long-range rocket artillery, but also for large support capabilities such as transport and tanker aircraft and naval support ships, countries such as Australia, the Netherlands, Oman and Israel partner particularly with the US and sometimes with France and the UK. In the context of these partnerships, they attempt to involve their national industries in either the development or production of components of such new capabilities, or in the resulting follow-up maintenance phase. Developing national industrial capabilities to address technological advances is also visible in Canada, Finland, Israel, Turkey, Indonesia and the UAE, but none of them solely engage in this strategy. The cases of Turkey, with its TB2 drones, and Israel, with its Iron Dome system, demonstrate that SMPs can indeed be innovators and frontrunners in the development of niche capabilities. Building their national defence industries, or niche developer and supplier capacity, SMPs such as Canada, Indonesia, Turkey and the UAE also foster signing tech transfer agreements.

*Military innovation outcomes: conceptual paths, technological patterns and magnitude of change*

As explained in the introduction, authors of the book chapters plotted their respective case studies on the innovation pathways framework developed by Michael Raska, which lays out the type of strategies deployed by SMPs in terms of the conceptual and technological patterns they pursue and the magnitude of organisational change this results in (see Figure 13.2).

Different clusters emerge in terms of the different innovation modes pursued by SMPs as revealed by Figure 13.2. First, it appears that very few of the SMPs are transformative innovators, although some SMPs do stand out. The most innovative SMPs, capitalising on continuous experimentation with a keen focus on the actual implementation, are, unsurprisingly, Israel, followed by Singapore. Israel in particular has developed innovative technologies and modes of operations in response to pressing operational problems. The missile threat from Hezbollah and Hamas resulted in the development, and following the Second Lebanon War, subsequent fielding of the Iron Dome defence system. The need to find, track and engage fleeting small targets such as mobile missile launchers fed the development of advanced C4ISR arrangements linking sensors, C2 nodes and shooters (mimicking US network-centric warfare concepts) to conduct time-sensitive-targeting operations.

Second, most SMPs go beyond mere emulation and speculation in that they do incrementally adjust existing military means and methods based on experiments by official organisations charged with experimentation. Most other SMPs, in particular the Western ones investigated here, have tended towards a policy of retaining

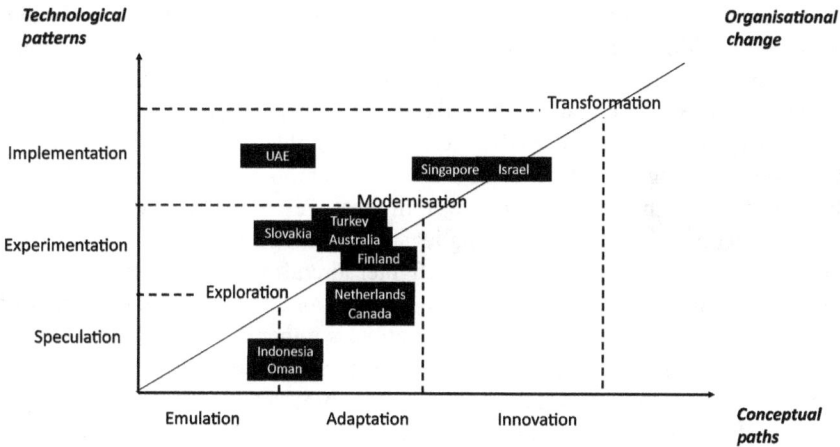

*Figure 13.2* Military innovation trajectories in small and middle powers: comparison of all case studies.[8]

*Source:* Figure created by authors based on Michael Raska, Military Innovation of Small States— Creating a Reverse Asymmetry (Routledge, 2016)[9]

existing systems, modernising those, and while doing so adapting to evolving threat contexts. Most new technological solutions are used to deal with existing tactical and technical challenges. When new weapon systems emerge on the horizon, they will experiment with those, and, budget permitting, may acquire those that have been developed, produced and demonstrated to be effective by partner nations, in doing so emulating a great power. A variety of SMPs, such as Singapore, the Netherlands and Finland, are involved in speculation and also experimentation with AI, robotics and cyberwarfare capabilities. Such developments, however, are mostly not driven by a new vision of future warfare or novel operational concepts that may be inspired by promises of new technologies or demanded by pressing threat developments.

Third, SMPs such as Oman, the UAE but also Indonesia, largely rely on emulation by importing new weapon systems that they buy off the shelf and try to fit in their force structures. Oman's decision to acquire advanced fighter aircraft, tanks, corvettes and air defence systems from several different countries, a variety which does not improve efficiency, is informed by its desire not just to obtain world-class systems but also to maintain strong relations with a number of important countries—the US, the UK, France and Turkey—as part of a multi-hedged insurance policy.

For most SMPs, the lack of an economy of scale precludes the construction of self-sufficient industries. Yet, some SMPs have substantial expertise and capacity in certain military niche technologies—artillery, shipbuilding, drones, radar systems and missile defence—that enable them to develop and produce specific modern military systems and afford a measure of national modernisation capacity for a service or a specific military function, which they also export to other states. The Netherlands, for instance, exploits its R&D infrastructure in maritime and radar technology, while Israel has developed modern missile defence systems, tanks, drones and even autonomous loitering weapon systems. Turkey produces tanks and advanced drones, and Finland state of the art artillery systems.

Overall, the introduction of advanced systems in the militaries of SMPs sometimes generates significant novel operational capabilities but these tend to be at the tactical rather than the strategic level. In most Western militaries, for instance, the introduction of the F-35, replacing legacy systems such as the F-16, results in more than just the modernisation of the fighter aircraft fleet, and instead expands the range of tactical missions a military can perform. Yet, with the exception of Singapore and Israel, the degree of innovation is rather limited in character and restricted to the technical, tactical and service-specific levels, rather than resulting in entirely novel force structures and new modes of warfare.

In most SMPs, innovation pertains to service-oriented changes that are mostly inspired by the regular need to modernise existing portfolios of capabilities in light of technical wear and new operational threat systems, rather than by a particular view on the character of future war, a national perspective on joint operations, or an overarching operational concept. Israel, Singapore, Finland and Australia are exceptions in this, as these countries have articulated a coherent strategic concept undergirding defence policy and military investment choices. But such a strategic or operational level rationale was absent in NATO SMPs here investigated in recent years.

This contrasts with the final two decades of the Cold War during which the trajectory of military innovation in NATO militaries was in no small measure informed by emerging operational concepts such as AirLand Battle and Follow-on-Forces concepts. Both concepts endeavoured to undermine the Warsaw Pact strategy which relied on multiple echelons of forces, the first of which would destroy the line of NATO army corps in Germany, and the second and third echelons would steamroll over the remnants and drive towards the ports in Western Europe to prevent US reinforcement units reaching those ports from the US mainland. NATO's military operational concepts targeted massed armour formations of the second and third echelons, and this provided the logic of European investments in modern tanks (Leopards), modern fighter aircraft (F-16, Tornado), long-range Multiple Launch Rocket Systems, and a dense belt of advanced air defence systems (Patriot) which helped to maintain air superiority and hence freedom of manoeuvre in Germany for ground troops.

The immediate post–Cold War era peace operations in the Balkans suggested a different logic, seemingly opening up policy options for drastic defence cuts in high-intensity warfare capabilities. After 9/11, and in the wake of military successes of Desert Storm and Iraqi Freedom, and in light of the need to conduct interventions at strategic distances, the US attempted to coax European militaries into buying into an ambitious NATO transformation agenda, aimed at accelerating European military modernisation and maintaining interoperability with US forces and visions of network-centric war (NCW). European SMPs restructured and downsized their militaries to carry out expeditionary operations but adopted NCW-style concepts only in name. After all, this required substantial investments in sensors, datalinks, precision-guided munitions, cruise missiles, electronic warfare systems and strategic enablers such as strategic transport capabilities, and it would shift the investment balance towards air- and sea-power capabilities. In the era of wars of choice, European governments were reluctant to fully adopt such a force transformation. Moreover, peace and stabilisation operations suggested to policymakers in SMPs that such high-end capabilities would not necessarily be required, provided the US and partners such as France and the UK would be willing to contribute such capabilities in coalition operations. This forms the domestic political logic explaining the reduction of the numbers of tanks, fighter squadrons, artillery units, and frigates and the complete disestablishment of categories of capabilities such as anti-submarine war platforms and multiple launch rocket systems in countries such as Canada and the Netherlands.[10]

From the mid-2010s onwards, the US introduced a new operational concept—Multi-Domain Operations—as a response to Russia's (and China's) aggression and the need to synchronise novel capabilities such as space assets, swarms of drones and cyber operations with more traditional capabilities. But similar to the previous wave of technological advances, whilst paying lip service to MDO in their (acquisition) plans, SMPs are only very reluctantly using MDO as a guiding beacon for national defence planning efforts. Instead, SMPs such as Australia, Slovakia, the

Netherlands and Turkey, but also Singapore, must, because of limited budgets and industrial backbone, balance what alliance obligations or emerging threats require, and what is feasible and affordable. Russia's ongoing war against Ukraine, and China's increasing regional assertiveness, are also affecting the defence planning priorities of SMPs as they refocus on the present—rather than on the future threat. For the smaller among the SMPs, another important factor is the limited availability of manpower. This affects for instance Oman, UAE, Slovakia, Israel, the Netherlands and Singapore. The latter two therefore have opted for exploring personnel-extensive solutions to military problems (i.e. technology), increasingly experimenting with AI applications, robots and autonomous systems to compensate for their lack of human capital.

Last but not least, innovation in SMPs such as Australia, Canada, and the Netherlands is also guided by a preference for multiple-role capabilities that are relevant against a variety of threats (such as F-35, strategic transport aircraft, air and missile defence systems and cyber defence capabilities), for a variety of objectives in a variety of strategic contexts, through which they can make credible contributions in most coalition operations, or by investing in maintaining strategically scarce and relevant capabilities such as submarines or air-to-air refuelling. Although the current threat environment is refocusing defence efforts, the inclination to multifunctional systems persists amongst SMPs.

### Types of defence planners: a theoretical discussion starter

The case studies under review in this volume each demonstrate the uniqueness of SMPs, highlighting the importance of variables such as history and strategic culture, the domestic political system and geography. At the same time, from the case studies a three-pronged typology clearly emerges based on a shared distinct set of characteristics (see Table 13.1).

First, a set of SMPs exhibits a stronger and much more focused threat perception than others, based on a combination of history and immediate security environment, and they take a much more rigorous and structured approach to defence planning. We have dubbed this cluster of countries the 'Strategic Defence Planners'. Finland and Israel are probably the best examples of how the continuous presence of external threats has shaped defence planning. Not coincidentally, they, although each in their own way, also represent fascinating examples of how the military is integrated into society, with a continuous focus on mobilisation and a total force/comprehensive security approach. The actual nature of the threat is of course very different. Finland is confronted with a territorial threat from its much larger neighbour Russia, while Israel faces the threat of militant groups as well as from other SMPs in the region. Singapore too, albeit to a lesser extent, has maintained a threat-oriented approach, in recognition of its vulnerability as a small island-state that lacks strategic depth and relies on technology to make up for what it lacks in manpower. It is in this category that the strongest relation between the three themes of national ambitions, defence planning approaches, and military innovation can be witnessed and where there appears to be the most logical cascading between 'setting the strategy' and the actual implementation and force posture outcomes.

In this cluster of SMPs, it is the intensity of the threat rather than only its nature that leaves a clear impact on the way defence planning is conducted and the approach taken to military innovation. The felt urgency has spurred both Finland and Israel, and to a degree, also Singapore, to put in place a highly strategic and, comparatively speaking, a more disciplined approach to defence planning than many of the others examined, with clear processes, institutions and an emphasis on long term planning, while keeping a good level of flexibility and pragmatism. This clear threat perception also trickles through in these countries' approach to military innovation, where there seems to be a link between the existence of a serious threat and decisions to invest in a focused domestic industrial base that is tailored to the type of direct threat it confronts. For example, Israel has developed cutting-edge missile defence and cyber capabilities, and Finland has become a producer of its own hardware (including armoured personnel vehicles and artillery) and ammunition, while specialising at the same time in (resilience against) hybrid threat actors.

These SMPs are also looking at their choice of alliances through the lens of threats. This is evident in the current changing geopolitical order, in which both these powers have adjusted their regional and global allegiances in an attempt to navigate the changes taking place. Finland, after the Russian annexation of Crimea, invested massively in building more cooperation with neighbouring Sweden and made a 'unanimous' domestic decision to join NATO after Russia decided to invade Ukraine. Israel for its part, was long able to survive in its hostile environment relying on US and European support, but has in recent years co-opted important powers in the region through the US-facilitated Abraham Accords, which saw it open diplomatic relations with the UAE, Bahrain, Morocco and Sudan, while talks were underway with the Kingdom of Saudi Arabia. The Israeli response to the large-scale attack by Hamas on Israel in October 2023 has seen these talks collapse, while also, at the time of writing, having derailed relations with some of the signatories of the Abraham Accords. Singapore, in turn, has close ties to the US but also takes care not to antagonise other powers trying to maintain a peaceful relationship with China, its largest trading partner.

A second cluster brings together the 'Transactional Defence Planners'. Indonesia, Oman, the UAE and Turkey seem to fall in this cluster. Here, the nature of the threat plays less of a role than other considerations. These SMPs display a larger tendency towards hedging at various levels, a dynamic that has been amplified in the current changing geopolitical context, as they incline more to building and maintaining relations with the various great powers while using great power competition to extract additional benefits for themselves. This dynamic can, for example, be witnessed in the procurement of defence or dual-use materiel, as well as in technology transfer requests and joint exercises and projects with, for example, the US, Russia, and/or China. In this cluster, defence planning is often used as a quid pro quo tool both internationally and domestically. Indeed, internationally, defence planning including acquisition processes can be used to establish and strengthen ties with third parties such as is the case for Oman, and/or to play off great powers against one another. Domestically, defence planning can be and often is a way to serve the interests of particular constituencies that are relevant for the support of the regime and/or the defence establishment. In some countries, such as Indonesia, the historical civil-military tension remains at the heart of this practice, as the regime, as a coup-prevention measure, prefers to dilute resources for the military and ensure that certain units will not become strong enough to consider a coup. In Turkey, a similar dynamic can be witnessed, in which defence procurement practices are used by the incumbent regime to consolidate domestic power.

Those practising a more transactional variant of defence planning, tend to have defence planning processes that are less linear and predominantly based on political considerations. This sometimes means that strategic planning and decision-making are being reversed—starting with what the desired procurement outcome is, rather than what actual threat is supposed to be countered. In terms of alliances, these SMPs display a level of non- or semi-commitment or, in the case of Turkey, a willingness to challenge existing commitments to pursue others at the same time. Often, these states do not align themselves with one particular side but tend to hedge by establishing ties with different powers, for instance, through opting for a diversity of major defence platform suppliers. In the context of current geopolitical tensions, they also tend to position themselves as bridges and brokers between these players. There is a less clear pattern here when it comes to societal resilience and the integration of the military within society. The role of the military in Turkish society is, for instance, deliberately being diminished by its political leadership, following the failed coup in 2016. In the UAE, on the contrary, the military's role is increasing, with military service having been introduced in 2014, be it more for reasons of building a stronger national identity rather than in response to a specific threat.

A third and final cluster consists of SMPs that do not consider defence and security very prominently in their national ambitions and considerations. This cluster we call the 'Complacent Defence Planners'. In contrast to defence planners who have been on the alert at all times, carefully crafting their alliances, dependencies and ambitions, these complacent defence planners have taken a much more complacent approach to their defence and security. Here, the perception prevailed that in an era of 'wars of choice', combined with the absence of direct threats and the conviction that a great power (in these cases pretty much always the US) would reliably underwrite their security, defence could be relegated low on the political agenda. Despite the necessity to operate at strategic distances which would suggest substantial new investments, the almost guaranteed Western military superiority vis-à-vis most potential opponents also contributed to this until recently relaxed attitude in most European militaries, Canada and Australia. Although national think tanks often warned of the risk of irrevocable disbandment of types of capabilities and the growing inability to conduct high-intensity combat operations, this lack of urgency at the political level has translated into diminished discipline and a paucity of strategic thinking in defence planning processes and military innovation, even if their approaches can be considered 'strategic' from a different perspective (e.g. domestic industry, financial considerations). Here as well, the ultimate driver seems to be the threat perception, which in these cases has been rather low. Indeed, for most of the past four decades, these complacent defence planners experienced a period of relatively insignificant threats. As such, defence planning and military innovation seem to have been pushed to the back burner, while the military in many cases decoupled from civil society. Conscription was abandoned or suspended, the military disappeared from crowded urban spaces, and the military instrument tool decreased in importance compared to the development of cooperation and diplomacy in international affairs.

Amongst the countries under review, Canada, the Netherlands and Slovakia seem to be the most obvious representatives of this more 'complacent' cluster, while Australia falls somewhere in between the Complacent and the Strategic Defence Planner cluster. What connects complacent defence planners is that broadly, in particular since the 1990s, the importance of territorial defence slipped away. The more optional nature of participation in military missions 'out of area' made the military predominantly an instrument of diplomacy. The lack of a direct threat and this ability to choose from 'a la carte deployments' also meant that there was more room for domestic—political, economic or other—considerations in the shaping of defence planning. As a result, these planning processes are not always clearly institutionalised, and they tend to be reactive and focused on consensus-building. Within this cluster, the overarching objective of many missions was to be a good ally, or at least a good enough ally. The postwar (and for Slovakia post-Cold War) alliance with the US through NATO provided the necessary security guarantees, and an overall sense of security that came with belonging to the West. For example, in this cluster, when an economic downturn had governments scrambling to make cuts, the defence organisation turned out to be an easy target. Efficiency and discipline of defence planning processes suffered as well during 'peacetime', and became vulnerable to competition between the different military services, civil-military competition and inter-departmental rivalries. Overall, this has led to at best medium levels of military innovations, largely private sector driven, and a low readiness of military forces that are not well integrated into society at large. While for the first cluster of strategic defence planners, the relationship between set ambitions, strategies, the defence planning approaches and military innovation could be judged as logical steps in a linear process, the sheer number of possible intervening variables throughout the process, makes that, for this 'Complacent Defence Planners' cluster, such relationships can only be assessed on an individual basis and with a deeper knowledge of internal political and bureaucratic dynamics. Australia falls somewhere in between the two types, as it is making the transition from taking a complacent approach to a much more strategic one, in the context of China being perceived as a threat to Australia's national security. Today, Australia is not just strengthening ties with the US but it is building up greater self-reliance in the pursuit of robust and innovative military forces. At the same time, a major defence review in 2023 was conducted by external experts rather than by the defence ministry itself, reflecting considerable institutional shortcomings in its defence planning processes.

Table 13.1 provides an overview of the three types of defence planners based on the sample in this volume and the structured focused comparison applied throughout the chapters.

*Table 13.1* A three-pronged typology of SMP's defence planning

| TYPE | Strategic Defence Planners | Transactional Defence Planners | Complacent Defence Planners |
|---|---|---|---|
| **Threat environment** | Direct and intense threats | Not a key driver | Indirect and not prominent |
| **Overarching role** | Reduce strategic risk | Quid pro quo tool; ensuring domestic stability and/ or creating international leverage | Tool of diplomacy |
| **Defence planning process** | Robust and linear; institutionalised; proactive; focused on threats | Non-linear; predominantly based on political considerations. | Haphazard; institutionalised planning processes not strictly followed; reactive; focused on consensus-building |
| **View on alliances and partnerships** | Security guarantees; strategic considerations | Non-committal | Security guarantees; financial considerations |
| **Defence industry** | Focused on building self-sufficiency | Uses procurement and contracting as a relations-building tool | Predominantly focused on financial considerations |
| **Level of innovation** | High, based on countering a clear threat | Varies | Medium, less focused on specific threats |
| **Societal resilience** | High readiness; Mobilisation-driven, force mix of conscription and volunteers; closer civil-military integration | Readiness varies; Not considered essential for force posture; military can play a role in national identity formation. | Low readiness; conscription-based force; military perceived as outside of society |
| **Who?** | Israel, Finland, Singapore, (Australia) | Indonesia, Oman, UAE and Turkey | Netherlands, Slovakia, Canada, (Australia) |

*Source:* Authors

## Final thoughts and future research avenues

In the more stable geopolitical climate that prevailed at least until the early 2000s, defence planning of SMPs was an arguably rather tedious and little exciting subject for research and analysis. In today's world, as the overarching security environment is getting more fluid, complex and less stable, the topic rightly deserves more attention. In many SMPs, the relative lack of knowledge about defence planning is rendered more urgent by the fact that continuous underinvestment has led to poor readiness of the armed forces, a situation that will take years to reverse. For some SMPs, the constraints brought by existing, costly legacy forces or sustaining force structures geared towards manpower-intensive stabilisation operations have produced an imbalance in capabilities for high intensity warfare that are more relevant in the emerging era of great power competition.

Many of the SMPs are currently adjusting to the changing situation. In that process, they will inevitably have to revisit their defence planning approaches if they are to respond to new dynamics as best as possible. For European SMPs, this holds distinct risks. The Russian invasion of Ukraine has removed any uncertainty about the nature of the security environment that Western SMPs may have perceived until 2022. The war has rapidly reprioritised defence planning towards preparing for a major conventional war. Most European militaries are now scrambling to purchase tanks, artillery, drones, air and missile defence systems and modern fifth-generation fighter aircraft, and rebuild depleted stockpiles of ammunition—all geared towards the new NATO deterrence by denial strategy. They may, as is feared, only enjoy a window of opportunity of a couple of years before Russia has rebuilt its heavily damaged conventional armed forces with which it can pose a direct threat to, for instance, the Baltic states.[11] The legacy of three decades of relative neglect and unwarranted optimism amongst complacent defence planners offers no grounds for optimism.

Other SMPs, especially in the Middle East and Asia, develop their defence capabilities, some of them less constrained by legacy forces; most of them face a whole different set of challenges related both to the availability of the required skills and expertise, the training of personnel, as well as the development of organisational structures and procedures to leverage old and novel military technologies for defence purposes. This situation harnesses opportunities as well as risks for these emerging powers. On the one hand, it offers the chance to achieve step change and create punctuated advantages. On the other hand, it means that they do not possess the planning experience that older powers have built up over decades, and have limited institutional memory to learn from past experiences. This may make them more agile, but it may also increase the risk of them committing strategic mistakes. In this context all SMPs struggle with striking the right balance between incremental force renewal and force modernisation through military innovation in order to reap the benefits of what some analysts prospect as a new revolution in military affairs.[12]

The starting premise of our book was that most literature on defence planning and military innovation has been based on studies of great powers. Seminal works that led to theory development by Murray et al., Rosen and Posen focused pre-dominantly for instance on Russia, the UK, Germany and the US. The case studies in this book demonstrate first and foremost that those factors that help explain defence policy and military innovation in great powers, including the presence of external threats and the availability of new technologies, may only be one among many and most likely not the most relevant ones for understanding the trajectories of SMP defence planning and concomitant force postures. In particular, domestic politics, strategic culture, inter- and intra-service rivalry and alliances need to be brought in as analytical lenses to understand the different types of logics at play within the defence planning and military innovation processes of SMPs. While the chapters in this book offered a kaleidoscopic set of observations, all convincingly demonstrated the complex interplay of factors in those processes.

In this book, we have analysed these SMP-idiosyncratic pathways using a struc-tured focused comparison framework which has proven useful for mapping and explaining the defence planning of SMPs. In addition, the framework developed by Raska allowed for a comparative assessment of military innovation paths and patterns of SMPs. Both frameworks will benefit from future research into a wider set of case studies to expand the knowledge base of defence planning practices in SMPs and to gauge the different weight of factors shaping defence planning and military innovation, which fell outside the scope of this volume. In addition, further work can focus on refining, expanding, and/or refuting the typology based on an assessment of other SMPs along the structured focused comparison framework. For example, this volume did not include any case studies based in the Americas other than the northernmost one Canada, while it also does not contain any SMPs in central Asia and Africa.

The findings of this book serve not just scholarly theory development but are also relevant for practitioners involved in real-world defence planning. The struc-tured focused framework presented here can help guide discussions both on how to structure national defence planning processes and what choices to make in the setting of ambitions, the formulation of priorities, the design of processes, and the selection of partners. Moreover, the evidence from the array of case studies can be useful in reflecting on the peculiarities of the national defence planning process, as well as offering a rich repertoire of good—as well as bad—practices. As case studies fall across different geographies, strategic cultures, and governance sys-tems, awareness of the different storylines will also help create an understanding of patterns in the modus operandi of other defence organisations. In particular for Western defence establishments, learning from what is going on in the 'rest of the world', will provide a healthy antidote to often internally focused strategic dis-cussions. With a clearly deteriorating security environment, those case studies of countries featuring a well-developed strategic approach to defence planning may prove informative for those who recently discovered that they have to take defence policy more seriously. If one thing is clear, in today's security environment, SMPs ignore the perils of inattentive defence planning at their own risk. We therefore hope that this volume will lead to better-informed and more deliberate defence planning in the SMPs covered in this book.

# Notes

1  Henry Bartlett, G. Holman, and Timothy Somes, 'The Art of Strategy and Force Planning', *Naval War College Review* 48, no. 2 (1995), https://digital-commons.usnwc.edu/
nwc-review/vol48/iss2/9.

2  Henrik Breitenbauch and André Ken Jakobsson, 'Coda: Exploring Defence Planning in
Future Research', *Defence Studies* 18, no. 3 (3 July 2018): 391, https://doi.org/10.1080/
14702436.2018.1497447.

3  See Jasen J. Castillo and Alexander B. Downes, 'Loyalty, Hedging, or Exit: How
Weaker Alliance Partners Respond to the Rise of New Threats', *Journal of Strategic
Studies* 46, no. 2 (23 February 2023): 227–68, https://doi.org/10.1080/01402390.202
0.1797690; Peter Viggo Jakobsen and Sten Rynning, 'Denmark: Happy to Fight, Will
Travel', *International Affairs* 95, no. 4 (1 July 2019): 877–95, https://doi.org/10.1093/
ia/iiz052; David A. Lake, 'Escape from the State of Nature: Authority and Hierarchy in
World Politics', *International Security* 32, no. 1 (2007): 47–79.

4  For an overview of classic and some more recent work (as cited in the introduction),
see Annette Baker Fox, *The Power of Small States: Diplomacy in World War II* (Chi-
cago, IL: University of Chicago Press, 1959), https://press.uchicago.edu/ucp/books/
book/chicago/P/bo8928706.html; David Vital, *The Inequality of States: A Study of the
Small in International Relations* (Oxford: Clarendon Press, 1967); Robert L. Rothstein,
*Alliances and Small Powers* (New York: Columbia University Press, 1968); Robert O.
Keohane, 'Lilliputians' Dilemmas: Small States in International Politics', *International
Organization* 23, no. 2 (1969): 291–310; Trygve Mathisen, *The Functions of Small
States in the Strategies of the Great Powers*, Scandinavian University Books (Oslo: Uni-
versitetsforlaget, 1971); Peter R. Baehr, 'Small States: A Tool for Analysis', ed. Edward
E. Azar and Marshall R. Singer, *World Politics* 27, no. 3 (1975): 456–66, https://doi.
org/10.2307/2010129; Håkan Wiberg, 'The Security of Small Nations: Challenges and
Defences', *Journal of Peace Research* 24, no. 4 (1 December 1987): 339–63, https://
doi.org/10.1177/002234338702400403; Tim Sweijs, 'The Role of Small Powers in
the Outbreak of Great Power War', (The Centre of Small State Studies, The Institute
of International Affairs, University of Island, 2010), www.semanticscholar.org/paper/
The-role-of-small-powers-in-the-outbreak-of-great-Sweijs/2f2db7d8fedf06c829891
b0a1e58e279343e741a; Håkan Edström, Dennis Gyllensporre, and Jacob Westberg,
*Military Strategy of Small States: Responding to External Shocks of the 21st Century*
(Abingdon, OX: Routledge, 2019), www.routledge.com/Military-Strategy-of-Small-
States-Responding-to-External-Shocks-of-the/Edstrom-Gyllensporre-Westberg/p/
book/9780367529598; Håkan Edström and Jacob Westberg, *Military Strategy of Middle
Powers: Competing for Security, Influence, and Status in the 21st Century* (Abingdon,
OX: Routledge, 2020).

5  For the dependency of European SMPs on the US in this regard for instance, Hugo
Meijer and Stephen G. Brooks, 'Illusions of Autonomy: Why Europe Cannot Provide
for Its Security If the United States Pulls Back', *International Security* 45, no. 4 (20
April 2021): 7–43, https://doi.org/10.1162/isec_a_00405. Tim Sweijs and Frans Osinga,
'VIII. Maintaining NATO's Technological Edge', *Whitehall Papers* 95, no. 1 (2 Janu-
ary 2019): 104–18, https://doi.org/10.1080/02681307.2019.1731216.

6  Michael E. O'Hanlon, *Forecasting Change in Military Technology, 2020–2040* (Washington, D.C.: Brookings Institution, September 2018), www.brookings.edu/articles/
forecasting-change-in-military-technology-2020-2040/.

7  Daniel Fiott, 'A Revolution Too Far? US Defence Innovation, Europe and NATO's
Military-Technological Gap', *Journal of Strategic Studies* 40, no. 3 (2017): 417–37,
DOI: 10.1080/01402390.2016.1176565

8  Indonesia's position is the result of the assessment of the authors of the concluding
chapter.

9  Michael Raska, 'A Structured-Phased Evolution: The Third Generation Force Transformation of the Singapore Armed Forces', in *Military Innovation in Small States: Creating a Reverse Asymmetry*, ed. M. Raska (New York: Routledge, 2016), 130–62.
10 See on this Theo Farrell, Terriff Terry, and Frans Osinga, eds., *A Transformation Gap?: American Innovations and European Military Change* (Stanford, CA: Stanford University Press, 2010).
11 Christian Mölling and Torben Schütz, 'Preventing the Next War: Germany and NATO Are in a Race Against Time', DGAP Policy Brief (German Council on Foreign Relations, November 2023), https://dgap.org/system/files/article_pdfs/34-DGAP-Policy-Brief-2023.pdf.
12 Time will tell whether 'this time is different', but if a Revolution in Military Affairs (RMA) comes to transpire, defence organisations surely want to come prepared. For a proponent of a coming RMA, see Christian Brose, 'The New Revolution in Military Affairs', *Foreign Affairs*, 16 April 2019, www.foreignaffairs.com/united-states/new-revolution-military-affairs.

## Bibliography

Baehr, Peter R. 'Small States: A Tool for Analysis'. Edited by Edward E. Azar and Marshall R. Singer. *World Politics* 27, no. 3 (1975): 456–66. https://doi.org/10.2307/2010129.
Breitenbauch, Henrik, and André Ken Jakobsson. 'Coda: Exploring Defence Planning in Future Research'. *Defence Studies* 18, no. 3 (3 July 2018): 391–94. https://doi.org/10.1080/14702436.2018.1497447.
Brose, Christian. 'The New Revolution in Military Affairs'. *Foreign Affairs*, 16 April 2019. www.foreignaffairs.com/united-states/new-revolution-military-affairs.
Castillo, Jasen J., and Alexander B. Downes. 'Loyalty, Hedging, or Exit: How Weaker Alliance Partners Respond to the Rise of New Threats'. *Journal of Strategic Studies* 46, no. 2 (23 February 2023): 227–68. https://doi.org/10.1080/01402390.2020.1797690.
Edström, Håkan, Dennis Gyllensporre, and Jacob Westberg. *Military Strategy of Small States: Responding to External Shocks of the 21st Century*. Routledge, 2019. www.routledge.com/Military-Strategy-of-Small-States-Responding-to-External-Shocks-of-the/Edstrom-Gyllensporre-Westberg/p/book/9780367529598.
Edström, Håkan, and Jacob Westberg. *Military Strategy of Middle Powers: Competing for Security, Influence, and Status in the 21st Century*. Abingdon, OX: Routledge, 2020.
Farrell, Theo, Terriff Terry, and Frans Osinga, eds. *A Transformation Gap? American Innovations and European Military Change*. Stanford, CA: Stanford University Press, 2010.
Fox, Annette Baker. *The Power of Small States: Diplomacy in World War II*. Chicago, IL: University of Chicago Press, 1959. https://press.uchicago.edu/ucp/books/book/chicago/P/bo8928706.html.
Jakobsen, Peter Viggo, and Sten Rynning. 'Denmark: Happy to Fight, Will Travel'. *International Affairs* 95, no. 4 (1 July 2019): 877–95. https://doi.org/10.1093/ia/iiz052.
Keohane, Robert O. 'Lilliputians' Dilemmas: Small States in International Politics'. *International Organization* 23, no. 2 (1969): 291–310.
Lake, David A. 'Escape from the State of Nature: Authority and Hierarchy in World Politics'. *International Security* 32, no. 1 (2007): 47–79.
Mathisen, Trygve. *The Functions of Small States in the Strategies of the Great Powers*. Scandinavian University Books. Oslo: Universitetsforlaget, 1971.
Meijer, Hugo, and Stephen G. Brooks. 'Illusions of Autonomy: Why Europe Cannot Provide for Its Security If the United States Pulls Back'. *International Security* 45, no. 4 (20 April 2021): 7–43. https://doi.org/10.1162/isec_a_00405.
Mölling, Christian, and Torben Schütz. 'Preventing the Next War: Germany and NATO Are in a Race Against Time'. DGAP Policy Brief. German Council on Foreign Relations, November 2023. https://dgap.org/system/files/article_pdfs/34-DGAP-Policy-Brief-2023.pdf.

O'Hanlon, Michael E. 'Forecasting Change in Military Technology, 2020–2040'. Washington, DC: Brookings Institution, September 2018. www.brookings.edu/articles/forecasting-change-in-military-technology-2020-2040/.

Rothstein, Robert L. *Alliances and Small Powers*. New York: Columbia University Press, 1968.

Sweijs, Tim. 'The Role of Small Powers in the Outbreak of Great Power War', (The Centre of Small State Studies, The Institute of International Affairs, University of Island, 2010), www.semanticscholar.org/paper/The-role-of-small-powers-in-the-outbreak-of-great-Sweijs/2f2db7d8fedf06c829891b0a1e58e279343e741a.

———, and Frans Osinga. 'VIII. Maintaining NATO's Technological Edge'. *Whitehall Papers* 95, no. 1 (2 January 2019): 104–18. https://doi.org/10.1080/02681307.2019.1731216.

Vital, David. *The Inequality of States: A Study of the Small in International Relations*. Oxford: Clarendon P, 1967.

Wiberg, Håkan. 'The Security of Small Nations: Challenges and Defences'. *Journal of Peace Research* 24, no. 4 (1 December 1987): 339–63. https://doi.org/10.1177/002234338702400403.

# Index

For Product Safety Concerns and Information please contact our EU
representative GPSR@taylorandfrancis.com
Taylor & Francis Verlag GmbH, Kaufingerstraße 24, 80331 München, Germany

www.ingramcontent.com/pod-product-compliance
Lightning Source LLC
Chambersburg PA
CBHW052120230326
41598CB00080B/3916